S0-BNR-132

 W9-BGL-511

Ecological Studies

Analysis and Synthesis

Edited by

W. D. Billings, Durham (USA) F. Golley, Athens (USA)

O. L. Lange, Würzburg (FRG) J. S. Olson, Oak Ridge (USA)

Volume 18

Remote Sensing
for Environmental Sciences

Edited by
Erwin Schanda

Contributors
R. T. H. Collis · D. J. Creasey · R. L. Grasty
Ph. Hartl · G. P. de Loor · P. B. Russell
A. E. Salerno · E. Schanda · P. W. Schaper

With 178 Figures

Springer-Verlag Berlin Heidelberg New York 1976

The picture on the cover is part of Fig. 19 on page 334.

ISBN 3-540-07465-1 Springer-Verlag Berlin · Heidelberg · New York
ISBN 0-387-07465-1 Springer-Verlag New York · Heidelberg · Berlin

Distributed in the British Commonwealth Market by Chapman & Hall Limited, London.

Library of Congress Cataloging in Publication Data. Main entry under title: Remote sensing for environmental sciences. (Ecological studies; v. 18). Includes bibliographies and index. 1. Remote sensing systems. I. Schanda, Erwin. II. Collis, Ronald Thomas, 1920— . III. Series. QH541.15.R4R39. 621.36′7. 75—37553.

Typesetting, printing. and binding: Brühlsche Universitätsdruckerei, Gießen.

Preface

The public's serious concern about the uncertainties and dangers of the consequences of human activities on environmental quality demands policies to control the situation and to prevent its deterioration. But far-reaching decisions on the environmental policy are impaired or even made impossible as long as the relevant ecological relations are not sufficiently understood and large-scale quantitative information on the most important parameters is not available in sufficient quality and quantity.

The techniques of remote sensing offer new ways of procuring data on natural phenomena with three main advantages

— the large distance between sensor and object prevents interference with the environmental conditions to be measured,
— the potentiality for large-scale and even global surveys yields a new dimension for the investigations of the environmental parameters,
— the extremely wide, spectral range covered by the whole diversity of sensors discloses many properties of the environmental media not detectable within a single wave band (as e.g. the visible).

These significant additions to the conventional methods of environmental studies and the particular qualification of several remote sensing methods for quantitative determination of the natural parameters makes this new investigation technique an important tool both to the scientists studying the ecological relationship and the administration in charge of the environmental planning and protection.

The aim of the present volume is to disseminate some knowledge on recent methods of remote sensing and their applicability. This book is an introduction for natural scientists and graduate students wishing to become acquainted with these methods and to take over this new tool in their specific investigations. Particular emphasis is put on the new methods which, though still in a stage of development offer considerable potentialities as operational systems in the near future. Information on the technology applied and still more on the basic investigations and large-scale applications are of recent date and represent therefore a valuable review even to the specialists. We have not striven for completeness, which is impossible within the available space, but have preferred to produce a readable textbook rather than a many-volumed handbook. Some established sensing methods (mainly those in the optical and near-infrared range) are

therefore presented in a considerably reduced form, in favor of more recent developments. The selection of topics for this book and of material treated within the various chapters has been made with the aim of illustrating the variety of methods which are representative and important for future applications in the widely spread natural disciplines; the examples of application in the natural sciences are given essentially to demonstrate their feasibility, while their ecological discussion has been strongly limited.

Of the total of nine chapters seven are on specific sensor methods and one is on image processing, which is important to all types of sensors and all spectral ranges as soon as some imagery is involved. Each of the sensor chapters contains

— a description of the basic techniques and the sensor systems employed,
— a treatment of the physical fundamentals of the object-sensor relation (i.e. a translation of earth-science parameters into sensor-specific observables),
— a discussion of a number of illustrative applications of the particular method.

The emphasis and length of each of these topics is very much dependent on the type and stage of development of a sensor. The authors of the various chapters are highly experienced specialists in their respective technical fields (physicists and electronics engineers) and are deeply involved in earth-science applications. This background makes them particularly qualified to present their methods to readers interested in earth- and life-science as well as to those more technically oriented.

I should like to thank my author colleagues for collecting all pertinent material for their respective chapters, for their consideration of the common purpose and for delivering their manuscripts within a very short period. I acknowledge also the valuable advice of the publishing company and the series editor Prof. O. L. Lange. My special thanks are due to my wife Gabriele and my children Susanne, Christine and Rüdiger for their patience with the burden of my additional work during this time.

Bern, October 1975   ERWIN SCHANDA

Contents

Contributors

COLLIS, R. T. H. Atmospheric Sciences Laboratory, Stanford Research Institute, Menlo Park, CA 94025, USA

CREASEY, D. J. Department of Electronic and Electrical Engineering, University of Birmingham, Birmingham 15, Great Britain

GRASTY, R. L. Ocean and Aquatic Affairs, Department of Environment, Victoria, B. C., V8W-1Y4, Canada

HARTL, PH. Institut für Luft- und Raumfahrt, Technische Universität Berlin, Salzufer 17—19, Geb. 12/IV, D-1000 Berlin 10, FRG

DE LOOR, G. P. Physics Laboratory, TNO, Den Haag, The Netherlands

RUSSELL, P. B. Atmospheric Sciences Laboratory, Stanford Research Institute, Menlo Park, CA 94024, USA

SALERNO, A. E. US Department of Interior, Geological Survey, 522, National Center, Reston, VA 22092, USA

SCHANDA, E. Institut für Angewandte Physik, Universität Bern, Sidlerstraße 5, CH-3012 Bern, Switzerland

SCHAPER, P. W. Jet Propulsion Laboratory, California Institute of Technology, Pasadena, CA 91103, USA

1. Introductory Remarks on Remote Sensing

1.1 Application Areas of Remote Sensing

The increased potentialities of the remote sensing techniques—in particular the new methods outside the visible part of the electromagnetic spectrum—and the growing experience in a variety of civil applications during the last decade demonstrated new ways to the solutions of many problems of the natural sciences and opened a broader view on the relations between various large scale natural phenomena. But the potential utilization of the remote sensing technology to many public services and managements, which is not yet fully realized, may be expected to bring even more important achievements during the next decades by the qualitative and quantitative improvement of the remotely sensed information and due to the increased effectiveness of their procurement. This can be achieved by the possibilities of quick, small to medium scale (e.g. ground-based or air-borne) but detailed, multi-sensor observation of special features within small areas or by less detailed, large to global scale (space borne) but repetitive and synoptic surveying.

The following list is only a modest and somewhat arbitrary sample of application areas and topics for which immediate gain may be expected:

Water

— Water resources inventory and management for agricultural and industrial use.
— Storage of water in the snow and ice cover and run-off forecast.
— Distribution of the humidity of the soil.
— Water quality (chemical, thermal and biological waste discharge control).
— Sea-ice boundaries and sea state (shipping).
— Thermal streams and salinity distribution in the oceans (marine resources, fish farming).
— Coastal zone, estuary and harbour activities.

Land and Vegetation

— Land use (inventory and planning).
— Soil classification and conservation (agricultural production, irrigation).
— Mineral inventory and exploration planning.

— Control of plant diseases (crop and forest protection).
— Disaster assessment and prediction (volcanos, earthquakes, landslides, avalanches).
— Land pollution control (waste disposal, contamination).
— Cartography.

Atmosphere

— Global weather mapping (cloud distribution, short-range warning).
— Horizontal and vertical temperature and water vapor distribution (all-range numerical weather forecast).
— Control of contents and distribution of minor constituents of the upper atmosphere on global basis.
— Pollution control of the lower atmosphere within limited areas (industrial and automobile emission).
— Survey of the earth's radiation belt (interaction with non-terrestrial phenomena).

Others

— Wild life control and protection.
— Survey of radiation hazards from nuclear power plants.
— Updating of the growth of urban areas.
— Traffic surveying and control.

There may be very different priorities for the utilization of remotely sensed environmental data in different regions of the world e.g. in extremely densely populated parts of Europe as compared to nearly unexplored polar or equatorial regions. But obviously there exists an urgent need for the availability of remote sensing tools as an aid for developing nations in fields such as

— basic land use surveys and inventories of resources.
— location of areas for mineral and energy prospection.
— establishing and updating of maps.

1.2 Sensing Systems

Remote sensing systems have to offer various levels of sophistication of the sensing capabilities, of the various degrees of reliability and of the various stages of the information processing, dependent on the areas of application and on the degree of development from experimental to operational utilization:

— Unambiguous identification of a feature to be observed (relation between the observables and the natural parameters of the object).
— Continuous or intermittent monitoring of temporal changes of the observed features.
— Mapping of the spatial distribution of the observed natural parameters and thematic data and image processing.

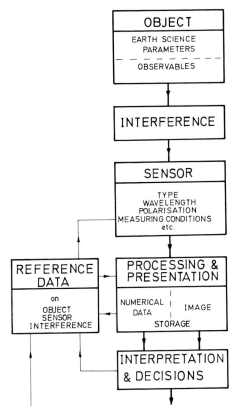

Fig. 1. Block diagram of a remote sensing system with optional automatic interpretation and decision. A complete system will comprise of several complementary types of sensors. The reference data will be generated by previous knowledge and by automatic adjustment

— Thematic interpretation of the observed features (relevant to the specific fields of the users) and automatic prediction.
— Judgement of the information, decision finding, eventually execution of action.

The last two—the highest—levels of performance are to some extent interfering with optional or compulsory intervention of man.

A block diagram of a complete sensing system including the object to be sensed, the interference by the medium carrying the information from the object to the sensor, and finally the—either human or automatic—interpretation and action, is presented in Fig. 1. In most cases there is no identity between the natural parameters and the observables (e.g. the distribution of an atmospheric constituent versus intensity and shape of a spectral line). The sensor itself, however, is nothing more than a transformer of the received information about the observables into an image or into an electric output voltage to be used in the data processing. Images are therefore in most cases only presentations artifically made in a format familiar to the user. Thus the output of a sensing system provides a new type of information in a "language" which is different from the language of the conventional description of the environmental parameters, and the untrained user will need time to learn the meaning of the new information and to readjust his concepts. This not only because of the unconventional kind of presentation, but even more because of the above-mentioned difference between the observables and the natural features themselves. Therefore the main problem of utilizing modern remote sensing methods is not so much the sensor technology or the techniques of data processing and presentation, but the translation of the observables to be sensed into the environmental parameters to be actually controlled. The prime aim of many applied research programs today is the attempt to understand what the sensor output data mean and how they can be used. The existence of reference data on the environmental phenomena (e.g. "ground truth" experience and instantaneous adjustment) and its correct use in the information processing is of eminent importance for the reliability of the interpretation. In a singlewavelength sensing situation there is almost always a smaller number of observables than there are parameters characterizing a natural phenomenon, and therefore an unambiguous identification is strictly impossible. A typical situation of this kind is the interpretation of black and white photography by the exclusive use of the albedo of the objects (no shape recognition). Table 1 gives a selection of the ranges of the albedo of various natural surfaces (as given in various handbooks and textbooks). There is a wide range of overlapping albedos around 30% in the visible spectrum. From this it becomes evident that a remote sensing

Table 1. Albedo of various surfaces (integral over the visible spectrum)

Surface	Percent of reflected light intensity
General albedo of the earth	
total spectrum	~ 35
visible spectrum	~ 39
Clouds (stratus) < 200 meters thick	5–65
200–1000 meters thick	30–85
Snow, fresh fallen	75–90
Snow, old	45–70
Sand, "white"	35–40 (increasing towards red)
Soil, light (deserts)	25–30 (increasing towards red)
Soil, dark (arable)	5–15 (increasing towards red)
Grass fields	5–30 (peaked at green)
Crops, green	5–15 (peaked at green)
Forest	5–10 (peaked at green)
Limestone	~ 36
Granite	~ 31
Volcano lava (Aetna)	~ 16
Water: sun's elevation (degrees)	
90	2
60	2,2
30	6
20	13,4
10	35,8
5	~ 60
< 3	> 90

system—even for a rather limited range of tasks—has to consist of sensors of various types and in various spectral ranges.

1.3 Remote Sensing and Spectral Constraints

The methods to be used for the remote probing of environmental parameters have to be selected due to their ability of transmitting the information of the observable over sufficiently long ranges to the sensor without unrestorable loss. The best suited principles are those based on the transmission of electromagnetic waves, which exhibit an excellent long-range effect over a spectrum of many decades of wavelengths.

But acoustic waves are also well-suited for many sensing problems and are superior to the electromagnetic methods, particularly where transmission media (e.g. sea water) are involved which attenuate electromagnetic waves very strongly.

The electromagnetic spectrum which is in use for the various sensing methods—as partly described in the following chapters—is presented in Fig. 2 with the usual units and the usual designations of the various spectral ranges. Under the heading sensor types a representative sample of instrumental methods is enumerated.

Many sensor types are based on analysing the reflection of an artificial signal transmitted to the remote scene. These methods are often comprized under the designation "active methods". "Passive methods", on the other hand, are based on the emission of a type of radiation by the observed objects characterizing the media or their environmental states.

The natural radiation of the media can be of very different origin. Atomic and molecular gases at normal environmental temperatures e.g. exhibit characteristic line spectra due to the transitions between different quantum states in the visible, infrared and microwave parts of the spectrum. A very much different source of electromagnetic radiation is the natural radioactivity of e.g. Uranium, resulting in the gamma-radiation spectrum characteristic for minerals containing this element. A source of very low frequency radio-emission is the electric discharge during lightning in a thunderstorm.

A very broad spectral region is covered by the electromagnetic radiation due to thermal agitation in any medium. Regarding a non-transparent, non-reflecting ("black") body, the radiation behavior can be described by the Planck radiation law. Figure 3 represents the radiance of an ideal black body in units of power per Hertz bandwidth, which is emitted per square meter of radiating surface into one unit of solid angle. The effect of the temperature of the radiating surface is shown to be very pronounced in the infrared region (wavelength 10 microns or shorter at environmental temperatures), while the radiance is linearly dependent on the temperature in the microwave range. Most media encountered in nature are not "black" but—within the limited spectral widths of the particular sensors—gray i.e. their radiance is reduced by a factor smaller than unity. The spectral behavior (Fig. 3) enables small temperature differences of the various types of the terrestrial surface to be detected very sensitively by measuring the infrared (approximately 10 microns) radiance, but this sensitivity to temperature changes masks the radia-

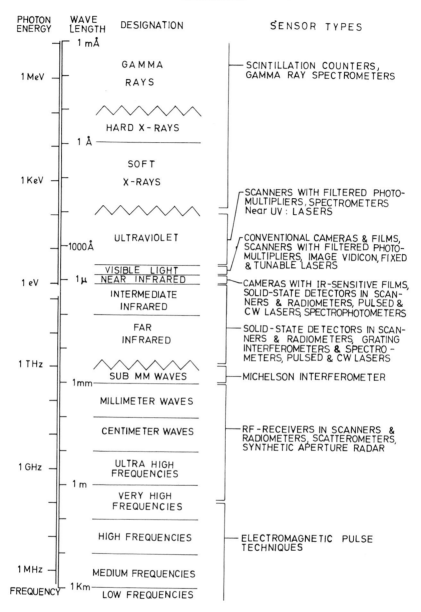

Fig. 2. The electromagnetic spectrum in units of wavelength, photon energy or frequency, whatever is appropriate. The various ranges of the continuous spectrum are named with their widely used designations. Different sensor types are required for the remote probing in the various ranges of the spectrum

tion differences due to the different emissivities, characterizing the type and composition of the media. In the microwave range (wavelength longer than 1 mm) on the other hand, slight changes of the emissivities (or albedos) are recognized very easily, but the sensitivity due to changes of the physical temperature is much less.

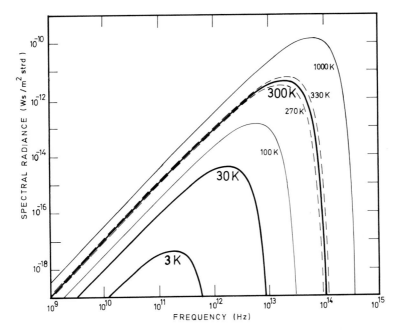

Fig. 3. The "black-body" radiation intensity per unit bandwidth and unit solid angle in the ultra high frequency (wavelength 30 cm) to the near ultraviolet (wavelength 0.3 μ) spectrum for various temperatures of the radiating body

An optimum utilization of the spectral range of Fig. 3 for probing remote objects is, however, prevented by the atmospheric attenuation, caused by molecular absorption lines of important constituents as O_2, H_2O, and CO_2. Figure 4 represents the horizontal path attenuation in the same spectral range under clear weather standard conditions. The absorption in units of decibel per kilometer is given by ten times the logarithm of the intensity ratio before and after having passed the distance of one kilometer. Low values as 0.1 db/km or less (upper part of the diagram) mean almost perfect transmission, high values (more than 20 db/km) mean almost perfect opaqueness of the atmosphere, and the sensing of remote objects (distance more than, say 1 km) is almost impossible. The graph (Fig. 4) makes clear that a considerable part of the far infrared and submillimeter wavelength range is inaccessible for environmental remote sensing—even with active methods—with one exception: the probing of the atmosphere itself.

Figure 5 presents the attenuation due to haze and fog, again for the spectrum between the visual and the microwave region. The "visibility" of an object through haze and fog is given as a comparison between the visual impression and the measured values of the transmission loss.

More details concerning the effects of the transmission medium on the feasibility of the various sensing methods are given in the respective chapters.

The ideas and proposals of a representative part of the European earth science community of future utilization of a variety of sensing methods in the ultraviolet to microwave spectrum for earth resources survey on board of the Space-Lab

Table 2. Typical examples of the utilization of remote sensing techniques for earth resources (ER) surveys by Space-Lab [ESRO, Study Report on definition of the technical requirements for an earth resources payload for the Space-Lab, ESRO contract SC/3/73/HQ (1973)]

Major application area	Experiment-Group	UVAS	BRUV	LIDAR	LH	ACCS	RaPo	MC	MSC	MRC	TT	MSS	IRS	IRR	IRSP	RBSP	SCR	SAR	MWR	MWSR	MWSC
Atmospheric measurements	Air pollution radiometry and spectrometry	×	×	×		×	×						×	×		×	×				×
	Atmospheric measurements	×	×	×		×							×	×	(×)	×			×	×	
	Atmospheric measurement supporting ER data evaluation	×					×						(×)	×	×		×				
	Combined laser altimetry and scatterometry		×	×	×																
Water measurements	Multispectral water imagery and radiometry—Land (incl. hydrology)	×						×	×	×		×	×	×				(×)	×		
	Multispectral water imagery and radiometry—Ocean				×				×				×	×				×	×	×	×
	Maritime traffic investigations				×			×			×		×	×				×	×	×	
	Multifrequency MW ocean imagery/radiom.																	×	×	×	×

Sensor abbreviations

UVAS = UV absorption spectrometer
BRUV = Backscatter radiometer in UV
LIDAR = Laser scatterometer
LH = Laser altimeter
ACCS = Atmosph. composition correlation spectrophotometer
RaPo = Radiometer polarimeter
MC = Metric camera
MSC = Multispectral camera
MRC = Multiresolution camera
MSS = Multispectral scanner
IRS = Infrared scanner
IRR = Infrared radiometer
IRSP = Infrared spectro-photometer
SAR = Synthetic aperture radar
MWR = Microwave radiometer
MWSR = MW scan. radiometer
MWSC = MW scatterometer
TT = Tracking telescope
RBSP = Radiation budget spectro-photometer
SCR = Selective chopper radiometer
UVPH = UV photometer

Land and surface measurements																	
Multisp. vegetation imagery and spectrom.	×	×	×	×			×	×	×	×	×	×					
Multisp. surface imagery radiom./geology, mineral resources	×	×	×	×			×	×	×	×	×	×					
Multisp. regional and environm. planning investigations	×		×	×		×	×	×	×	(×)	×	×					
Multifrequency land imagery and radiometry (MW)	×	×	×	×			×										
Investigation of urban and industrial areas	×		×	×		×	×	×	×	×	×	×	×		×		×	×

E. SCHANDA

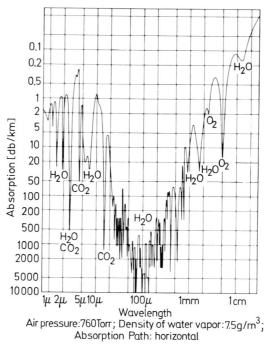

Air pressure: 760 Torr; Density of water vapor: 7.5 g/m³;
Absorption Path: horizontal

Fig. 4. Transmission loss in air under clear weather and standard atmospheric conditions in the near infrared to the microwave spectral range. (Personal communication by F. KNEU-BÜHL, ETH-Zürich, 1974)

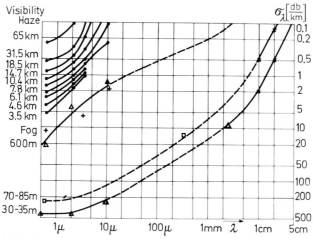

Fig. 5. Transmission loss in a horizontal atmospheric path with haze or fog; for comparison the optical visibility. (Personal communication by F. KNEUBÜHL, ETH-Zürich, 1974)

have been collected and evaluated by the European Space Research Organization (ESRO). Table 2 indicates the methods which are estimated to produce the most useful results on an experimental manned mission for the given earth science investigations.

2. Aerospace Photography

A. E. SALERNO

2.1 Introduction

Aerospace photography is being used extensively as a Remote Sensor to map and detect the natural resources of the Earth. The types of natural resources being surveyed are many and varied; ranging from timberland inventory to water pollution detection. The use of aerospace photography to survey Earth resources has provided a method of rapidly inspecting large areas. Investigators in agriculture, geography, geology, hydrology and oceanography are able to use aerospace photography as a valuable tool for the collection of data. These investigators are able to interpret remote views and translate the photographic images into data regarding ground features of interest. They are familiar with the types of water, terrain, and cultural features which are considered important. In many cases, specific characteristics of target areas are known by the investigator, a fact which can help the photographic community in choosing the proper materials and conditions.

Although not expert in photographic science, most investigators have become familiar with photographic materials and processes. He is then in a better position to know target characteristics and their anticipated photographic results in order to optimize a mission.

Remote sensing of Earth resources involves the discrimination of a target (natural or man-made) from the surround.

The basic problem is one of determining the amount of reflected and/or emitted energy from objects in each portion of the optical spectrum by measuring the recorded densities on exposed and processed film. The photographic and operative factors can be briefly summarized as follows:

1. Illumination (natural or artificial).
2. Spectral reflectance or emittance.
3. Sensor spectral sensitivity (film).
4. Spectral transmission of optical train (lens, filter, atmosphere).
5. Processing control (contrast).
6. Image Quality (Tone Reproduction, Granularity).
7. Geometry (Resolution, Distortion).
8. Interpretation.

To discriminate the target from its background, the film must be able to record density differences due to exposure differences. These exposure differences are the result of combining spectral reflectance, average energy level and difference in size.

Film can be made sensitive to the optical spectrum of 380 nm to 1350 nm. In a Photographic System, the sensitivities of a film are limited by low film speeds for wavelengths above 900 nm and by glass absorption of wavelength below 380 nm. Once a spectral region or band has been selected, a film with a high spectral sensitivity in that region should be used in conjunction with any necessary cutoff filters.

For target discrimination, the gamma should be as high as possible and still record the exposure range of the scene. Most medium-altitude (2.1 km or 7000 feet) aerial scenes have a $\log E$ scene range of 1.2. At higher altitude, the contrast reduction by the atmosphere decreases this range (see Atmosphere HAZE, Section 2.4.4).

The detail rendition required in an aerospace scene will be dependent upon several factors. Certainly, the target contrast and the resolution capability of the photographic system will be of primary importance. The scale of the photograph will be particularly important if there is a known ground resolution. In most Earth resources studies, broad patterns are of more interest than specific identification of objects.

Finally, interpretation becomes possible. By combining the above requirement and records and orchestrating each in its respective role and its relation to the whole, the disciplinarians can interpret the terminal record in terms of the original Earth resource. It must be remembered that most of the photographic records were obtained through narrow spectral windows and recorded on B/W film, which necessitates processing control (gamma or contrast uniformity) between bands so that the original spectral ratios are maintained. Color film is a special case involving simultaneous recording in several bands, and since the tripack material is processed as one, integrity of processing is maintained.

To assist in any photographic mission planning for a given Earth Resource discipline, this chapter will deal with all the required photographic parameters.

For convenience, the chapter is divided into several sections, namely:

Photographic Parameters, which describes the theory of the photographic process via a graphical representation known as the characteristic curve.

Camera, which briefly describes its role and common features of a well made cartographic camera.

Film, which is the photo sensitive recording medium on which much depends.

Photographic Attributes of Earth Resources, which uses several disciplines to describe the interplay between selected spectral reflectances, the intervening anomalies and the possible enhancement of the record, all in hopes of acquiring data remotely.

2.2 Characteristics of Photographic Process

2.2.1 Sensitometry (General)

Sensitometry is primarily the accurate measurement of the sensitivity or the response to light of a photographic emulsion. However, since the effect of an exposure is apparent only after development, it can be said that sensitometry deals with both exposure and development and their relation to one another.

Some of the characteristics of an emulsion that may be determined by sensitometry are the speed, color sensitivity, filter factor, effect of varying conditions of processing, graininess, fogging tendencies and latitude or exposure range of the film.

Although sensitometry is most important for the manufacturer of emulsions, it is advisable for the Earth scientist to have a basic understanding of it so that he can obtain consistently good results under varying conditions. HURTER and DRIFFIELD, in about 1890, were the first to carry out a systematic study of the relationship between exposure and development. Extensive studies of the subject have been made since then, but the basic principles advanced by HURTER and DRIFFIELD still apply. In general, a sensitometric test is made by giving an emulsion a series of known and controlled exposures ranging from a very thin (shadow) to a dense (highlight). The emulsion is then processed under very exacting conditions and the resultant densities measured, plotted and studied.

A practical method would be one in which the intensity and quality of the light could be accurately controlled and where each different density could be arranged in progressive sequence. Such a method is accomplished by a machine known as a sensitometer. The light source for a sensitometer is usually a tungsten bulb with a means of controlling the intensity and the color output.

2.2.2 The Characteristic Curve

In order to analyze and understand the results of the sensitometric test, it is necessary that we plot the density of each step in a test strip in relation to the exposure which was required to produce that density. We draw a characteristic curve on a graph to plot the densities of the exposed and processed strip against the logarithms of their corresponding exposures. In honor of Messrs. HURTER and DRIFFIELD, the characteristic curve that is plotted is sometimes referred to as the "H and D" curve. It is also referred to as the "$D \log E$" curve, where $D \log E$ stands for Density (ordinate of graph) and Logarithm of Exposure (abscissa of graph), a "sensitometric curve" or "characteristic curve".

2.2.2.1 Plotting a Characteristic D logE Curve

The H and D terminology does not describe the type of curve which is being plotted. If the values of density, as measured in a test strip, are placed in a column beside the exposures which produced them, the two columns would represent the relationship between exposures and densities for that particular emulsion (Table 1). Since the relationship between exposure and density is difficult to visualize from a table of this sort, it is much better to use a graphic representation as shown in Fig. 1.

In this graph, the density ($\log O$) is plotted against the logarithm of the exposure ($\log E$) of a hypothetical emulsion. It should be noted that the exposures and densities are plotted in geometric progression by powers of 10 or in a logarithmic progression.

Table 1. Relation between exposure, log exposure, density
and opacity

$E(MSC)$	$\log E$	Density $(\log O)$	Opacity
.001	$\bar{3}$	0.11	1.29
.010	$\bar{2}.0$	0.20	1.58
.10	$\bar{1}.0$	0.42	2.63
1.00	0.0	0.85	7.08
10.00	1.0	1.35	22.40
100.00	2.0	1.85	70.80
1000.00	3.0	2.23	170.00

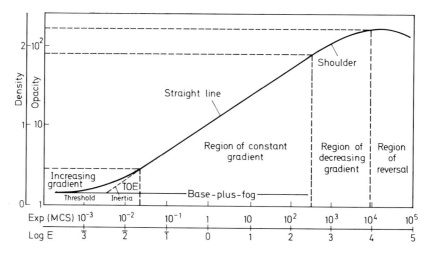

Fig. 1. Graphic representation of density versus log exposure

The exposure E is the product of the light intensity I and the time t. This is expressed by the reciprocity law:

$$E = It.$$

From this relationship, it can be seen that the exposure can be held constant with varying time or intensity as long as the product of the two is constant. The developed density is measured in terms of its ability to block the passage of light. The light-blocking ability can be defined as transmission T, opacity O or density D. These are defined as follows:

$$T = I_t/I_i,$$
$$O = I_i/I_t = 1/T,$$
$$D = \log O = -\log T,$$

where I_i is the light incident on the film, and I_t is the light transmitted by the film.

Density is used for photographic purposes because of the large range of transmissions which are encountered and the relationship which exists between density and the mass of developed silver per unit area.

To construct a $D \log E$ curve, we must determine the log exposures which produced the densities in the strips to be plotted. Knowing that exposure is the product of illumination multiplied by time, we can compute the $\log E$ simply by finding the log of the illumination and the log of the time and adding them. All pertinent data are supplied by manufactures of sensitometers and can usually be applied directly. However, we can calculate the log exposure range of a sensitometer by first measuring the output of the light source in meter candles, finding the logarithm (to the base 10) of the light value, and adding it to the logarithm of the exposure time in seconds. If filters are used in the system, we merely subtract their density values from the log exposure to get the new log exposure. In timescale sensitometers, the time-intensity value must be calculated for each step. But in intensity-scale instruments, the constant log exposure is attenuated by the density of each step in the step wedge, and we can find the $\log E$ for each step by subtracting the density of each step from the log exposure of the sensitometer. Quite often, instead of reading the densities of each step of the exposure modulator wedge, we read just the minimum density steps (to obtain the maximum $\log E$) and assume density increments of 0.15 for a 21-step wedge or increments of 0.30 for an 11-step wedge.

The $D \log E$ curve is a graphic representation of the effects of exposure and development. Changes in either cause the curves to shift. Increasing the exposure moves the curve to the right, while decreasing the exposure moves it to the left. Increasing development causes the slope of the curve to become steeper, while decreasing development makes it flatter.

Since the curve represents a continuous condition, you can locate any intermediate density and its corresponding log exposure on it. You are not confined to the steps of the wedge.

2.2.2.2 Interpreting Sensitometric Curves

The principal function of sensitometry is to measure, record, and represent graphically the reaction of photosensitive emulsions to varying conditions of exposure and development and to analyze and interpret results. Through sensitometric measurement, we can not only determine the effective film speed, latitude and useful exposure range of emulsions, but we can also exercise contrast control. Virtually all sensitometric determinations include the analysis of $D \log E$ curves as part of the procedure.

2.2.2.3 Structure of the Curve

Figure 1 illustrates a graph of a hypothetical curve with all of the essential parts identified. The lower portion, or toe, of the curve is the region of increasing gradient where increases in density are greater than proportional to their corresponding exposure increases. The straight-line portion is the region of constant gradient where density increases are proportional to their corresponding expo-

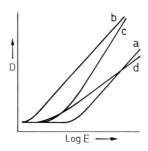

Fig. 2. Effect of exposure and development upon D-log$E$ curve

sure increases. The upper portion, or shoulder, is the region of decreasing gradient where density increases are less than proportional to their corresponding increases in exposure. Finally, if the exposure continues to increase, the density values will decrease even more, and we have the region of reversal. In practice, most exposures will fall on the toe and straight-line gradients. It is seldom that an exposure will produce densities up on the shoulder and probably never in the reversal region. It takes extremely long exposures to produce the reversal effect.

The characteristic curve shown in Fig. 1 could be that of a typical panchromatic emulsion. Notice that the curve is placed on the graph so that some density above zero is shown as the least density. This is the density of the base-plus-fog level of the film. The length of this flat portion of the toe is affected by the amount of overall exposure the film has received. If the exposure is low, the flat portion of the toe will be longer and the straight-line portion will be moved to the right (*a* of Fig. 2). On the other hand, increasing the exposure will shorten the flat portion of the toe and move the straight line portion to the left (*b* of Fig. 2).

Where the toe is flat, the silver halides have not been exposed long enough to render them developable under the developing conditions used in the test. It is possible that a more energetic developer or increased developing time will shorten the flat part of the toe. However, it will also steepen the straight-line gradient and will not appreciably change the extreme end of the toe (*c* of Fig. 2). Shortening development in an attempt to compensate for over-exposure may produce a toe but will also produce a flatter curve (*d* of Fig. 2).

The *threshold* is the point on the toe of the curve where the density first becomes perceptible. This is variable. No two observers see exactly alike and will disagree upon the location of a threshold density. This is why some procedures for calculating film speed (including the new ASA) set a specific density above base fog as a basis for measurement.

On the straight-line gradient, equal changes in log$E$ produce equal (or almost equal) changes in density. Any deviations from a straight line occurring in normally exposed and developed sensitometric strips will average out, and these can be ignored in most of the calculations. Since we believe this portion to be linear, having density differences which change proportionally with the log$E$, we can select an exposure for any desired density (on the straight line) or predict a density for any preselected exposure.

Normally, the shoulder of a characteristic curve is somewhat convex. The density changes represented in it are less than corresponding $\log E$ changes. Density is still increasing, but not as rapidly as it is in the straight line or toe. It continues to increase until maximum density (Dmax) is reached. If the exposure continues to increase beyond this point, the reversal effect is produced. Most camera exposures avoid the shoulder. In this area, detail is lost because the density differences are not great enough to allow the viewer to distinquish between them, and the result is blocked up highlights.

2.2.2.4 Log*E* Axis

If you photograph a uniformly lighted gray card, the exposure you give it can be represented as a single point somewhere on the $\log E$ axis. If you increase the light and make a second photograph using the same camera settings, this exposure can also be a point on the $\log E$ axis. It will, however, occur to the *right* of the first point. The distance separating the two points will depend upon how much more light is used for the second exposure. If you place a second card having a different reflectance beside the first, and photograph the two together, the exposure can be represented as two points on the $\log E$ axis. The separation between them will depend upon the ratio of light intensities as they are reflected by the two cards. Increasing the exposure (for instance by slowing the shutter or opening the aperture) will move both points to the right but will not change the distance between them. Decreasing the camera exposure will, of course, move the points to the left on the $\log E$ axis. This is an oversimplified description of what takes place whenever camera exposures are made. The number of different exposures produced is equal to the number of different luminance levels in the scene. Each can be represented by some point on the $\log E$ axis, and changes in the camera setting simply shift all of the points either to the right or to the left.

2.2.2.5 Density Axis

If you process the exposure you made of the two cards mentioned above, two densities will result, and these can also be plotted on the graph. Each density will occupy a position directly above its corresponding log exposure. The plot on the right (having received the greater exposure) will be higher than the one on the left. In this manner the densities produced by all of the brightnesses of a scene (if you could measure them) could be plotted, and they could be connected by a smooth curve having a toe, a straight-line portion, and a shoulder. The degree of development given is reflected in the shape of the curve and the slope of the straight-line portion. An increase in development causes the straight-line portion to become steeper and increases the difference between the densities lying upon it and in the toe. The base fog level and the areas of minimum density are affected very little by increases in development, and the changes in the toe shape are largely the result of the straight line becoming steeper. With increasing development, the straight line continues to steepen until a point is reached where additional development no longer causes an increase in the gradient. This point is referred to as *gamma* infinity (Fig. 3), and will be discussed in more detail later in this chapter.

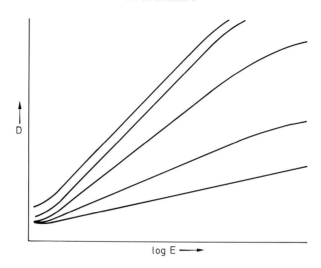

Fig. 3. Effect of development to infinity

2.2.2.6 Gamma

The numerical description of the effect of development upon the characteristic curve is designated by the Greek letter gamma (γ). It is frequently defined as the ratio of the difference between any two densities on the straight-line portion of the curve and the difference between their corresponding log exposures and is shown by the equation

$$\gamma = \frac{\varDelta D}{\varDelta \log E} \tag{1}$$

where $\varDelta D$ is the interval between straight-line densities, and $\varDelta \log E$ is the interval between the log exposures producing these densities (Fig. 4).

Equation (1) is an expression of the average slope or gradient (G) for the interval $\varDelta \log E$. The gradient of the curve, except for the straight-line portion, is not constant and must be expressed in the differential form

$$G = \frac{dD}{d \log E} \tag{2}$$

However, the constant gradient of the straight-line portion can also be stated as the *tangent* of the angle occurring between the straight-line portion of the curve and the log$E$ axis (Fig. 4).

The tangent of this angle was designated by Hurter and Driffield as gamma (γ). Therefore, the straight-line portion is

$$G = \frac{dD}{d \log E} = \text{constant} = \tan \alpha = \gamma \tag{3}$$

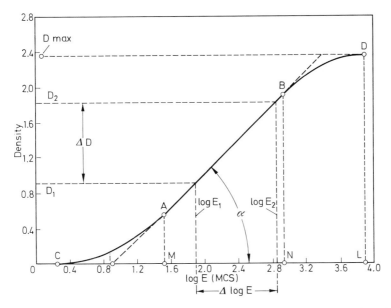

Fig.4. *D*-log*E* curve showing ΔD versus Δ log*E*, tangent ratio, inertia, and location of pertinent density and log*E* point

As an example, let us assume that two areas in an object being photographed have brightnesses of 10 units and 80 units. The Δlog*E* value of this object would be log 80 − log 10, or 0.90. Let us assume further that the densities produced by these areas are both located on the straight-line portion of the curve and that the $\gamma = 0.8$.
Then,

$$\frac{\Delta D}{0.90} = 0.8$$

Therefore,

$$\Delta D = 0.90 \times 0.8 = 0.72$$

When $\alpha = 45°$, than α, or γ, becomes unity (1.0); consequently, any log *E* increment is rendered in the negative by an identical density difference. If γ is less than unity, the proportional reproduction will be correct, but the object brightness scale will be compressed. And if γ is greater than unity, correct proportional reproduction will also be obtained, but the brightness scale will be expanded.

2.2.2.7 *The Significance of Gamma*

Gamma is often referred to as development contrast, but this is not an accurate description. It is a useful measure of degree of development and, as such, is a valuable processing control. An increase in gamma indicates an increase in development, all other parameters being unchanged. An increase in time, temperature, agitation, or developer activity will result in increased gamma, and a decrease in any of these factors will result in decreased gamma. A moderately energetic developer can produce high gamma if the time, temperature, or both are increased. Moreover, the reverse is true; high energy developer can be made to produce low gamma if the other factors are held back.

It is important to remember that gamma relates only to development and not to exposure. A camera negative developed to a predetermined gamma will possess that gamma regardless of the exposure. The densities in the negative, which correspond to straight line densities in the $D \log E$ curve, will have the same density differences. Of course, sensitometric curves made at different exposures would occupy correspondingly different positions on a graph.

2.2.3 Development Control

We have seen that changes in development alter the character of the H and D curve; therefore, it is logical that the curve should be used to monitor and control development.

As mentioned earlier, development can be extended until gamma infinity is obtained; and when this happens, the effect of further development is seen chiefly in the fog level. The point at which gamma infinity occurs varies mainly with the emulsion, and this provides valuable information concerning films. However, since gamma infinity is also affected to some degree by the developer used, it is usually considered in terms of film-developer combinations. But more important than the fact that an emulsion reaches gamma-infinity is that intermediate positions of its slope correlate to the development given. By plotting the changes which occur against the factors causing them, we can establish some very useful processing controls.

2.2.3.1 Time-gamma and Time-fog Curves

Since processing results are quite consistent for a given filmdevelopment combination, it is a simple matter to construct a chart which will allow you to select a processing time to produce a desired gamma. To construct such a chart, you would first plot a family of curves developed at varying times with constant agitation and temperature. Figure 5 illustrates a family of five curves developed in a given developer for 3, 6, 9, 12, and 20 min. The values of gamma obtained are 0.5, 0.75, 0.9, 0.98, and 1.0 respectively. These gamma values are then plotted against their development times on a graph which is usually located (for convenience) in one corner of the larger graph. The points are then connected with a smooth curve extending a short distance beyond each of the outermost points.

The families of curves you plot for your time-gamma chart will generally show slight differences in base fog levels at the various developing times, and these differences can be plotted on the time-gamma chart as shown in Fig. 5.

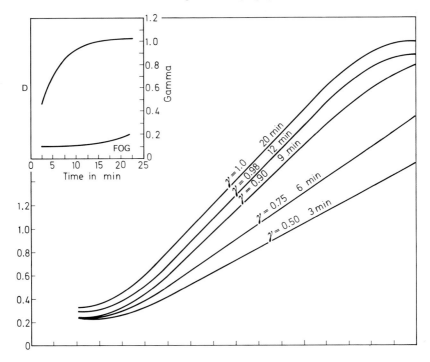

Fig. 5. Plotting time-gamma and time-fog curve

2.2.3.2 Time-temperature Charts

Another chart which is extremely useful in process control is the time-temperature chart. This is actually a time-temperature-gamma chart, since its purpose is to indicate time-temperature combinations to produce a given gamma. The charts published by manufacturers of films indicate recommended gammas for ordinary photographic applications. However, remote sensing requirements of precision laboratories sometimes call for radical departures from the ordinary, and you would need to plot time-temperature charts for your particular project.

To construct a time-temperature chart, start with two time-gamma curves plotted from families of curves processed at different temperatures. Any reasonable temperature interval may be used but should not be less than 10° F. The selected gamma is read on both time-gamma curves, and the time is plotted against the temperature. A straight line is then drawn through the two plots as shown in Fig. 6. This step is repeated for each gamma value you desire to plot.

2.2.4 Exposure Control

In some photographic applications, it is advisable to confine exposures to the straight-line region of the curve. However, there are good reasons why it is frequently advisable to use exposures beginning somewhere on the toe. For example, while the density differences in the toe are less than those in the straight-line

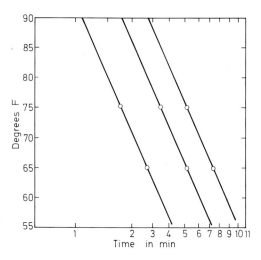

Fig. 6. Time-temperature chart

gradient, the rate at which they increase is greater in the toe than anywhere else on the curve. With the proper printing exposure, the shadow details occurring at or above the minimum useful gradient will be quite apparent. Also, a range of scene luminances with the minimum brightness recorded on the toe rather than on the straight line will produce a negative with an overall low density. This will permit shorter printing time. In addition, placing the minimum exposure on the toe increases the exposure latitude (in the direction of increased exposure) over that available on the straight-line gradient. This practice is so prevalent that many photographic technicians favor using, instead of gamma, a numerical value of some average gradient originating on the toe. This is particularly true when contrast is the prime consideration. This is an additional tool and does not replace the very useful gamma. Contrast Index, (which see) as the new value is sometimes called, is employed mainly in sensitometric analysis of subject luminances and exposure, while gamma is used chiefly in process control, evaluation of density differences and tone reproduction control.

2.2.4.1 Latitude

In Fig. 4 the points M and N on the $\log E$ axis correspond to the limits of the straight-line portion, and the distance between these points is known as *latitude*. Latitude can be expressed either in $\log E$ units or in exposure units shown thus

$$\text{Latitude} = \log E_n - \log E_m \, (\log E \text{ units})$$
or
$$\text{Latitude} = E_n/E_m \, (\text{exposure units}) \, .$$

The latitude value of a given $D \log E$ curve determines the maximum object contrast (ratio of maximum to minimum object luminance) which may be proportionally rendered between density and log exposure on a specific photographic material processed under specific conditions. Latitude is not constant for a given photographic material since it depends primarily upon the degree of development given and, to some extent, upon the spectral composition of the radiation to which the material was exposed.

2.2.4.2 Contrast Index

Because γ is equal to the ratio of the negative density difference to the corresponding log exposure difference, it is used by many persons to express the contrast of a photographic material. But we should never forget that γ describes only the straight-line portion of the curve and gives no information concerning the contrast characteristics of other parts of the $D \log E$ curve.

Some research scientists at Eastman Kodak Company have suggested an alternate basis for describing contrast. Reasoning that a density of 0.20 above base-plus-fog density can be a useful threshold density, that point on the $D \log E$ curve was selected as a starting point. Then, assuming that a $\log E$ range ($\Delta \log E$) of 2.00 will provide a suitable range of tones, it was selected as a limiting gradient point. A value equal to the tangent of an imaginary line drawn between these two points is determined and called (by EK) contrast index.

Figure 7 illustrates this. Point A is 0.20 above base-plus-fog, and point B is separated from it by 2.0 $\log E$ increments. Contrast index (C.I.) is derived by the equation

$$C.I. = \tan\alpha.$$

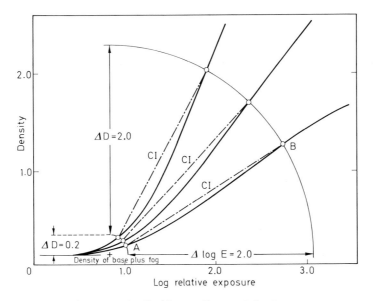

Fig. 7. Locating limiting gradient on D-$\log E$ curve

2.2.5 Film Speed

Speed, as applied to photographic emulsions, indicates a measure of the emulsions' sensitivity to an exposing radiation. By definition, it is the reciprocal of the exposure needed to produce a desired result. There have been many approaches to the problem of rating film sensitivity, the more important of which are namely:

H and D (inertia speed)
Scheiner speed
D.I.N. speed
Weston Speed
Average limiting gradient (ALG)
A.S.A. speed
Federal Specification speeds (FSS).

The determination of speed points for a photographic emulsion is one of the more disputed areas in photography. The controversy concerns whether a fixed density or a fractional-gamma speed point is the best indicator. No matter how one juggles the numbers, the speed point calculation always returns to using an exposure value which produces the best picture or one which contains the most information. Therefore, a speed point should suit the particular situation or application for which the film is to be used.

It is imperative that the sensitometric conditions of the test duplicate, in all respects, the actual conditions of the photographic mission. In other words, the tests should conform to practice in the following areas: time of exposure, intensity and spectral characteristics of light source, development or process conditions, pre and post mission storage conditions, etc. Due to this and other situations, there have been developed the many different speed criteria listed above.

Any photographic reference book listed in the bibliography can be consulted for the different speed criteria. We will simply and briefly discuss the two basic methods, namely, the fixed density and fractional-gradient.

The fixed-density speed point is a fairly straightforward technique. The D-$\log E$ curve shown in Fig. 8 indicates how the fixed-density point is found. The figure shows point a at a fixed density of 0.1 (this value varies) above base-plus-fog. The speed point is found by dropping a perpendicular to the $\log E$ axis. The speed value (S) is then computed as the reciprocal of the exposure:

$$S = K/E$$

where K is a constant, and E is the exposure in either photometric or radiometric units.

The fractional-gradient technique as shown uses a 0.6 gamma speed point. The constant for the gamma fraction varies from 0.2 to 0.6.

A line of the proper slope $(0.6 \times \gamma)$ is made tangent in the toe of the curve. The point of tangency is the speed point. A perpendicular is then dropped to the $\log E$ axis, and the speed value is calculated in the same manner as the fixed-density speed.

From this example, it can be seen that a fairly large difference can occur in speed-point determination. For this particular curve, the fractional-gradient

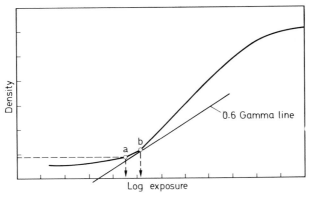

Fig. 8. Fixed density and fractional-gradient speed determination methods

Table 2. Camera film speed systems (historical)

Speed system	Speed point	Formula
H and D (Inertia)	Inertia = intersection of log$E$ axis with extension of straight line	$S = \dfrac{34}{E}$
SCHEINER	First perceptible density above base fog	$S = \dfrac{10}{E}$
D.I.N.	0.10 above base fog	$S = \log_{10} \dfrac{0.49}{E}$
WESTON	Sum of base fog, and gamma	$S = \dfrac{4}{E}$

method placed the point at a higher exposure, giving the film a lower speed value. The type of speed calculation used will be dependent upon the application. Tables 2 and 3 summarize some of the more common speed points and their applications.

2.2.5.1 ASA Film Speed

Aerospace photography does not use any of the aforementioned speed criteria. In order to better understand aerial film speed and ground photography film speed, a somewhat detailed explanation of ASA speed is considered advisable.

The former ASA speed system was based upon the exposure required to produce a "good" print of a scene having a luminance ratio of 1:32 or a Δ log$E$ of 1.50.

L. A. JONES (1935) after extensive investigation and statistical study, found that a speed point which corresponded to a point on the characteristic curve where the minimum gradient is equal to three tenths (0.3) of the average gradient (G) of the log$E$ range of 1.50 provided an exposure which would prod-

Table 3. Camera film speed systems (modern)

Application	Speed point [a]	Source
Pictorial black-and-white (a) negative	0.10 density 0.30 density	USA PH2.2-1963 Federal standard 170a, Method A
(b) positive and reversal	1.00 density	Federal standard 170a, method B
Aerial black-and-white (Negative)	Slope of 0.5 × gamma	Federal specification L-F-330, method B
	Slope of 0.6 × gamma	Eastman Kodak Co., kodak aerial exposure computer publication No. R-10
Pictorial color (a) Negative	$0.2 \times \Delta D$ between toe and specified $\log E$	USA PH2.27-1965
(b) Reversal	Midpoint between gradient shadow point and fixed highlight density	USA PH2.21-1961
Aerial color (a) Negative (b) Positive	Practical test Practical test	

[a] Fixed density levels are over base-plus-fog.

uce a satisfactory print. The exposure index became

$$S = \frac{1}{1/4 \log E}.$$

This calculation provided a safety factor which, in effect, placed the exposure higher on the curve, and most photographers found that this produced excessive overall negative density. In practice, most photographers doubled the film speed thereby adjusting the rated speed to their particular needs. It naturally followed that a new ASA speed was needed, and consideration must be given to the adjusted ratings and inherent safety factor.

Therefore, the American Standard Method for Determining Speed of Photographic Negative Materials (Monochrome, Continuous Tone), Number PH 2.5-1960, was approved and adopted by the American Standards Association, April 1, 1960.

This system fixes a starting point of 0.10 density units above base fog. The curve characteristic is controlled by requiring a 0.80 density difference (ΔD) to a corresponding 1.30 $\log E$ differences ($\Delta \log E$).

The formula for computing the arithmetic speed is

$$S_x = \frac{0.8}{E}.$$

Where S_x is the arithmetic speed, and E is the exposure in meter candle seconds required to produce the 0.1 density above base fog.

2.2.5.2 Aerial Exposure Index

In aerospace photography, the speed value that is used is Aerial Film Speed. Prior to its establishment, Aerial Exposure Index was used.

The Aerial Exposure Index for black and white aerial films is defined as $\frac{1}{2}E$, where E is the exposure in meter candle seconds, at the point on the characteristic curve where the slope is 0.6 of the measured gamma. The table of Aerial Film Characteristics show both Aerial Exposure Index as well as Aerial Film Speed. The Aerial Exposure Index in this table was determined for the same processing conditions as those for the corresponding Aerial Film Speed or Effective Aerial Film Speed for that product.

2.2.5.3 Aerial Film Speed

Aerial Film Speeds or Effective Aerial Film Speed are for use with the latest Kodak Aerial Exposure Computer, Kodak Publication R-10 (12/70 edition), in determining the correct camera exposure for aerial photography. Aerial Film Speeds and Effective Aerial Film Speeds are not equivalent to, and should not be confused with, conventional film speeds (see ASA Speeds above) used in pictorial photography.

Aerial Film Speeds as defined by ANSI Standard PH 2.34-1969 for black and white negative aerial film is $3/2E$; E is the exposure (in meter candle seconds) at the point on the characteristic curve where the density is 0.3 above gross fog.

In case of infra red sensitive and color aerial films, the Effective Aerial Film Speeds are derived empirically by practical aerial photographs. Usually the film speeds values found in data sheets or charts are obtained by rounding the calculated value to the nearest $\sqrt[3]{2}$ step (equivalent to 1/3 f-stop).

2.2.6 Reciprocity Law Anomalies

The conditions of exposure for sensitometric testing must conform to those encountered in actual practice; the response characteristics change with variations in exposure time and intensity and the spectral distribution of the exposing source. The differences in response due to spectral distribution are discussed in the section on spectral sensitivity. Differences due to variations in exposure time and intensity fall under the heading of "reciprocity law failure", which is actually a misnomer. The law itself ($E = It$) does not actually fail; the film fails to respond equally to exposures which follow the reciprocity law. A more appropriate name

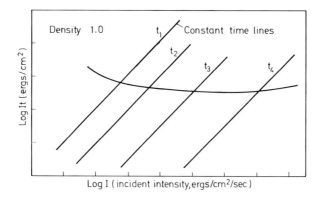

Fig. 9. Reciprocity curve

for the phenomenon is "reciprocity law anomaly". Reciprocity law anomalies are usually represented graphically. The total exposure *(It)* required to produce a fixed density is determined for various values of *I* and *t*. Log*It* is then plotted against log*I*, as shown in Fig. 9. The anomalies are present when the plot departs from a horizontal line. The optimum range of times and intensities can be determined from the graph. The point where the curve has no slope change represents no anomaly.

2.2.7 Density Measurement

The apparent density of a photographic image varies, depending upon the geometry of the measuring system and the spectral response of the receptor. Therefore, it is important to choose measuring techniques which conform to the application of the density data. For example, if the imagery will be viewed by the eye, the density should be a measurement which will reflect the response of the eye. If the imagery will be viewed by an electronic scanner, the density measurement should reflect the response of the electronic system. A third case would be when imagery is to be used for printing onto another light-sensitive material. The density should again represent how the light-sensitive material will "see" the density.

This introduction leads into the three basic types of density. These are: Visual, printing and photoelectric. The distinction between the different types of density measurements is not very important if the density being measured is spectrally neutral and nonscattering. This is rarely the case, since even a neutral silver image has unequal absorption across the visual spectrum and exhibits scattering properties. Due to this, the sensitivity and geometry of the receptor measuring the density must be considered with regard to how the densities will be used.

The two broad classes of density measurement geometry are diffuse and specular. Generally, diffuse measurement gives a lower density than the specular type. By looking at Figs. 10 a and 12 a, the difference between specular and diffuse can readily be seen [1].

---

[1] Diagrams from PH 2.19-1959.

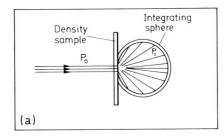

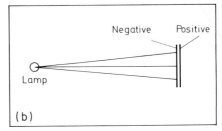

Fig. 10 a and b. Totally diffuse density measurement

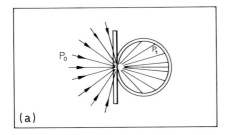

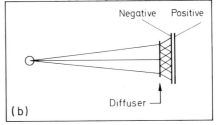

Fig. 11 a and b. Doubly diffuse density measurement

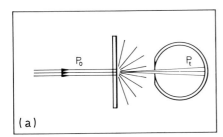

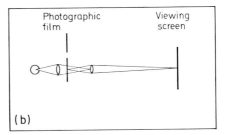

Fig. 12 a and b. Specular density measurement

The diagrams show that the diffuse measurement, shown in Fig. 10 a, collects all of the transmitted light, while the specular measurement of Fig. 12 a only measures the specular component.

A set of standards has been established for which types of geometries are representative of a given practical condition[2]. Figure 10 a shows the geometry for totally diffuse density measurement, which simulates the geometry of a material when it is being contact printed. Figure 10 b shows the corresponding geometry of a common contact printing setup. Figure 11 a indicates the geometry for doubly diffuse density measurement. This geometry represents what is encountered if a contact print has a diffuser between the light source and the photographic materials. The corresponding printer geometry for doubly diffuse density is shown in Fig. 11 b. Figure 12 a illustrates the density geometry of specular density. This

---

[2] PH 2.19-1957.

geometry corresponds to the case shown in Fig. 12b which illustrates projection. Many projectors are, however, only semispecular, so that the effective density of the film used in them is somewhere between specular and totally diffuse. Transparencies which are viewed on an opal glass illuminator, instead of being projected, approach totally diffuse density. Small discrepancies, however, do occur because of the interreflection between the opal glass and the film.

2.2.8 Measurement of Color

The opening statements concerning density measurements indicated that when the density being measured was neutral, the spectral response of the receptor was not important. On the other hand, the cyan, magenta, and yellow dye layers of a color film are not neutral. As a result, several types of density measurements are presently being used in color sensitometry. The classification shown in Fig. 13 represents the classical manner of outlining color density measurements.

1. Integral densitometry is the measurement of the integrated effect of the combined color images. Integral density measurements of color films correspond to the same type of density measurements made on black-and-white film.

Integral densities of color tripack materials are obtained from three broadband regions of the spectrum (red, green, and blue). Each of the three dye layers contributes differently to the density in each of the three regions. Therefore, integral densities reflect the total contributions of all of the dye layers. The red, green, and blue filters chosen for integral densitometry must correspond to the application.

Printing densities of color material are an important tool. The printing density of a color material will specify the characteristics of this intermediate image in terms of the response of a duplicating material. Due to the difficulty of specifying the response of a color material upon exposure, printing densities are operationally defined.

Because of interimage effects (action of one dye layer upon another) in print materials, the density in each layer is dependent upon the exposure it received and

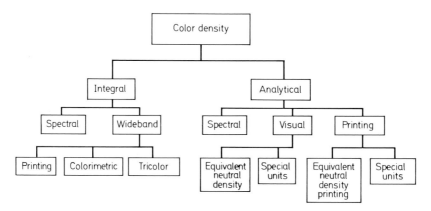

Fig. 13. Color density schematic

upon what exposure other layers received. Because of this, no single spectral sensitivity response can be determined. As a result, MEES and JAMES (1967) define printing density in the following way:

> "Consider two color film images operating in a printer; let the absorption difference between their images produce a difference, ΔD, in a print dye concentration. If, without the color-film image, the printing source is adjusted to produce a specified neutral, and the insertion into the printing beam of a nonselective absorber of density N results in a print which differs from the neutral by the same ΔD in dye concentration, then the printing density difference between the two color film images is equal to N."

Stated another way, the printing densities of a color film to the three dye colors are defined as silver densities. When a silver wedge produces the same effect as the dye image, the dye image is said to have the density of that silver image. Printing densities can be approximated with limited accuracy by adjusting the response of the densitometer to that of the printing material. As mentioned earlier, interimage effects limit the accuracy of this technique.

Colorimetric densities represent the integrated effect the eye sees from a color image. This is in contrast to printing densities, which represent response of the duplicating stock. The densitometer response is adjusted to correspond to that of the X, Y, and Z CIE primaries. When properly manipulated, the colorimetric densities indicate how the colors of a film appear to a standard observer, as defined by the CIE. In addition, the Y primary is the photopic luminous efficiency function. The colorimetric density D_y is called visual density. The D_y density relates to the brightness or tone of a color.

Spectral integral densities are measured at specific wavelengths. Both colorimetric and printing densities are broadband. The American Standard PH 2.1-1952 specifies that spectral densities be measured at the 4358 A (blue) and 5461 A (green) lines of the mercury arc spectrum and the 6438 A (red) line of the cadmium arc spectrum. Narrow-band optical filters are also used with tungsten sources as a substitute.

2. Analytical density is an expression of the amount of dye in each of the layers of a color material. This is contrasted by the integral measurement, which gives the combined effect of all dye layers present. The use of analytical densities enables researchers to perform sensitometric studies of cause-and-effect relationships in color image formation. Analytical densities are also useful in the manufacture of color materials.

A great deal of work is being done with black-and-white multispectral photography. Black-and-white photographs are being made through selected bands in the red, green, and blue and near-infrared. A color film could accomplish the same results as the multiband black-and-white photography, since the image within each dye layer represents a given spectrally sensitive range. Information within each layer would then be separated through the use of analytical densities. The color film also has the advantage of having all three layers in register. The disadvantage lies in the fact that each dye record cannot be easily manipulated to change its contrast or sensitivity. Two extensions of the analytical density concept are equivalent neutral density and equivalent neutral printing density. The difference between the two is analogous to the differences between colorimetric and integral printing densities. "The equivalent density of a color in any subtractive color process is the visual density it would have if it were converted to a neutral gray by superimposing the just-required amounts of the fundamental colors of the process." This is a quote from EVANS et al. (1953) as they defined the concept of gray equivalent density.

From the above definition, it can be seen that equivalent neutral densities *(END)* which have equal values produce visual neutrals when combined. Equivalent neutral printing densities *(ENPD)* are similar to *END*, except that *ENPD*'s having equal values produce a visual neutral in the print stock.

Several different techniques have been used to measure analytical and *END* densities. A complete description of the techniques to generate analytical densities is given in EVANS et al. (1953), Principles of Color Photography.

Analytical densities are converted from integral density measurements using matrix algebra. The integral values are converted to *END*, or analytical densities, by the following set of general equations:

$$C = A_1 D_r + A_2 D_g + A_3 D_b$$

$$M = A_4 D_r + A_5 D_g + A_6 D_b$$

$$Y = A_7 D_r + A_8 D_g + A_9 D_b$$

where C, M, Y = cyan, magenta, yellow dye densities:

D_r, D_g, D_b = red, green, and blue integral densities; and

A_n = conversion coefficients.

2.2.9 Tone Reproduction Theory

The reproduction of a scene by a black-and-white material is accomplished through a scale of tones in the imagery which represent brightness, luminance or radiance of the original scene. The ability of a film or photographic system to reproduce the original scene is evaluated by the use of tone reproduction theory.

The terms "brightness", "luminance", and "radiance" are not to be used interchangeably. "Brightness", as defined by the Optical Society of America, is the magnitude of the subjective sensation produced by light, and "luminance" is the magnitude of the stimulus expressed in photometric units. "Radiance" is an unweighted measure of absolute energy. The relation between brightness and luminance is complex. Brightness is determined only subjectively or by means of psychological scaling procedures. Luminances are determined empirically by weighting the radiance with the luminosity function of the eye.

The theory of tone reproduction is divided into objective and subjective areas. Objective tone reproduction deals with the photographic reproduction of scene luminance or radiance of each point in a photograph corresponding to points in the original scene. The subjective theory of tone reproduction deals with the reproduction of brightness of the original scene. If the brightness of the reproduction match the original, then the photograph, illuminant, and associated surroundings are considered to be a combination that provides exact reproduction of the visual effect. Objective tone reproduction formed the basis for the extension of the reproduction theory into subjective measurements. Subjective tone reproduction considers the state of the eye as it adapts itself to viewing the reproduction of the original scene. The treatment of subjective tone reproduction becomes a psychological problem. In order to keep the treatment of tone reproduction reasonable, the subjective aspects will not be dealt with in a detailed manner here. The selected references which appear in the bibliography, however, do contain a thorough treatment of the theory.

It should also be pointed out that the treatment of tone reproduction of color materials is a relatively unexplored area. Equivalent neutral density *(END)* measurements of color materials have been made for tone reproduction studies in the past. The important aspect of color reproduction is not considered with this approach. The use of colorimetric densities have been used to relate the original scene color reproduction. The color differences are usually very large, but the reproduction may be considered acceptable. For example, the reproduction of true flesh color is not at all desirable. The state-of-the-art has not advanced sufficiently far in color and color tone reproduction of original scenes to warrant other than discussion on objective tone reproduction.

The objective tone reproduction characteristics are displayed in a graphical manner. This display is in the form of a curve of densities of the reproduction versus the logarithm of the corresponding scene luminances or radiances. Exact tone reproduction is represented by a straight-line plot having a slope of 1.0. Exact luminance reproduction is possible by viewing the photograph at the proper illuminance level. If more or less light is used on the photograph, the luminances in the reproduction would be proportional rather than equal to the original scene.

The tone reproduction diagram developed by L. A. JONES (1921) represents a way to graphically display the various component effects on the final tone reproduction. The elements which should be considered in a tone reproduction are many. The following is a brief listing of the major components: Original camera film characteristics, flare characteristic of the camera system, flare introduced by the atmosphere for scenes over long path lengths, characteristics of the duplicating film, and the flare introduced by the equipment to print the original film.

The preferred objective tone reproduction characteristic does not necessarily agree with an exact match of the ideal reproduction. The difference is due to the manner in which the reproductions are normally viewed. Reflection-type prints are commonly viewed under a room illuminance of 30–100 foot-candles. Sheet film transparency illuminators that have a luminance of 300–500 foot-lamberts are normally used in rooms where the general light level is the same as is used for reflection print viewing. Slide transparencies and motion picture films are viewed in a dark room with a screen luminance of 10–50 foot-lamberts.

Figure 14 is a tone reproduction plot of a preferred reflection print, motion picture and slide, and a transparency on a viewer. Notice that the plot of the reflection print falls below the ideal reproduction line. According to the plot, the preferred print is lighter than the original scene. Also, the midtone reproduction

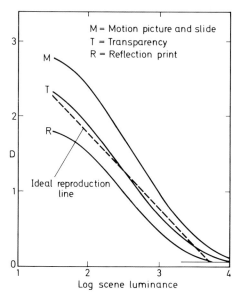

Fig. 14. Preferred objective tone reproduction

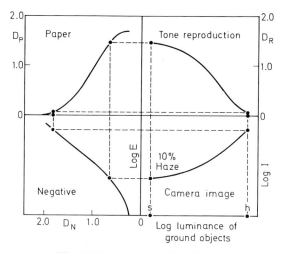

Fig. 15. Tone reproduction diagram

has a slope of 1.1–1.2. Both of these aspects are due to the reduced light levels under which the reflection print is viewed. If the print had higher densities, it would appear too dark. The transparency fits the ideal curve more closely than the reflection print. This is again true because of the viewing conditions. Due to the combination of the dark surroundings and low light levels at which slides and motion pictures are viewed, their preferred reproductions are quite different from the other two curves.

As a matter of demonstration, Fig. 15 shows how the various components of the tone reproduction cycle can be cascaded graphically. The camera image quadrant shows the original scene range and the effect a 10-percent haze level has on the camera image. The resulting radiances are then transferred to the D-log$E$ curve of the negative material. The densities of the negative material then modulate the exposure to the paper material. In the tone reproduction quadrant, the densities of the paper are then compared to the original scene radiances.

The tone reproduction diagram is useful to determine the negative characteristics necessary to approximate a preferred reproduction or ideal reproduction. The characteristics of both the duplicating material and the atmosphere are nearly constant. Most of the available manipulation is in the negative quadrant. By plotting backward through the duplicating material and the desired tone reproduction curve, a new negative curve can be generated. This new curve will then provide the sensitometric properties best suited for that reproduction.

2.2.10 Spectral Sensitivity

Knowledge of the spectral response characteristics of a sensitive material enables prediction of its response to a variety of exposing sources. Determination of spectral sensitivity of photographic materials involves exposing the film specimen to a spectrum of known energy distribution. Analysis of resultant density data permits describing the response as a function of wavelength of the exposing radiation.

The spectral sensitivity of photographic materials is usually defined as the reciprocal of the energy at the given wavelength required to produce a given response upon development. The data for spectral sensitivity are presented either in graphical or tabular form. The spectral sensitivity function does change shape, depending upon the response level from which it is derived. Generally, spectral sensitivity data supplied by the manufacturer of photographic materials are referenced to a density level of 1.00 above base-plus-fog density. As supplied, these data reflect the general trend for the material and are adequate for comparing one film with another for selection purposes.

However, when one is conducting remote sensing experiments involving precision photographic radiometry, consideration must be given to determination of the spectral response characteristics at the density levels of interest. This may be ignored, if the spectral energy distribution of the exposing light is known, by determining the characteristic curve of the material under the conditions of the experiment.

The spectral sensitivity function is determined experimentally by exposing film samples to nonselectively modulated monochromatic radiation and generating a family of D-log$E$ curves from which the absolute sensitivity is calculated. There are several pieces of commercially available equipment, usually termed "spectral sensitometers" or "wedge spectrographs", to aid in generating the spectral sensitivity data. All are equipped with either prisms or a diffraction grating to separate the radiation of a lamp into its spectrum. Exposures of varying wavelength radiation are then made sequentially across a film sample. Modulation of the exposure is usually accomplished by means of a rapidly rotating sector wheel. Since there is very little energy available in the monochromatic exposures, long exposure times are necessary to produce usable data. The long exposure times lead to reciprocity law anomalies. The effect of the reciprocity law anomalies can be considered constant (independent of wavelength); therefore, as long as the exposure time is held constant throughout the spectrum, a constant correction factor can be applied to adjust the data for reciprocity law failure.

In all cases, density measurements used in spectral sensitivity studies must conform to standard practices. Black-and-white density measurement techniques are described in USASI Standard number PH 2.19-1959. Density measurement of color materials becomes quite involved, since the independent response of each of the three dye-forming layers must be determined. In order to extract these data from a tripack material, one must resort to conversion of integral densities to analytical densities. Since some assumptions that are necessary for this conversion may or may not be valid, the analytical densities may be subject to error. Most spectral sensitivity data supplied by the color film manufactures are generated from special single-layer coatings of each of the three dye-forming emulsions. The type of density measurement normally used for color spectral sensitivity determinations is the equivalent neutral density *(END)* of the dye.

2.2.11 Resolution

Resolution, expressed as the minimum detectable spatial frequency, has been used to describe the ability of a photographic material to record retrievable detail. The measurement of resolution is, in part, subjective and involves the complexities of visual perception. Visual perception of detail is dependent upon the spatial frequency (or size) and modulation (or contrast) of the image being viewed. Combined with the complexities of visual perception, the microscopic structure of the photographic image is nonhomogeneous. The image (signal) is superimposed on a background of grain (noise). The noise tends to obscure the signal, making detection of the signal more difficult. Higher signal-to-noise ratios (signal/noise) result in images that are more easily perceived. Ideally, one desires to record an image under conditions which permit a high signal-to-noise ratio. Evaluation of areospace photographs is often done at or near the resolution limit where the signal-to-noise ratio is low. It is desirable to describe the ability of a photographic material to record information under the adverse conditions of "aerial reconnaissance" as practiced in remote sensing.

A generalized Modulation/Resolution *(M/R)* curve (see Fig. 16) represents the threshold modulation the image of a tribar target must have in order to be resolvable with a given photographic material.

The usual method used to determine resolution involves imaging a target consisting of a series of progressively smaller three-bar elements which have equally wide lines and spaces. Resolution is then expressed as the maximum spatial frequency, in lines-per-millimeter, at which a representative sample of human observers can preceive all three bars. Resolution is dependent upon the target contrast, which is defined as the exposure modulation M as follows:

$$M = (E_{max} - E_{min})/(E_{max} + E_{min}) .$$

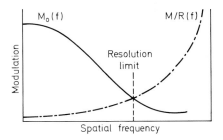

Fig. 16. Resolvable information, between the curves of the output modulation function and the input modulation-versus-resolution curve. The limiting resolution is defined by the intersection of the system modulation transfer function and the M/R curve

E_{max} and E_{min} are the maximum and minimum exposures received by the film from the light and dark bars. In general, resolution increases nonlinearly with modulation. This relationship is presented as a plot of input modulation versus resolution and is termed an "M/R curve." The term "AIM curve" (Aerial Image Modulation detectability) is used when the targets have been optically printed onto the film. Since the M/R relationship is generally nonlinear, it is apparent that comparison of the capabilities of two films on the basis of "one number" resolutions is dangerous, if not meaningless.

The data are normally generated in the same manner for both black-and-white and color films. In the case of color films, an attempt is made to obtain neutral images of the target. One must be careful when interpreting resolution data for color films, since the resolution usually varies with image color.

The area above the M/R curve in Fig. 16 represents the region of tribar target detectibility. The area under the curve represents images of tribar targets that cannot be resolved by the photographic material. Resolution data presented in the form of an M/R curve are useful in predicting the limiting resolution of a system/film combination. The limiting resolution is determined by the intersection of the modulation transfer function (MTF) of the system and the M/R curve.

Given the modulation transfer function for a system and the input modulation distribution, one can determine the output modulation by the equation

$$M_I(f) \times T_S(f) = M_O(f)$$

where M_I = input modulation,
T_S = system MTF, and
M_O = output modulation.

The resolution limit of the system/input combination is reached at the point where output modulation is equal to the M/R, e.e., when the following conditions are met:

$$M_O(f) = M/R(f) = M_I(f) \times T_S(f).$$

This is shown graphically in Fig. 16.

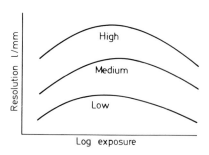

Fig. 17. Resolution vs log exposure for three levels of input modulation

The area bounded by $M_o(f)$ and $M/R(f)$ left of the resolution limit represents resolvable information.

The M/R curve is to be regarded as an absolute threshold. If insufficient input modulation is present at a given spatial frequency, that frequency will not be resolved. There are some philosophical and some physical inconsistencies connected with the M/R concept. If one simply accepts this curve as a useful threshold reflecting a likely trend, then its use can be accepted in good conscience.

Generation of the M/R curve is accomplished experimentally by exposing the test film to images of tribar targets of varying modulation and by determining the maximum resolution obtainable at each modulation level. The response variable (resolution) varies from a low value, through a maximum, then a lower value as the exposure level is increased. Consequently, an exposure series must be run at each modulation level.

Figure 17 illustrates the dependence of the response (resolution) of a photographic material upon exposure for three general cases of input modulation. The resolution versus log exposure curves are usually quadratic in nature, passing through a maximum. The maximum resolution may be determined analytically by applying the first derivative of the least-squares model,

$$\hat{y} = a + bx + cx^2 + e$$

which has been fit to the experimental data; y is the predicted resolution for the input log exposure x, and e represents the error involved in applying the model. The first derivative is set equal to zero to determine the log exposure at which the maximum occurs:

$$b + 2cx = 0$$

thus, the maximum resolution is predicted by the least-squares model when

$$x = -b/2c \, .$$

The maximum predicted resolution is

$$\hat{y} = a - (b^2/4c)$$

where a, b, and c are the least-squares coefficients.

The least-squares technique is applied to experimental data gathered at several input modulations to obtain an array of maximum resolution. Input modulation is plotted versus maximum resolution to determine the M/R curve.

The exposure levels to be used in the experiment are chosen to cover the dynamic range of the test film so that the difference between successive exposures is a constant logarithmic value. For example, suppose 10 exposure levels are to be used with a film having a scale of 2.0 log exposure units. The difference between successive exposures will be

$$\Delta \log E = 2.0/10 = 0.20 \,.$$

2.2.12 Granularity [3]

Root-mean-square (RMS) granularity is a measurement that may be used to a certain extent to compare the imaging capabilities of photographic materials. The parameter computed as RMS granularity is the standard deviation of densities σ_D computed from a stationary record of independent density samples. The density record is generated by scanning a uniformly exposed and processed film sample with a microdensitometer or Micro-Analyzer. The deviations of density are caused by the granular structure of the film. RMS granularity generally increases nonlinearly with density level, and is dependent upon the area of the scanning aperture. When the scanning aperture is large in comparison with the grain size of the film, RMS granularity is approximately inversely proportional to the square root of the aperture area.

Since RMS granularity varies with both density level and aperture area, it is apparent that some set of standard conditions should be adopted. An industry-wide standard has not been set; however, at Eastman Kodak, a major film manufacturer has chosen an average net density of 1.00 and a circular aperture 48 nm in diameter as its standard conditions. These conditions produce granularity measurements which closely correlate with the visual sensation of "graininess" when the film is viewed at 12X magnification.

The record from which granularity calculations are made must consist of independent samples; it must also be stationary. Independence of sampling is achieved by generating the record such that no two adjacent densities are determined from overlapping areas on the film. The conditions for stationarity are achieved if the average density level does not vary within the record. If nonstationarity exists in the data, the measurement of RMS granularity will be exaggerated; however, a small linear trend in density can be tolerated if σ_D does not vary significantly along the record.

Computation of σ_D and the test statistics for stationarity are simplified by normalizing the record by subtracting the least-squares line-of-best-fit from the record. Then, an adequate assurance of stationarity is the chi-squared goodness-of-fit test.

---

[3] ERB, Keith; Kelch, David; Contract No. NAS 9-7003, Prepared for Photographic Technology Laboratory, M.S.C., Houston, Texas, prepared by Data Corporation, May 1960.

Assurance of a stationary record requires that the film sample chosen for granularity measurement be uniform in density and free of mottle, dust, dirt and scratches. The exposure must not be made through a carbon or other types of step tablet which are inherently grainy, since the grain pattern of the stip tablet will be overprinted on the grain of the test sample.

RMS granularity generally varies for small sample sizes. It is therefore necessary to use two to three thousand densities to calculate reliable RMS granularity values. The calculations are accomplished as follows:

1. *Least-Squares Line-of-Best-Fit:*

$$\hat{D} = b_0 + b_1 X_i + \text{error}$$

where

$$b_0 = (\textstyle\sum_{i=1}^{n} D_i - b_1 \sum_{i=1}^{n} X_i)/n$$
$$b_1 = [n\textstyle\sum_{i=1}^{n} X_i D_i - (\sum_{i=1}^{n} X_i)(\sum_{i=1}^{n} D_i)]/[n\sum_{i=1}^{n} X_i^2 - (\sum_{i=1}^{n} X_i)^2]$$

and D_i is the i-th density sample,
    X_i is the position of the i-th sample,
    $\hat{D}_i$ is the average density at the i-th position as predicted by the line-of-best-fit,
    n is the sample size, and
    b_0, b_1 are the least-squares coefficients of the line-of-best-fit.

2. *Normalized Record:*

$$D_i' = D_i - \hat{D}_i$$

where D_i' is the value of the i-th sample after normalization.

3. *Chi-Squared Goodness-of-Fit:* The record is divided into N equal segments consisting of n/N samples each. The average of each segment is then computed and placed in appropriate positions in a frequency histogram of J cells. The frequency of occurrence (F_j) of the segment averages within the j-th cell is compared with the expected frequency E_j for that cell, based on the Gaussian distribution having a standard deviation equal to that of the segment averages and a mean of zero. The test statistic x^2 is calculated by the equation:

$$x^2 = \textstyle\sum_{j=1}^{J} [(F_j - E_j)^2 / E_j].$$

The test statistic is then compared with the tabulated chi-squared distribution for $(J-1)$ degrees of freedom. If the test statistic exceeds the tabulated value, the record cannot be considered stationary. The tabulated distribution can be found in most elementary statistics text books.

4. *RMS Granularity:* Having verified that the record of density samples is stationary, *RMS* granularity is calculated as follows:

$$RMS \text{ granularity} = \{[\textstyle\sum_{i=1}^{n}(D_i')^2 - (\sum_{i=1}^{n} D_i')^2/n]/(n-1)\}^{\frac{1}{2}}.$$

The simplified equations most seen in the literature is

$$RMS \text{ granularity} = \sigma \big/ \sqrt{a}$$

where $\sigma = \sqrt{\sum \frac{(\bar{x} - x)^2}{n}}$.

and $\bar{x}$ = average density,
        x = deviation of density from average,
        a = aperture diameter in nanometers.

2.2.13 Dimensional Stability and Mechanical Properties

The final film parameter that must be considered is dimensional stability.

Dimensional stability of photographic materials is the ability of the material to retain its size and shape under varying environmental conditions. Either permanent or temporary dimensional changes are caused by processing, aging and changes in temperature and humidity. The changes affect the relative distances between image points and introduce error in mensuration of target relationships. The dimensional changes can be ignored in cases in which a high degree of mensuration accuracy and precision is not required; however, they may be significant in the mapping and astronomy fields. Although it is possible to define a portion of the changes under carefully controlled conditions, it is difficult and inconvenient to correct mensuration data for dimensional changes with a high degree of certainty. When precision measurement of photographic images is a major consideration, it is advisable to choose a film processing high dimensional stability.

The dimensional stability of photographic materials is governed primarily by the base or support material. Common support materials in increasing order of dimensional stability are paper, cellulose triacetate, cellulose triacetate butyrate, polystyrene, polyethylene, polyesters and glass.

Humidity and thermal changes of dimensions most often are temporary or reversible changes and may exhibit hysteresis. Dimensional changes caused by processing and aging are usually permanent or irreversible. All the previously mentioned effects are due to inherent characteristics of the photographic film. Additional causes of dimensional changes are traceable to physical handling of the material. The support material may become stretched by tension applied during manufacture, and by transport in the camera and processing machine. Mechanical relaxation of the base may occur during storage.

Methods for testing dimensional stability characteristics are given in USA Standard PH 1.32-1959, "Methods for Determining the Dimensional Change Characteristics of Photographic Films and Papers". This and other useful USA Standards are listed in the bibliography.

The mechanical properties of photographic films are also primarily determined by the base material. Unless extreme environmental conditions or mechanical stress and strain are anticipated during use of the film, comparison of mechanical properties can usually be ignored when choosing one film over another.

The mechanical properties must be considered when low temperature and humidity conditions will be encountered. The high extremes of temperature and humidity affect the sensitometric properties of the film adversely before the mechanical properties are affected detrimentally. Methods for testing brittleness, scratch resistance, and curl are given in USA standards PH 1.31-1965, PH 1.37-1963, and PH 1.29-1958, respectively.

2.3 Cameras and Films

2.3.1 Cameras (General)

Aerospace photography generally means taking photographs of the ground from the air or space. It has two main purposes: to provide maps and economic information.

Cartographic cameras tend to resemble one another and show little departure from the basic essentials, while reconnaissance cameras take varied forms according to their role. Special cameras are also required for missile recording.

2.3.2 Aerial Cameras—Common Features

All aerospace cameras used in Remote Sensing must be very robust in construction to stand relatively rough handling, and they are generally much larger and heavier than ground cameras. Success has been achieved in a few novel applications where by two or more ground cameras (Mitchell Vintens, Hasselblads, Leicas, etc.) were assembled on a common mount, boresighted and synchronized for multispectral photography. While some experiments in Remote Sensing have used small hand held cameras, the majority of aerospace cameras are designed for fixed mounting in the platform.

In principle, the aerospace camera is a rigid fixed-focus high precision instrument designed to take roll films. Operation may be manual but is usually electrically powered, with fast recycling rates and varying degress of automation. See Fig. 18 which portrays a specially designed camera for aerospace photography. The Earth Terrain Camera (ETC) as shown in Fig. 18 was designed to take high resolution photographs of the Earth's surface while mounted in the Orbital Workshop (OWS) of Skylab. With the Camera Control Box, astronauts operated the ETC in an automatic mode for overlapping topographic coverage, or in a manual mode for single frame capability. Other controls provided versatility while indicators provided essential monitoring.

The Earth Terrain Camera typifies a RS camera in that it consists of five main assemblies; a body, a magazine, a lens cone and camera mount, a Camera Control Box, and a window assembly. The camera body, in addition to housing the magazine, contains the focal plane shutter assembly. The magazine holds (61 m) 200 feet of thin base film for extended topographic coverage. The lens cone and camera mount contains the major electronics and the 0.450 m (18 in.), f/4.0 lens which provides high photographic resolution. Interchangeable lens filters are provided to optimize the exposure on various films. To enhance image quality,

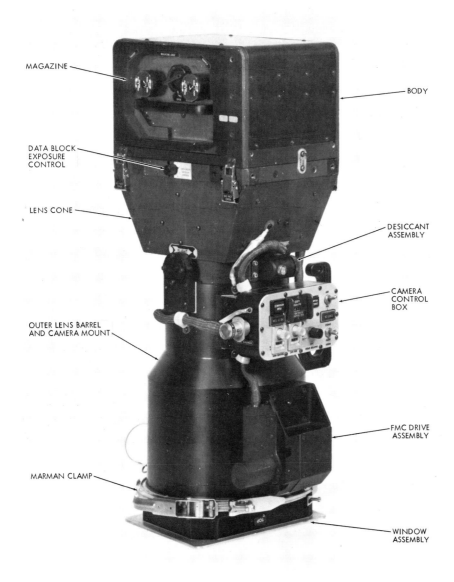

MAGAZINE

BODY

DATA BLOCK
EXPOSURE
CONTROL

LENS CONE

DESICCANT
ASSEMBLY

CAMERA
CONTROL
BOX

OUTER LENS BARREL
AND CAMERA MOUNT

FMC DRIVE
ASSEMBLY

MARMAN CLAMP

WINDOW
ASSEMBLY

Fig. 18. Earth Terrain Camera (ETC) system (front view). (With permission of Actron,
Division of McDonnell Douglas Corporation, Monrovia, USA)

the camera is rocked in the line of flight at controlled rates for forward motion compensation.

Formats in America range from 5.72×5.72 cms (2-$1/4 \times 2$-$1/4$ ins.) to 45.72×45.72 cms (18×18 ins.), the most common being 22.86×22.86 cms (9×9 ins.) and in Europe, 18×18 cms (7×7 ins.). Films, both perforated and unperforated, from 70 mm to 469.9 mm (2.75 to 18.5 ins.) wide are used. These are carried in interchangeable magazines holding up to 304.8 m (1000 feet).

Fast panchromatic emulsions, are most common, used in conjunction with a minus blue (yellow, green or red) filter, but slower types and infra-red are employed for special purposes. Color is also proving increasingly valuable. These emulsions are usually coated on dimensionally stable "topographic" bases (usually polyester film).

2.3.3 Optical System

Aerospace photographs must provide the best possible definition, so highly corrected lenses are mandatory. In particular, the lens must be free from distor-.tion as far as possible. Since the subject is always at infinity, no focusing is required. The lens is fixed permanently at one end of a cone or optical unit; a register glass if used defines the focal plane being likewise fixed permanently at the other end. A survey camera is required to measure angles between objects on the ground with the precision of a theodolite, and for this to be possible, the optical unit must be rigid with no possibility of relative movement between lens and focal plane once assembled.

The register glass against which the film is located by the pressure plate is a feature of British survey cameras, but American and Continental designers prefer a different system in which the film is held by vacuum against a rigid flat metal plate while pressed down against a metal frame defining the focal plane. The film should be held flat to 0.0127 mm (0.0005 in.) or better because small departures from flatness which have a negligible effect on definition cause appreciable errors in the geometry required to make accurate maps. Remote Sensing Applications do not require this precision in geometry. However, in most cases, available survey cameras are used rather than design and fabricate tailor made cameras which may cost just as much as the more precise versions.

The cone of a good survey (or RS) camera is an alloy casting, the locating faces for lens mount and focal plane being machined flat and parallel within fine and specified limits. Diaphragms are often cast in the inside of the cone to keep flare light to a minimum.

In those cameras-fitted with a register glass, this is often engraved with an accurately calibrated grid. By reference to the grid pattern on each picture, it is possible to correct for small residual distortions which occur even with modern polyester base film.

2.3.4 Calibration

Modern aerospace lenses have strikingly low distortion figures; in the best examples no image point is displaced from its theoretical position by more than 0.01 mm.

Aircraft cameras used in Remote Sensing are very commonly used at heights of 4500–6100 m (15000–20000 feet). A 3 m (10 feet) object is reproduced by a 153 mm (6 in.) lens as a patch 0.10 mm (0.004 in.) diameter. At a 12200 m (40000 foot) flying height, the same lens records the same target with a 0.05 mm

(0.002 in.) diameter. Thus the need for good resolution can be appreciated, not only for "normal" flying altitudes, but more so for the "synoptic" coverage obtained by high altitude and space photography and highly favored by many disciplinarians. Formerly the lens resolution was the limiting factor, even with the somewhat grainy high-speed film normally used. With the best modern lenses this is no longer true, except in the outer zones of the field.

2.3.5 Image Motion

The exposure time used depends on the speed of the platform and the scale of the photograph. Focal plane or diaphragm shutters may be fitted, the former speeded up to 1/2000 sec. Focal plane shutters introduce distortion however, due to the platform's forward motion or movement during the shutter blind transit.

Diaphragm shutters circumvent this distortion and are therefore preferable. They are basically similar to those in ground cameras but much larger. The stresses involved in fast recycling over long periods make efficiency and reliability difficult to attain, especially at high speed and large diameters. Exposure times range between 1/250–1/800 sec. Certain types of electrically driven disc shutters overcome these difficulties and also give shorter exposures.

Fast shutter speeds freeze image movement on the film. But where this is not feasible, the alternative is to move the film in step with the image. This is the principle of Image Motion Compensation (IMC), embodied in some aerospace cameras or separately in the magazines. The film is moved during exposure, usually by the transport mechanism or by moving the film aperature or pressure platen. Speed of image movement (v) is calculated from the platform's height (H) and speed (V) and lens focal length (f).

According to the equation: $$\frac{v}{V} = \frac{f}{H}.$$

The image thus stays stationary on the film and longer exposures are possible.

2.3.6 Hand Held Cameras

Most of what is written today about Remote Sensing unfortunately deals with large platforms, large expensive cameras, large investments for data reduction, etc. Attendance at symposia in RS has revealed a deluge of requests from disciplinarians who did not have large organizational backing, government sponsorship, nor large amounts of private monies. For these dedicated scientists, hand held cameras are a boon and any platform from a helicopter to a commercial airplane is used.

Hand held cameras have few if any of the above mentioned refinements. Used mainly for taking oblique "pin points" of single features, they are designed as compactly as possible, protectively cowled and with double hand grips and trigger release for maximum steadiness. Other characteristics are fast lever-wind operation and open frame sights. Maximum film size is about 175 mm (7 ins.), the commonest being 70 mm (2.8 in.), usually held in magazines.

Some fixed type "air" cameras are convertible to hand operation simply by fitting a special mount. Others may be rigid, fixed-focus versions of standard ground cameras and may even take plates.

Most precision ground cameras can still be used for this type of photography, provided they are protected against slipstream and rough handling. Miniature cameras, although reported in the literature, are less suitable, owing to the inherent low information capacity of 35 mm film.

2.3.7 Special Cameras

Although it is desirable for a RS camera to do many jobs, the search for greater speed and coverage as well as increased specialization has led to a wide variety of types and techniques.

A logical development of IMC is the continuous strip camera. This moves the film continuously past a fixed focal plane slit at the calculated image motion compensation speed, making an unbroken record of the ground along the film. Both shutter and intermittent film movement are thus eliminated, but a fairly complex control system is needed. Another development is the panoramic camera, in which a rotating prism scans from horizon to horizon across the flight path and records on film moving past a fixed slit in step with it.

To save time, or transmit in real time, in flight processing is sometimes used. In this type of system the film is carried in special magazines in which it is

Fig. 19. I²S Mark I multispectral camera with 150 mm lenses. (With permission of International Imaging System, Mountain View, California, USA)

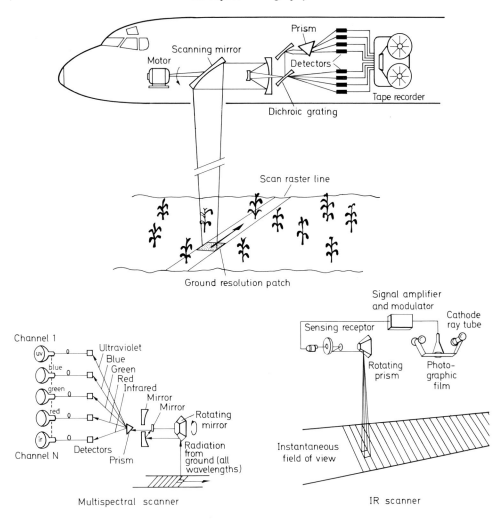

Fig. 20. Schematic of multispectral and infrared scanners

developed and stabilized immediately after exposure. The chemicals are held in thixotropic pastes or on rolls of absorbent material. The developed negative can be examined on landing or may be scanned by on board television camera (Lunar Orbiter Program) and relayed to the ground. Alternatively, the television camera (ERTS' RBV) or a scanner (ERTS' MSS) may itself record the ground, and the transmitted pictures being then photographed.

Another extension of this principle is used with satellite cameras, which record on magnetic, electrostatic or thermoplastic tape. This is played back through a ground data reduction instrument which records in photographic form for visual analysis or digital form for computer analysis of the data.

Figures 19 and 20 as well as Tables 4 through 6 are self-explanatory.

Table 4. Specifications of I^2S multispectral camera

Specifications	
Type, Camera Body	K-22 (I^2S modified configuration)
Cycle Rate	2 seconds
Power Requirement	24 volts d.c., 5 amps. Electrical Connector: AN3102A-16-11S
Shutter	Focal plane: One supplied, specify either Type A or B. Type A 1/150–1/350 second; Type B 1/350–1/800 second
Lenses	4 each, Schneider Xenotar, 150 mm or 100 mm, f/2.8. Specify type desired
Filters	1 each #47B (blue); #57A (green); #25 (red); #88A (infrared). IR blocking filters are included with #47B, #57A and #25 filters
Masked Format	9 by 9 inches with four each 3.5 by 3.5 inch images
Size, Overall	15 by 15 by 8 inches (without magazine) 15 by 15 by 15 inches (with A-5A magazine)
Weight, Camera	35.4 pounds (without A-5A magazine) 57 pounds (with A-5A magazine, no film)
Mounting	Trunnions compatible with types A-8, A-11, A-11A, A-27, and A-27A camera mounts
Intervalometer	Compatible with B-3B, B-5A or equivalent. Electrical Connector: AN3102A-16S-6P
Magazine Type	A-5A (standard configuration)
Focal Plane	Vacuum: 1 inch of mercury, at $1^1/_2$ cubic feet per minute airflow, at all operating altitudes
Weight, Magazine	21.6 pounds (with empty take-up spool)
Film Capacity	$9^1/_2$ inches wide by 250 feet, 4.0 mil Estar base, Kodak Spec. No. 952
Number of Exposures	300 multispectral sets per 250 foot film load

Standard filter configuration for
I^2S multispectral camera, Mark I

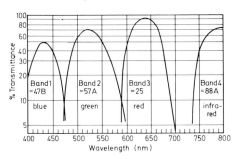

Typical Wratten filters used in the I^2S Multispectral Camera, Mark I. Interference filters block the longer wavelength transmission of the blue, green and red filters. Other filters and passbands may be selected, such as an 89B on Band 4 which has 11 percent transmission at 700 nm and opens rapidly into the infrared region. Other filter combinations available on request.

A/C Mounting Requirements

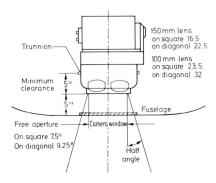

Designation	Special characteristics	Input film type	Model
Mark I	150 mm Lenses	$9^1/_2''$ Roll	00500-01
Mark I	100 mm Lenses	$9^1/_2''$ Roll	00500-02
Mark II	150 mm Lens	2-70 mm Rolls	00500-03
Mark II	100 mm Lens	2-70 mm Rolls	00500-04

Also specify type A (1/150–1/350 sec.) or type B (1/350–1/800 sec.) shutter.

2.3.8 Film (General)

Photographic film is a radiation sensitized material in the form of an emulsion coated on a flexible support (base)—i.e., celluloid plastic, etc. Film is inherently sensitive to the shorter wavelengths of EM spectrum. With the addition of sensitizing dyes during manufacture, photographic films are made sensitive over a broader wavelength interval than the eye.

Humans can see or perceive only a small portion of the spectrum. Light, therefore, is that form of EM energy which can stimulate vision, Fig. 21. The normal observer sees wavelengths of 400–700 nm and as Fig. 22 shows, the sensitivity is not uniform.

Film, therefore, as currently manufactured for Remote Sensing are typified by Fig. 23, which shows the spectral sensitivities of four commonly used aerial films.

These aerial films, used for Remote Sensing, differ from the films normally used for ground photography. The two main important differences are: (1) they have extended red sensitivity which allows short wavelength cut-off filters to be used to eliminate much of the scattered (haze) light without sacrificing speed, and (2) they are normally developed to a higher gamma, i.e., they are more "contrasty" in order to counteract the reduced contrast apparent from high altitudes caused by the scatter of image light out of the recorded beam and extraneous light being scattered into the recorded beam. Figure 24 show that light scatter is a function of wavelength, and reported by MIDDLETON (1952) and CURCIO (1961) to be proportional to $\lambda_{-1.3+0.6}$. Therefore, as the figures show, the scattered light is greater at the shorter wavelengths.

Table 5. Universal cartographic camera specifications

A. Camera body
    1. Name   Camera, Cartographic, Universal (KC-)
    2. Mission: High and low Altitude, with or without IMC
    3. Lens: Geocon I, 6″ f/5.6
    4. Format: 9 inch × 9 inch
    5. Lens filters: Clear vignetting, No. 12 vignetting, No. 25
                                   vignetting
    6. Mode of operation: Pulse with and without IMC:
                                   Auto cycle with IMC.
    7. IMC   0.1 in. per sec. to 2.0 inches per sec.
    8. Cycling interval: 5-second pulse mode for non-IMC
                                   2-second maximum for auto cycle with IMC.
    9. Exposure control: Automatic self contained for ASA 20 thru 200;
                                   Manual.
   10. Exposure time: 1/50 to 1/750 second effective, continuously
                                   variable by AEC; Manual −1/50, 1/100, 1/200,
                                   1/400 second.
   11. Aperture: f/5.6 to f/16, continuously variable by AEC;
                                   Manual −f/5.6, f/8, f/11, f/16.
   12. Resolution: 40 lines/mm AWAR on EK 2402 High Contrast.
   13. Distortion: Radial − 10 microns maximum with no more
                                   than ±5 microns asymmetry from the average,
                                   on 12 semi-diagonals at 30° increments when
                                   referred to the Point of Symmetry.
   14. Fiducials: 4 primary, one at the center of each side;
                                   4 secondary, one at each corner.
   15. Reseau: Peripheral, 8 points, one at mid-point between
                                   each fiducial.
   16. Data:
        a) Binary Code:
        b) Cone Temperature:
        c) Platen Temperature:
        d) Camera Serial No.:
        e) Lens CFL, Serial No.:

B. Camera magazine (without IMC) type KC-( ) A
    1. Reseau: 5 × 5 on 2-inch centers*
    2. Film capacity: 9$\frac{1}{2}$ inch wide, 390 ft. 440 exposures
    3. Platen flatness: ±.0001 inch
    4. Mode of operation: Pulse, without IMC
    5. Data:
        a) Platen serial no.
        b) Vacuum indicator

C. Camera magazine (with IMC) type KC-( ) B
    1. Reseau: 5 × 5 on 2-inch centers*
    2. Film capacity: 9$\frac{1}{2}$ inch wide, 500 ft., 550 exposures
    3. Platen flatness: ±.0001 inch
    4. Mode of operation: Pulse with IMC; Autocycle with IMC
    5. IMC rate: 0.1 to 2.0 inch per second
    6. Data:
        a) Platen serial no.
        b) Vacuum indicator

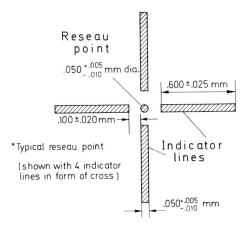

Reseau
point
.050 $^{+.005}_{-.010}$ mm dia.

.600 ±.025 mm

.100±.020mm

*Typical reseau point

(shown with 4 indicator
lines in form of cross)

Indicator
lines

.050$^{+.005}_{-.010}$ mm

Panchromatic film is sensitive to about the same part of the EM spectrum as the human eye, however, its response is more uniform, i.e., it is more nearly equal in sensitivity throughout its effective range than is the eye.

Color film is comprised of a sandwich of three emulsions on one substrate. Each of these emulsions has its primary sensitivity in a different part of the visible spectrum from the other two.

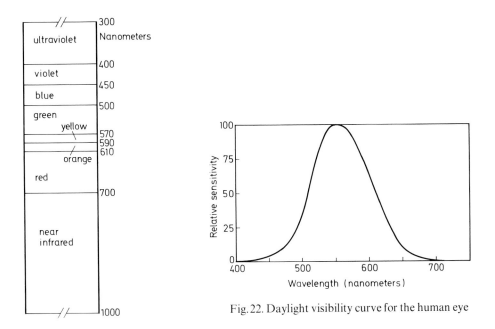

Fig. 22. Daylight visibility curve for the human eye

Fig. 21. Visible and photographic portion of the electromagnetic spectrum. This small portion is approximately that part which can be utilized in photography; only the part between roughly 400 and 700 nanometers is visible to the human eye

Table 6. European mapping cameras

Camera type	Use	Basic design	Cycling time	Focal length	Lens apert.	Format size	Shutter type	Shutter speed (sec.)	Weight (lbs.)
RC-8	Mapping	Frame	3.5 sec.	6″	f/3.6	9″ × 9″	Between lens	1/100–1/700 continuously variable	190
RC-9	Mapping	Frame	3.5 sec.	3.5″	f/5.6	9″ × 9″	Between lens	1/300	180
Galileo Santoni Mod. VI	Mapping	Frame	2.5 sec.	6″	f/5.6	9″ × 9″	Between lens	1/125 1/200 1/300 1/400	80
SOM Film Camera	Mapping	Frame	4 sec.	125 mm	f/6.2	18 × 18 cm	Between lens	1/75 and 1/100; 1/125 and 1/150	75
SOM Plate Camera 125 mm	Mapping	Frame	4 sec	125 cm	f/6.2	18 × 18 cm	Between lens	1/75 and 1/100; 1/125 and 1/150	200
SOM Plate Camera 210 mm	Mapping	Frame		210 mm	f/5	18 × 18 cm	Rotating between lens	1/75 and 1/100; 1/150 and 1/200 and	220
SOM Plate Camera 300 mm	Mapping	Frame		300 mm	f/5	18 × 18 cm	Rotating between lens	1/75 and 1/100; 1/150 and 1/200 and	225
RMK A 15/23	Mapping	Frame	2.5 sec minim.	6″	f/5.6	9″ × 9″	Between lens	1/100– 1/1000	57 kg
RMK AR 15/23	Triangulation	Frame	2.5 sec. minim.	6″	f/5.6	9″ × 9″	Between lens	1/100– 1/1000	59 kg
RMK A 30/23	Mapping	Frame	2.5 sec. minim.	12″	f/5.6	9″ × 9″	Between lens	1/100– 1/1000	57 kg
RMK A 21/23	Mapping	Frame	2.5 sec. minim.	8¼″	f/5.6	9″ × 9″	Between lens	1/100– 1/1000	55 kg
RMK A 60/23	Mapping	Frame	2.5 sec. minim.	24″	f/6.3	9″ × 9″	Between lens	1/100– 1/1000	54 kg
RMK 21/18	Mapping	Frame	2.5 sec. minim	8¼″	f/4	7″ × 7″	Between lens	1/100– 1/1000	35 kg
RMK 11.5/18	Mapping	Frame	2.5 sec. minim.	4¼″	f/5.6	7″ × 7″	Between lens	1/100– 1/1000	35 kg

The three spectral sensitivities are blue, green and red. During process, the positive portrayal of the scene by the three emulsions are colors which are complementary to the sensitivity of the particular emulsion layer, namely yellow, magenta and cyan. Since the color film "sees" the same spectrum as the eye, the portrayal with this film is conveniently called natural color (Fig. 25).

Infrared film is sensitive in the visible and near infrared portion of the spectrum to 900 nm. Under special cryogenic conditions, the spectral sensitivity can be extended to 1200 nm. Most IR B/W films also show a relative lack of green sensitivity which manufacturers hope to overcome.

Color IR film is analogous in structure to natural color film. However, the three emulsion layers are sensitive to green, red and near infrared. Notice that in

Table 6 (continued)

Film mag.	Film size	Film Lond	Film spools Core Dia.	Flange Dia.	Mount used	Current status	Mf's Name	General remarks	Camera type
Attached	$9^1/_2''$	200'	Standard	Standard	Integral	Commercially available	Wild Heerbrugg	Aviogon lens	RC-8
Integ. w/ detachable cassettes	$9^1/_2''$	200'	Standard	Standard	Integral	Commercially available	Wild Heerbrugg	Super aviogon lens	RC-9
Separate	$9^1/_2''$	180' 360'	$2^1/_2''$	$6^1/_4''$	Special	Std.	Officine Galileo, Florence	Orthogon	Galileo Santoni Mod. VI
Separate	19 cm	50 m	4.35 cm	13.5 cm			SOM, Paris	Aquilor	SOM Film Camera
Separate plate magazine	19 cm	96 Plates	Plate thickness	1.7 mm			SOM, Paris	Aquilor	SOM Plate Camera 125 mm
Separate plate magazine	19 cm	96 Plates	Plate thickness	1.7 mm			SOM, Paris	Orthor (orthosopic)	SOM Plate Camera 210 mm
Separate plate magazine	19 cm	96 Plates	Plate thickness	1.7 mm			SOM, Paris	Orthor (orthoscopic)	SOM Plate Camera 300 mm
Separate	$9^1/_2''$	120 m	55 mm	168 mm	AS II	24 V– 28 V	Zeiss, Oberkochen	Used for pan, color, infrared	RMK A 15/23
Separate	$9^1/_2''$	120 m	55 mm	168 mm	AS II	24 V– 28 V	Zeiss, Oberkochen	Used for pan, color, infrared	RMK AR 15/23
Separate	$9^1/_2''$	120 m	55 mm	168 mm	AS II	24 V– 28 V	Zeiss, Oberkochen	Used for pan, color, infrared	RMK A 30/23
Separate	$9^1/_2''$	120 m	55 mm	168 mm	AS II	24 V– 28 V	Zeiss, Oberkochen	Used for pan, color, infrared	RMK A 21/23
Separate	$9^1/_2''$	120 m	55 mm	168 mm	AS II	24 V– 28 V	Zeiss, Oberkochen	Used for pan, color, infrared	RMK A 60/23
Separate	$7^1/_2''$	120 m	55 mm	168 mm	AS I	24 V– 28 V	Zeiss, Oberkochen		RMK 21/18
Separate	$7^1/_2''$	120 m	55 mm	168 mm	AS I	24 V– 28 V	Zeiss, Oberkochen		RMK 11.5/18

both color films there is a considerable overlap between the sensitivities of the adjacent layers or emulsions. (See Fig. 26.)

The Return Beam Vidicons (RBV) on ERTS-1 approximate the spectral regions used by Kodak for their Infrared Aerographic film. The spectral sensitivities for the three filtered vidicons are shown in Fig. 27.

The Multispectral Scanner (MSS) aboard ERTS-1 operates in four spectral bands. Three of these bands approximately coincide with those of color IR film. The additional longer wavelength presently exceeds normal color IR film (Fig. 28).

2.3.9 Aerial Photographic Materials

Technological progress in emulsion making has made possible aerial film that cover a broad range of sensitivity, speed and definition. Furthermore, the strength

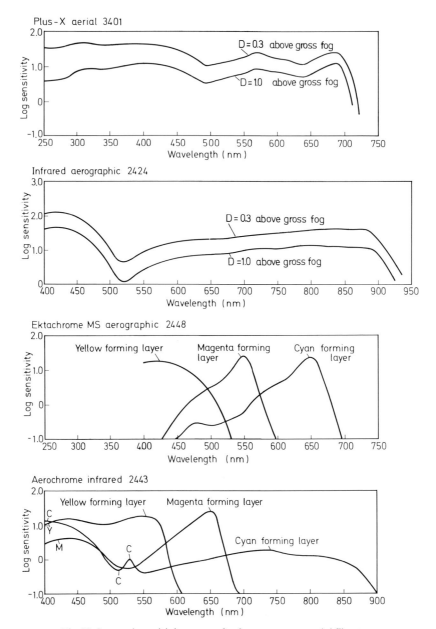

Fig. 23. Spectral sensitivity curves for four common aerial film types

and rigidity of recently adopted polyester support favors the manufacture of aerial films that are appreciably thinner than those available in the past. This has resulted in greater film lengths for a given roll diameter or, conversely, in a reduction in weight and size for a stipulated footage. The potential capability of in-flight processing and trends toward high temperature, rapid processing tech-

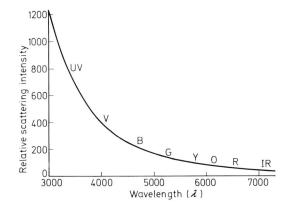

Fig. 24. Rayleigh, or small particle scattering, named for the man who first explained it theoretically, is shown graphically above. Rayleigh scattering exists when the particle size is smaller than the wavelength. Scattering intensity is proportional to the square of the volume of the scattering particle. For a given particle size, the relation can be stated as follows: The intensity of scattering increases with the inverse fourth power of the wavelength, as shown in the plot above. The blueness of the sky, caused by scattering of sunlight, is explained by Rayleigh's law. The short wavelengths which are blue light are more strongly scattered than are red. (From JENKENS and WHITE, 1957)

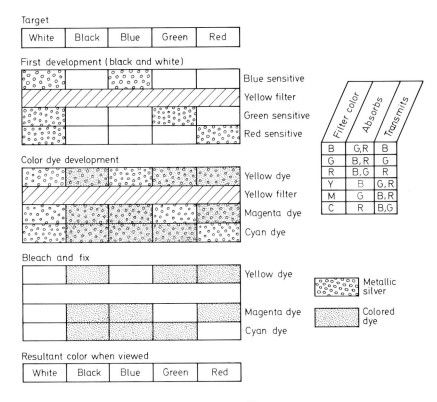

Fig. 25. Color film

A. E. SALERNO

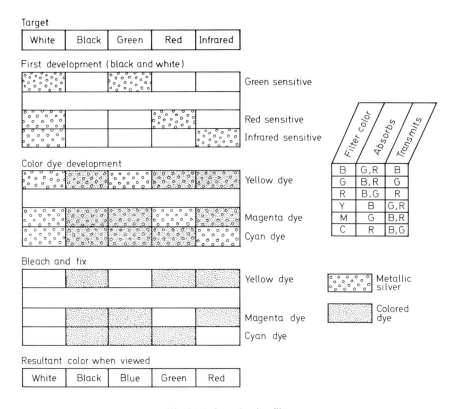

Target

| White | Black | Green | Red | Infrared |

First development (black and white)

Green sensitive

Red sensitive
Infrared sensitive

Filter color	Absorbs	Transmits
B	G,R	B
G	B,R	G
R	B,G	R
Y	B	G,R
M	G	B,R
C	R	B,G

Color dye development

Yellow dye

Magenta dye
Cyan dye

Bleach and fix

Yellow dye

Magenta dye
Cyan dye

Metallic silver

Colored dye

Resultant color when viewed

| White | Black | Blue | Green | Red |

Fig. 26. Infrared color film

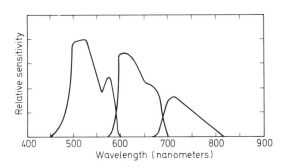

Fig. 27. Spectral bands of ERTS I return beam vidicons

niques can serve to present to the photo interpreter the information he needs in exceptionally short times.

The situation, being as dynamic as it is, makes it difficult for hardbound copy to keep abreast of the times. The changing environment very frequently evokes a new material while, at the same time, rendering another obsolete. In this respect, the particular films that are used in aerial photography can never be guaranteed a

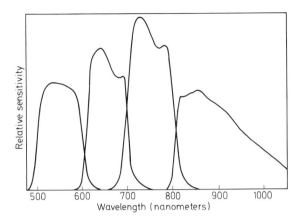

Fig. 28. Spectral bands of ERTS I multispectral scanners

predetermined tenure. Product lists fluctuate as response to the changing times, calling for the introduction of new materials as others phase out. However, despite the frequency of revisions in product listings, the specific product changes are most often small and are readily tolerated by existing systems.

As already mentioned, aerospace acquision films fall into two characteristic types: black and white or color. They may be subdivided further into panchromatic, infrared or color sensitized; high, medium and slow speed. Color products may be classified as negative or reversal types. These broad classes provide the outline for typical presentations of the available aerial films by the manufacturers. Table 7 shows all the available color films made throughout the world. This list was compiled by Mr. Norman L. Fritz of Eastman Kodak for the Color Committee of the American Society of Photogrammetry.

2.3.10 Black-and-white Camera Films

All Kodak black-and-white camera films feature ESTAR polyester base. This base is manufactured in four nominal thicknesses:

ESTAR Ultra-Thin Base	1.5 mils
ESTAR Thin Base	2.5 mils
ESTAR Base	4.0 mils
ESTAR Thick Base	7.0 mils.

Films coated on the two thinner ESTAR Bases are intended primarily for aerial reconnaissance, while those on 4.0 and 7.0 mil ESTAR Base find use in mapping and related applications.

The speed values of these products are currently expressed as *Aerial Film Speed* in accordance with ANSI Standard PH 2.34-1969, "A Method for Determining the Speeds of Monochrome Photographic Negative Films for Aerial Photography". Prior to the existence of this ANSI Standard, a fractional gradient system, known as *Aerial Exposure Index*, prevailed. In the latter system, the speed

Table 7. Available color aerial photographic materials

Film (camera)	Film number	Description	Type	Base Thick (mils)	Type	Backing
AGFA AVIPHOT Color	CN	Unmasked color negative film of medium speed	Neg	4	Acetate	Clear Gel
AGFA AVIPHOT Chrome	CT	Color-reversal highspeed film	Pos	4	Acetate	Clear Gel
GAF 1000 Blue Insensitive	7575	High-speed, special-purpose, color-reversal film for use over water to record bottom information	Pos	4	Polyester	
GAF 200 Aerial Color	7230	High-speed, color-reversal film for aerial photography	Pos	4.5	Acetate	None
KODAK AERO-COLOR Negative (ESTAR Base)	2445	High-speed, color-negative film for mapping and reconnaissance	Neg	4	Polyester	Fast Drying
KODAK EKTA-CHROME MS AEROGRAPHIC (ESTAR Base)	2448	Color reversal film for low-to-medium altitude aerial mapping and reconnaissance	Pos	4	Polyester	Fast Drying
KODAK AERO-CHROME Infrared (ESTAR Base)	2443	Falme-color for vegetation surveys and camouflage detection	Pos	4	Polyester	Fast Drying[c]
Kodak AERO-CHROME Infrared (ESTAR Thin Base)	3443	False-color; high spool capacity and minimum storage space	Pos	2.5	Polyester	Clear Gel
KODAK Aerial Color (ESTAR Thin Base)	SO-242	Slow-speed, high resolution film for high-altitude reconnaissance	Pos	2.5	Polyester	Clear Gel
KODAK Aerial Color (ESTAR Ultra-Thin Base)	SO-253	Similar to SO-242; ultra-thin base for maximum spool capacity	Pos	1.5	Polyester	Clear Gel
KODAK EKTA-CHROMS EF AEROGRAPHIC (ESTAR Base)	SO-307	High-speed, color-reversal film for aerial mapping and reconnaissance	Pos	4	Polyester	Fast Drying

Film (Duplicating and Printing)

Film (camera)	Film number	Description	Type	Base Thick (mils)	Type	Backing
AGFA DUPLICHROME	D 13	Cut sheet color-reversal duplicating film	Pos	7	Acetate	Gel
CIBACHROME Transparent type D	661	Dye-Bleach reversal film for duplicating aerial transparencies	Pos	7	Polyester	Dyed Gel

Table 7 (continued)

Illuminant	Aerial[a] exposure index	Effective[a] aerial-film speed	Resolving[a] power		Granu-larity[a]	Process[f]	Literature[f]
			1000:1	1.0:1			
Daylight			80	40		CN	Agfa-Gevaert Trade Publications
Daylight			80	40		CU	
Daylight	100		80	40	.045	AR-2C	
Daylight	20		100	50	.025	AR-1C or AR-2C	
Daylight	32	100	80	40	13	AERO-NEG Color	M-70, M-29
Daylight	6	32	80	40	12	EA-5	M-29
Daylight	10[d]	40[d]	63	32	17	EA-5	M-50, M-29
Daylight	10[d]	40[d]	63	32	17	EA-5	M-69, M-29
Daylight	2	6	200	100	11	ME-4 (Modified), EA-5	M-74
Daylight	2	6	200	100	11	ME-4 (Modified), EA-5	M-74
Daylight	12[e]	64	80	40	13	EA-5	M-78
	Exposure						
3200 °K	10 sec at 2 ft CNDL[g]		100			P-20 or P-10	CDPO1-6/72

Table 7 (continued)

Film (camera)	Film number	Description	Type	Base Thick (mils)	Type	Backing
GAF Color Duplicating	7470	Low-contrast, reversal color film for duplicating color transparencies	Pos	4.5	Acetate	None
KODAK EKTA-CHROME AERO-GRAPHIC Dupli-cating (ESTAR Base)	SO-360	Low-contrast, color-reversal film for making duplicate transparencies	Pos	4	Polyester	Fast Drying
KODAK EKTA-COLOR Print (ESTAR Thick Base)	4100	Cut sheets for making color aerial dia-positives from 2445 film	Neg	7	Polyester	Clear Gel

Reflection Print Materials

Film (camera)	Film number	Description	Type	Base Thick (mils)	Type	Backing
GAF (PRINTON®) Reversal Color on White Opaque Base	6410	Reversal color print material for making opaque reflection color prints from reversal color transparencies	Pos	7	Acetate	Clear Gel
KODAK EKTA-COLOR 37 RC Paper		Rolls and sheets for making paper prints from color negatives	Neg		RC Paper	
KODAK EKTA-CHROME RC Paper		Rolls and sheets for makimg paper prints from color positive transparencies	Pos		RC Paper	

b Aerial Film Speeds and Effective Aerial Film Speeds are for use with the new Kodak Aerial Exposure Computer, Kodak Publication No. R-10 (12/70 edition) in determining the correct camera exposure for aerial (air-to-ground) photography. Aerial Film Speeds and Effective Aerial Film Speeds are not equivalent to, and should not be confused with, conventional film speeds used in pictorial photography. Aerial Film Speed as defined in ANSI Standard PH2.34-1969 for black-and-white negative aerial film is $3/2 E$; E is the exposure (in meter-candle-seconds) at the point on the characteristic curve where the density is 0.3 above gross fog. Effective Aerial Film Speeds are values determined for films such as infrared-sensitive and color, and films not processed under the conditions specified in the

point is very sensitive to slight variations in the shape of the characteristic curve; as such, it is difficult to determine accurately and repeatably.

Aerial Exposure Index (AEI) for black-and-white negative aerial films is defined as $1/2 E$, where E is the exposure (in meter-candle-seconds) at the point on

Table 7 (continued)

Illuminant	Aerial[a] exposure index	Effective[a] aerial-film speed	Resolving[a] power 1000:1	1.0:1	Granu-larity[a]	Process[f]	Literature[f]
3000 °K	0.3 sec at 3 ft CNDL[g]		120	55	.022	AR-1C or AR-2C	
3000 °K	3 sec at 3 ft CNDL[g]		125	63	8	EA-5	M-72
3200 °K	10 sec at 1 ft CNDL[g]		125	63	16	AERO-NEC Color C-22 (Modified)	
Daylight	0.15 sec at 3 ft CNDL[g]		60	43		AR-1 (Modified)	
						EKTAPRINT 3 Chemicals	
						EA-5 (Modified)	

Standard. All the speed values given on this chart were obtained by rounding the calculated values to the nearest $\sqrt{2}$ step (equivalent to 1/3 f-stop).

[c] Some coations of this film have been manufactured with a clear-gel backing and a nominal total film thickness of 5.1 mils.

[d] With a Kodak Wratten Filter No. 12.

[e] SO-397 Film can be exposed at 2 times the normal Effective Aerial Film Speed with very little loss in image quality if processing is "pushed" or "forced" in the Kodak

[e] SO-397 Film can be exposed at 2 times the normal Effective Aerial Film Speed with very little loss in image quality if processing is "pushed" or "forced" in the Kodak Ektachrome RT Processor, Model 1811 (modified).

[f] Information on processing, and the literature listed can be obtained from the manufactures.

[g] Suggested trial exposure. Generally several filters are required, the specific ones depending on the product.

the toe of the characteristic curve where the slope is equal to 0.6 of the measured gamma with the processing conditions always specified.

The speed-rating method specified by the ANSI Standard is a fixed-density speed point system that is quick and easy. *Aerial Film Speed (AFS)* for black-

and-white negative aerial films is defined as $3/2\,E$, where E is the exposure (in meter-candle-seconds) at the point on the characteristic curve where the density is 0.3 above gross fog. The processing conditions are strictly defined in the Standard.

Effective Aerial Film Speed (EAFS) is the term prescribed for black-and-white aerial films processed under conditions other than that specified in the Standard. Although different processing conditions may be used, the basic speed criterion is still the same. The processing conditions must be specified along with the Effective Aerial Film Speed value.

2.3.11 Color Camera Films

Color films are usually comprised of several layers of which at least three are photosensitive emulsions. In normal color films, each emulsion is sensitized to a different primary color and is capable of producing a separate record of the brightness of a single primary color. In a color negative, the colors of the combined images will be complementary to those of the original scene. In reversal color products, the integrated images are a close visual reproduction of the subject.

Technically, any portion of the spectrum to which photographic materials are sensitive can be recorded in a color film if the individual emulsion layers are correspondingly sensitized. Furthermore, the color developed in a particular emulsion layer need bear no relationship to the color of light to which it is sensitive. Such a false-sensitized product can be used to emphasize differences between objects which are visually quite similar. To illustrate, Kodak Aerochrome Infrared Film 2443 (ESTAR Base) emphasizes differences in infrared reflection. Its three emulsion layers are sensitized respectively to green, red, and infrared radiation. They are exposed through a yellow filter to absorb the blue radiation to which all three layers are also sensitive. When the film is processed as recommended, the green-sensitive layer is developed to a yellow positive image; the red-sensitive layer to a magenta positive image; and the infrared-sensitive layer to a cyan positive image. The colors blue, green, and red appear in the final transparency, but with false rendition; the blue has resulted from green exposure, green from red exposure and red from infrared exposure.

Infrared-sensitive color films were originally designed for camouflage detection. A more recent use lies in forest survey work wherein diseased foliage can be distinguished from healthy foliage by interpretation of the infrared reflectance of the foliage as recorded on the film.

Although ANSI Standard PH 2.34-1969 does not apply to color or infrared-sensitive films, *Effective Aerial Film Speeds* can be assigned to such products. Speed values for Kodak color and infrared-sensitive films are determined empirically by the Kodak Aerial Photography Laboratory from actual controlled flight tests with aerial cameras.

2.4 Photographic Attributes of Earth Sciences

The Earth sciences have been broadly divided into several disciplines: agriculture, land-use, demography, mapping, eutriphication, geography, hydrology, geology, and oceanography. Scientists in each of these areas are interested in photographically recording different types of target characteristics under a variety of environmental conditions. The characteristics under scrutiny, for the most part, are those which enable distinction between phenomena of interest. As a result, there is a particular set of photographic parameters which can satisfy a given situation. Unfortunately, not all desired photographic parameters can be found nor incorporated into one material or system; it is therefore necessary to compromise when designing a single photo-optical system. An efficient compromise can be reached if the Earth science target characteristics can be ranked in order of importance. An efficient system can be designed if more than one sensor, system or mission is carried out.

In general, each aerospace photograph has different types of information for each of the Earth scientists. Target characteristics, therefore, must be considered for the specific areas within a scene from which the data are being collected. For example, a hydrologist may require a differentiation of moisture content. An agriculturalist may be interested in diseased crop detection within a growing field. A geologist may search for different minerals; whereas a land manager is trying to determine how the land was scarred from mining these minerals. For these reasons, there can be more than one type of target within the same scene, and the same target can have different interpretations according to its photographic attributes.

Earth science targets can be divided into four main categories of photographic characteristics. These are:

1. Spectral distribution.
2. Physical size and shape.
3. Illuminance.
4. Environment.

The interrelationships of these target characteristics also play an important role and must be considered for optimum systems design. For example, the contrast as well as the illumination level becomes increasingly important as the object size becomes smaller.

Before discussing a few photographic parameters which scientists must be aware of in order to design a meaningful aerospace remote sensing mission, a few examples of specific attributes in a selected group of Earth Resources Discipline will be described, such as in Agriculture, Geography and Geology and Oceanography.

2.4.1 Agriculture

The most important target characteristic for agricultural investigations through remote sensing is the spectral distribution of the target. Conventional color, false color (IR) and multispectral photography are used in this discipline. The best target discrimination is generally obtained from the false color and multispectral photography. Accurate color reproduction of the target area from high altitude or space using conventional color materials has not materialized, since the blue-green spectral region is scattered and attenuated. (See later Sections for further discussion.)

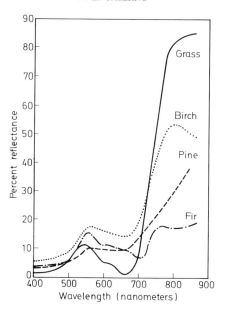

Fig. 29. Spectral reflectances of foliage

The agriculturist is interested in discriminating between types of vegetation; the ability to distinguish healthy plants from diseased plants; and/or the condition of the vegetation during its growth cycle. The best spectral region for detecting these differences in vegetation is the near-infrared region, where the subject contrast is high. The spectral reflectance curves for various typical foliage types indicate why the near-IR region is so useful (see Fig. 29). Note the differences in the curves in the green spectral band (500–600 nm). These differences are slight when compared to the differences in the near-IR (700–850 nm). It should also be pointed out that the curves in Fig. 29 are typical for a particular temporal situation. The reflectance curves of the vegetation will change in the visible and infrared wavelengths due to water content, climate, soil type, fertilization, etc. Each of these variables and its synergistic effects are of main interest to an agriculturalist.

Soil moisture can also be detected in the same manner as vegetation discrimination. Water has a high absorption coefficient to the infrared radiation. When added to soil (not standing water), water will also absorb the visible radiation and the soil appears darker.

Agriculturalists are also interested in detection of vegetation diseases, decay and damage assessment. The near-infrared spectral region has been found to be a good indication of vegetation vigor. Healthy vegetation has a very high reflectance of the IR energy.

Neither size and shape of the target nor illumination level have very much meaning in remote sensing and photointerpretation in agriculture. As mentioned, the spectral differences play the most important part.

By photographing vegetation from the air and more so from space, the slight spectral differences within a vegetation type become obscured. The very powerful discriminating band of IR radiation is absorbed by water vapor. As the intervening atmosphere becomes longer in the optical path, so too does the absorption of IR by water vapor increase. Therefore, the agriculturalists are instantly concerned with and pay close attention to humidity levels. At high humidity levels and increasing altitude, a marked reduction in the infrared exposure of the film is experienced unless exposure control is predicted and utilized.

In summary, the uses of aerospace photography in agriculture are:

1. Vegetation density determination.
2. Soil series and moisture content.
3. Grass/brush/timberland interface detection.
4. Plant species identification.
5. Damage assessment and prediction.
6. Irrigation water location.

The spectral distribution of the target is the most important parameter, and due concern is customary as the platform goes higher. As the spectral differences are attenuated, the scientists may resort to use of secondary target characteristics.

2.4.2 Geography and Geology

Aerospace photography has proven useful in geographic and geologic investigations in remote sensing.

The targets of interest in geography and geology fall into the general classification of surface features of the Earth, which can be further subdivided into cultural and terrain features.

Cultural features encompass such areas as land-use, transportation linkage and population movements. The surveying of cultural features utilizes a straight-forward approach involving assessment of man-made patterns such as roads, buildings etc. These cultural features, for the most part, are characterized by the presence of simple geometrical figures such as straight lines, or straight line combinations (parallels, perpendiculars, angles, etc.) and circles.

Cultural features are easily detected in aerospace photographs. A detection problem which can arise is due to the object size in relation to its contrast to the background. For example, two objects of the same size, but at different contrast with the background, are not detected to the same degree, and possibly one of them will not be discerned at all. The higher contrast object will be detected more easily.

The ability of the camera film to record objects of various sizes and contrast can be related to the resolution of the photographic system at different image modulation. From this relationship, the threshsold resolution is determined for a given contrast level or modulation. The scale at which a scene must be recorded can be predicted if the target size and contrast are known.

Natural terrain features, on the other hand, do not exhibit the well-defined geometric patterns found in cultural features. Terrain features, however, do have

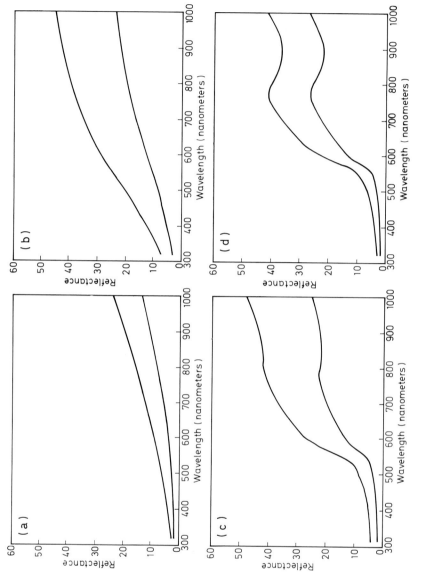

Fig. 30a–d. Spectral reflectance of various characteristic soil types; (a) soil type I; (b) soil type II; (c) soil type III; (d) soil type IV

some patterns. The most obvious of these patterns are the linear features of fault lines and stratification of sediment. In other aspects, such as boundaries between different soil types which are not sharply defined, terrain features are not always well defined.

True color and false color photographic techniques have been used to improve the discrimination of terrain features. The degree of improvement of one color system over the other is dependent upon the spectral characteristics of the terrain features. For example, H.R. Condit (1969), in "Spectral Reflectance of Soil and Sand" has characterized soil types into three general classes of spectral reflectances. Figure 30a shows a soil type with a gradual increase in reflectance with an increase in wavelength, whereas graph b shows a much greater reflectance change with wavelength. Another type of soil, as graph c shows, has a sharp break at 575 nm. This type soil was found to be most typical according to the study.

If conventional or true color film were used for remote sensing these three types of soils, good discrimination and detection is achieved. However, soil type IV, as seen from Fig. 30d, shows no difference than soil type III of Fig. 30c in the visible region. A greater difference does exist between these two soil types in the near-infrared region. For applications of this sort, increased discrimination can be obtained only in the near-IR region by using multispectral photographs or color IR film which will record images in this very powerful band.

Most natural colors of nature fall within a fairly restricted range. Figure 31 shows a C.I.E. Coordinate System according to the 1931 Standard Observer. This figure was taken from HENDLEY and HECHT (1949). Within the C.I.E. Coordinate

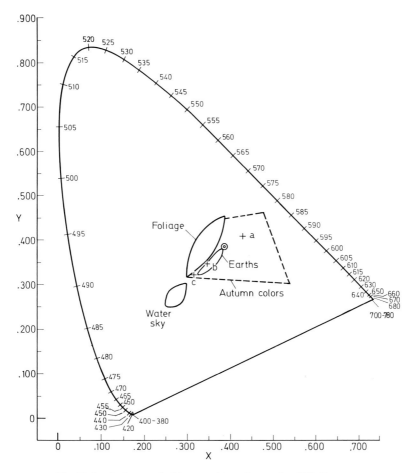

Fig. 31. Loci of natural objects and terrains on the CIE diagram

System is also plotted the gamut of colors for foliage, water and sky, earths and a large gamut of colors as seen in Autumn months of the year.

The color groups indicated are to a certain extent disaturated and limited in hue. From this C.I.E. plot of possible natural colors it can be seen that the infrared band adds another dimension and can greatly improve discrimination between natural features. Rumors abound that experiments in sensitizing materials beyond the present limit of 900 nm are currently under way, and much success is evident as well as predicted in further assisting discrimination of Earth features through spectral reflectances.

It is as difficult to distinguish the one most important target characteristic for geographic and geologic disciplines as it is for any other discipline. For cultural features, the relationship between the physical size and the contrast of the target is usually the most important consideration. Spectral differences or distribution and contrast of terrain features is more important for geologists.

2.4.3 Hydrology and Oceanography

Although quite distinctive, both of these disciplines, hydrology and oceanography, share a common objective. Each of these disciplines must contend with either photographing the water or through the water. An oceanographer may be interested in sea state coverage which requires no water penetration, or the detection of bottom detail which requires maximum water penetration.

These applications of remote sensing will serve as models for describing the types of problems associated with light attenuation by water. This attenuation of light by water varies, depending upon mineral and algae content. Generally, with increasing amounts of "pollutants", the peak transmission will shift toward the longer wavelengths; on the other hand, the overall transmission will decrease. Figure 32 shows the effect of varying concentrations of chlorophyll. Figure 33 shows the attenuation of light by actual samples of sea water. Both figures exhibit the shift toward longer wavelengths.

From these figures, it can be seen that the red spectral region from 600 to 700 nm is strongly attenuated by water. This spectral band is best for obtaining sea state information. Conversely, if water penetration is required, this spectral band should be avoided and the blue-green region of peak transmission should be utilized. Again from Fig. 32 it can be seen that these regions are variable depending on the pollutants in the water. In most cases, the spectral band width for water penetration should be confined to 450–600 nm.

As indicated earlier, an exposure increase is required when exposing for water penetration. Due to the absorption of light by pollutants, the amount of energy return from the bottom is greatly reduced, and aerospace photographic missions are further burdened by requiring the detection of the very subtle changes in color and detail. These low radiance values are further masked by light scatter (which see), which is explained below under haze and its effect. There is also present another contrast reduction due to the scatter of light at the air/water interface. In order to enhance these subtle changes or low contrast of bottom detail and other hydrological/oceanographic features, a high gamma photographic process should be used. The high gamma is achieved by selecting a high-contrast film and chem-

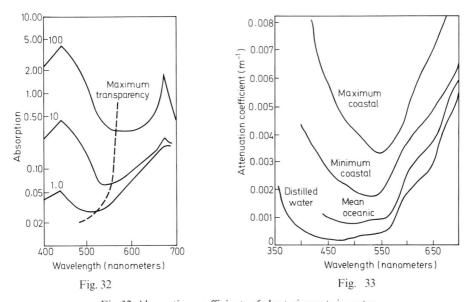

Fig. 32

Fig. 33

Fig. 32. Absorption coefficients of plant pigments in water

Fig. 33. Attenuation coefficient versus wavelength for distilled water and sea water

istry combination. This high gamma will give larger density differences in re-corded image for a given set of exposures and radiance levels.

Scientists in hydrology and oceanography are also interested in the broad patterns of effluent flow, in river currents, in subsurface structure detection or in sea color analysis. For these applications high-resolution is not required, al-though there will be some applications where fine detail is required for shoal and coastal mapping. In these cases there must be a compromise between the required film speed and resolution, since films with high speed have low resolution, and high resolution films normally have slow speeds.

2.4.4 Anomalies, Atmospheric

The preceding sections have shown that remote sensing of Earth resources using photographic media depends upon recording radiance values. Most Earth features, especially those encountered in Oceanography, will exhibit low radiance and little contrast with its background. Some oceanographic features such as currents, water masses, upwelling and water fertility must be recorded in the blue-green part of the optical spectrum where Rayleigh and Mie scattering of the light by the atmosphere is most pronounced. The atmospheric scatter of this spectral region degrades image contrast and distorts the scene contrast ratios. This situa-tion affects all sensors, whether photographic or electronic.

2.4.4.1 Haze Effects on Scene Contrast

The effect of atmospheric haze (air luminance) is to compress the contrast of low reflectance/radiance subjects more than the "brighter" or highly reflective surfaces. For example, if a dark substance has 5% reflectance, a light subject reflects 90% of the incident energy and the atmosphere contributes 50% back scatter in air luminance, then the relationship between dark and light becomes 0.55–1.40 or 1:2.6. In the absence of haze the ratio would be 1:18.

2.4.4.2 Haze and Light Attenuation Curves

TUPPER and NELSON (1955) developed a set of curves showing the relationship between various percentages of haze and the effect of compressing the luminance of ground subjects relative to image illuminance. The reader will note from Fig. 34 that the curves cross over at an intermediate point. This is due to the fact that atmospheric gases and suspended particles absorb some of the incident energy as well as reflect and scatter, thereby causing loss of potential image illumination from a beam of irradiance.

In an extension of this technique into multispectral remote sensing, Ross measured reflectances from a six-step grey tablet on the ground, imaged in aerial photographs for comparison of the relative effects of a given haze condition as separately affecting the blue, green, and red regions. The data is plotted in Fig. 35. Note that Ross also recorded data using an I²S Green filter which is similar to but not identical to a Wratten 57 A filter.

The curves in Fig. 36, derived from ELTERMAN's (1970) Vertical Attenuation Model, approximate the order of atmospheric attenuation of potential image forming energy to be expected during average atmospheric conditions suitable for "good" remote sensing. Little scattering or haze is found in the near-infrared regions, and absorption is the main effect of attenuation. As the wavelengths become shorter, both absorption and scatter increase. However, the scatter which

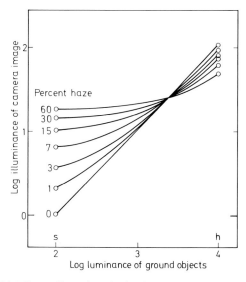

Fig. 34. Effects of haze in reducing image contrast in camera

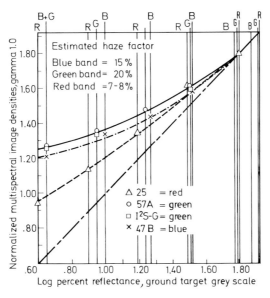

Fig. 35. Measurements of ground target greyscale steps in multispectral images, normalized. (After Ross, 1969)

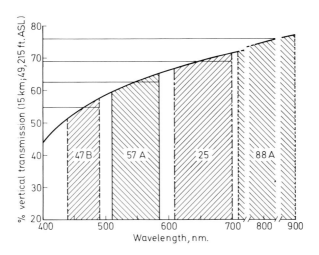

Fig. 36. Order of atmospheric absorption with vertical transmission through 15 km and one air mass, meterological range 13 km (Derived from ELTERMAN). The passbands of Wratten filters commonly used in multispectral cameras are shown

is a function of wavelength, and reported by MIDDLETON (1952) and CURCIO (1961) to be proportional to $\lambda - 1.3 \pm 0.6$, far exceeds the absorption thereby contributing to an increase in the apparent scene radiance. If the blue-green atmospheric attenuation is accounted for a set of curves (Fig. 37), may be produced which express the effects of various amounts of haze in distorting the tone

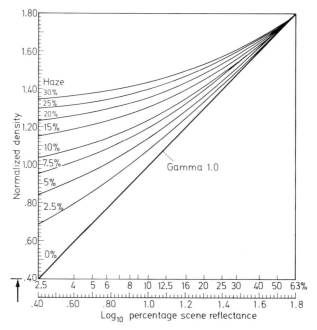

Fig. 37. Curves illustrating effects of different amounts of atmospheric haze in distorting the linear reproduction of scene radiances

reproduction curves of scene radiances. These curves originate at a 63% point in the peak 550 nm response of ERTS-1, MSS 4 (green) band; but to a first approximation can be considered useable for remote sensing in the 480–650 nm region.

The equation for the basic curves of Fig. 37 is

$\log I = \log (L + kL_{max})$ where
I = image illuminance
L = luminance of corresponding ground target
k = haze factor (e.g., $k = 0.1 = 10\%$)
L_{max} = maximum ground luminance.

While these curves can predict what happens to the image recorded through various amounts of haze, it is necessary to know the extent of the haze factor if a scene contrast correction is to be made during the reproduction phase; or if, in the case of multispectral imagery, density data are to be reduced to percent reflectance or radiance figures for target analysis and interpretation.

It is possible to determine the magnitude of the haze factor and its effect on the tone reproduction distortion by analyzing the light and dark portions of a scene. The accuracy with which this may be done depends on how accurately scene reflectances are known, and the sensitometry of the image acquisition and reproduction system. For example, BROCK has demonstrated that atmospheric transmission and haze effects can be found by measuring the image densities of two ground areas of known brightness, one of low and the other of high reflectances. The highly reflective area is essentially a reference of total transmission, whereas the density of the less reflective area is used to measure haze effects. When the image density data are normalized to the known ground subject values, it is found that the (negative) density of the highly reflective area is less than it should be because of atmospheric attenuation of the image light during transmission, while the area of low reflectance reaches a higher density than predicted because haze light has increased the apparent radiance, despite the inherent attenuation losses.

2.4.5 Factors[4] Affecting the Photo Image

The appearance of a plant on RS photography at a given scale is influenced by several factors (1) its morphological characteristics, (2) its spectral reflectance, (3) its seasonal state or maturity, (4) sun angle, (5) spectral sensitivity of sensor (film) and (6) spectral transmittance of lenses, filters, atmosphere etc. It is difficult to discuss each of these factors separately since the effect or relative importance of each is often determined by the particular combination of all factors. In many cases, the synergistic effect is more predominant than any single effect can be mathematically or intractively predicted. With these limitations in mind, the following brief summary of the effects of each factor is discussed.

1. The Morphological Characteristics of a plant include size, shape, orientation of the leaves, height and foliage density. These factors affect the textural appearance of the plant and also determine the proportion of the reflectance from leaf surface and stems, other plant parts and underlying litter and soil. As applied to the photographic image, "texture" may be defined as the frequency of tone change within the image.

2. The Spectral Reflectance Characteristics of a healthy green plant are governed primarily by its foliage. In a broader sense, however, a plant's reflectance represents an integration of the reflectance contribution from its leaves, stems, flowering parts and fruits and may even be influenced by spectral characteristics of underlying litter and soil.

3. Seasonal State or Maturity may change the spectral reflectance of a plant by altering the proportion or controlling the presence or absence of the plant parts mentioned above. Flowering, for instance, usually occurs during a short period of each year; deciduous plants, of course, are leafless during a period of each year and present a dramatically different appearance than during the leafy stage. Furthermore, the autumn coloration of some deciduous plants is distinctive enough to be an identifying characteristic. Because of these factors, the "spectral signature" of a plant species may vary during the season and during its life.

4. The Sun's Angle affects the amount of shadow present in the photographic image and the depth to which the plant canopy is illuminated. When the sun is at the zenith, the illumination of the plant interior and the underlying soil and litter is at its maximum. As the angle from the zenith increases, the proportion of reflectance from the outer leaves increases while that from the stems, litter and soil decreases as they are immersed in shadow.

---

[4] Some of these factors also affect the interpretation of other RS features. The discussion on plants is used as an illustration.

5. *Spectral Sensitivity of Film.* The "tonal" response with which plant is recorded on an aerial photograph is a function of the intensity of the reflected energy returned from the plant and the sensitivity of the film (or sensor) to this energy. Both the intensity of reflected energy and the film sensitivity vary according to wavelength; therefore, the appearance of a brush plant will vary according to the type of film used. Records of the same scene using Panchromatic film will be different than that of the tonal representation recorded on orthochromatic film.

6. *The Spectral Transmittance* of the optical system (filters, lens, windows etc.) governs the wavelengths of light allowed to strike the film. Filters used in the acquisition of the aerial photography preferentially absorb those wavelengths which the scientist decides to exclude from the photographic record. Many filters are available, each allowing passage of a different set of wavelengths, usually called bands. Since the energy reflected from the plant and the film sensitivity is wavelength (hue) dependent, it is possible by selecting the proper combination of filters and film to utilize those wavelength intervals which best discriminate or maximize response differences between species of plants, vitality of a specie, age etc. The spectral differences thus obtained assist discrimination and photo interpretation in all fields of Earth resources.

2.4.6 Image Formation on Color Film

As noted elsewhere, color films are manufactured by coating three emulsion layers and filter layer, separated by clear gelatin layers, on a film base.

Natural color film usually has the uppermost layer sensitive to blue light and forming a yellow dye image after processing. (EK Film SO 242 has a green sensitive, magenta dye layer as the uppermost layer, which is partly the reason for achieving high resolution.) This top layer is inherently blue sensitive, blind and completely transparent to red and green light. After forming its latent image, the blue light is absorbed by a yellow second layer. This layer does not record any light but merely removes blue light from exposing the other two layers.

The green sensitive third layer is not sensitive to red light. After processing, it will contain the magenta dye image.

Lastly, the fourth layer or third sensitive emulsion layer is sensitive to red and blue light. Because blue has been eliminated and does not reach this layer, exposure is made only by red light. The resultant dye image is a complementary cyan (blue green).

The net effect then is to have a three emulsion layer film with the first layer sensitive to blue (yellow forming dye), the second layer sensitive to green (magenta forming dye), and the third layer sensitive to red (cyan forming dye).

Color IR film also uses three emulsion layers. However, the green sensitive layer contains the yellow forming dye couplers; the red sensitive layer contains the magenta forming dye couplers, and the IR sensitive layer contains the cyan dye couplers. The minus blue or yellow filter is usually placed in the optical path of the camera, although experimental films were made with yellow absorbing gelatin as the uppermost layer. The usual order in which the layers are laid down in Color IR film is IR (cyan) first, green (yellow) second, and red (magenta) last.

On the processed film, the image of a blue target would appear black, a green target would appear blue; a red target would appear green; and infrared target would appear red.

Most Earth resource features reflect in more than one part of the visible spectrum. For example, viable green vegetation reflects infrared very much, and to a lesser degree the green band as well. The resulting color is therefore a combination of red from the IR reflectance and a little blue from the green reflectance, giving the characteristic bluish-red color of growing vegetation. On the other hand, diseased or dying vegetation does not reflect IR energy. The resulting color is therefore blue and easily detected among viable growth.

2.4.7 Processing of Color Film

The major difference between reversal and negative color is the mode of processing. Negative color requires a color forming first and only developer, and the colors formed are a negative (opposite polarity) and complementary to the hue of the image forming light. For example, blue light forms a latent image which subsequently causes the color coupler and developer molecules to combine into a yellow dye. However, in a reversal film, the latent image is reduced to metallic silver in a B/W first developer. Now, any residual silver halide is made to cause dye coupling in a second or positive working developer. The same blue light will now form a yellow positive image. Since yellow is a blue light modulator, the yellow color acts as a valve to permit or prohibit blue light causing a visual sensation. As a negative image, it holds back blue light and as a yellow positive image it permits passage of blue light.

The exposed reversal film is therefore first developed in a black and white developer, which produces a negative silver image in each of the three emulsion layers in proportion to their exposure. The developed silver is no longer light sensitive, but the residual unaffected silver halide remains light sensitive. The film is next completely re-exposed by passing it between white light photo floods to render the remaining silver halide developable. (An alternate used in many processes is to use a fogging agent in the color developer which produces the same net effect as light reexposure.)

The film is developed in a color developer which reduces the silver halide to metallic silver. The developer is oxidized in the process and combines with the couplers in the emulsion layers; the product of this chemical reaction is a color dye. Each emulsion layer carries a different coupler which forms a different colored dye; yellow in the blue sensitive layer; magenta in the green sensitive layer; and cyan in the red sensitive layer.

The film is next treated in a bleach which converts metallic silver to soluble salts which are washed out of the gelatin layers. The integral yellow filter layer is initially made of finely suspended metallic silver, so it too is chemically converted to a soluble salt and subsequently washed away.

The net result of a reversal color process is that yellow, magenta and cyan dyes, respectively, are formed in the blue, green, and red sensitive layers in an inverse proportion to the exposure.

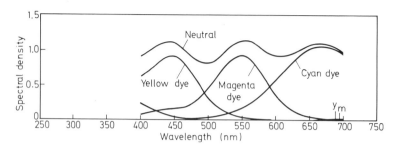

Fig. 38. Spectral density of Ektachrome dyes

The spectral absorption of the dyes used in E.K. Ektachrome are shown in Fig. 38. When the processed film or transparency is viewed over white light, the original colors of the scene are reproduced by subtraction from the white light of various color components of these dyes. For example, blue is obtained when magneta and cyan, but no yellow is present. The magenta dye absorbs green and transmits blue and red; the cyan dye absorbs red and transmits blue and green. The only color which these two dyes permit to pass through is blue, which then stimulates the eye and we see blue. Blue is the only color of light not absorbed by any dye layer in the area of the film representing imaging blue light from the target. The blue light caused an image in the yellow layer, but since this image removed any silver halide which would have promoted dye formation in the color developer, there is no yellow dye present to absorb the blue light.

Areas where image forming light (absence of light) radiated from a black target, no negative B/W silver is produced. All dyes are produced in the color developer, and the net result is that no light will be transmitted by the transparency and a black record is present.

2.4.8 Photo Systems, Choice

For any photographic remote sensing mission, the choice of camera film[5] involves a systems tradeoff between resolution, RMS granularity, speed, spectral sensitivity. The latter becomes significant for different disciplines, i.e., oceanography preferring to record in the 465–515 nm bandwidth, whereas agriculturalists would prefer the infrared for viable/diseased vegetation. In order to maximize tone reproduction and information content, any film must be exposed and processed so that all reflectances are recorded on the linear or straight line portion of the film characteristic curve. Laboratory experiments using the contemplated film and camera filter combinations can determine predicted success by separately exposing calibrated step tablets in a sensitometer (see Section 2.2, Characteristics of the Photographic Process).

---

[5] Kodak Data for Aerial Photography, Kodak Publication No. M 29, 1971.

2.4.9 Image Enhancement Procedures

Images of Earth resource phenomena may be recorded by black-and-white or color film, vidicons, multispectral scanners, radar and thermal imaging systems. Images may be produced in several spectral bands simultaneously, each band influenced differently by atmospheric effects, sun elevation, season, time of day etc. "Enhancement" techniques are frequently applied to an image for emphasizing spatial detail and/or spectral differences to improve detection, recognition and interpretation of the image subjects for visual or machine aided interpretation stations. Many enhancement processes have been published.

The system to be described below is a photooptical technique. The reader should be cautioned that there are a number of other photooptical procedures, each producing an "enhanced" image for better interpretation. However, most photointerpreters will argue that no single technique is a best or only one required. Instead, Earth Resource Photointerpreters require a combination of techniques, including analysis of raw data.

Most aerospace imagery is supplied to the analyst in transparency analog form, subjects being represented by a continuous black-and-white tone scale ranging from clear to opaque; or in color films, where subject reflectances are recorded through a continuous change of saturation, hues and chroma. In a majority of cases there will exist a very low contrast between the target or area of interest and its background. The low initial brightness values are such that changes in the analog black-and-white density or color information by which they are recorded often makes the detection and identification of significant phenomena impossible by normal visual means. The major contribution of any image enhancement process is to alter the analog color/grey scale in ways which make these small differences detectible.

The photooptical processes include color-separation of the aerospace natural color and color-IR images into black-and-white form to isolate one or more of the three spectral records; alter the contrast and densities through reproduction; density slice photographically and subsequently reproduce each density increment into symbolic contrasting colors; and spectral masking to suppress spectral reflectances common to two or more separate spectral records, such as water from land or urban from culture.

Additive color viewing, used for projecting ERTS or multispectral camera images (1^2S or Spectral Data Systems) taken in different spectral bands also provides a means of using strongly contrasting colors to emphasize spectral differences. Electronic image digitization (similar to 1^2S Digital or Spatial Data Systems video scanner) enables real-time conversion of image analog density differences into videodigital form, which may be displayed in contrasting colors on a video color monitor, or interfaced with image storage devices for computer processing.

Bibliography

Airborne Photographic Equipment, Vol. II: Controls, Accessories and Miscellaneous Equipment. RECON CENTRAL, U.S. Air Force Avionics Laboratory, Wright-Patterson AFB. Ohio (1961).

ALTMAN,J.H.: The Measurement of RMS Granularity. Appl. Opt. **3**, 35 (1964).

ANDERSON,P.N., DALKE,G.W., HARALICK,R.M., KELLY,G.L., MOORE,R.K.: Electronic Multi-Image Analog-Digital Processor and Color Display. IEEE Convention Digest, March 1972.

ATKINSON,J.H., JR., JONES,R.E.: Atmospheric Limitations on Ground Resolution from Space Photography. SPIE J. **1**, 1962.

BADGLEY,P.C.: Orbital Remote Sensing and Natural Resources. Photogrammetric Eng. **37**, 780–790 (1966).

BALL,G.H., HALL,D.J.: ISODATA. A Novel Method of Data Analysis and Pattern Classification. Stanford Research Institute, Menlo Park, California, Technical Report 1965.

BARNEA,D.I., SILVERMAN,H.F.: A Class of Algorithms for Fast Digital Image Registration. IEEE Transact. Computers **C-21**, No. 2, 179–186 (1972).

BARROWS,R.S.: Factors Affecting the Recognition of Small, Low-Contrast Photographic Images. Phot. Sci. Eng. **1** (1957).

BILLMEYER,F.: Optical Aspects of Color-XI: Color Scales and Chromaticity Diagrams. Opt. Spectra **3**, 74.

BIRD,G.R., JONES,R.C., AMES,A.E.: The Efficiency of Radiation Detection by Photographic Films: State of the Art and Methods of Improvement. Appl. Opt. December 1969.

BREWER,W.L., WILLIAMS,F.C.: An Objective Method for Determination of Equivalent Neutral Densities of Color Film Images-I: Definitions of Basic Concepts. J. Opt. Soc. Am. **44** (1954).

BROCK,G.C.: The Physical Aspects of Aerial Photography, pp. 130–135. Dover Publications, Inc. 1967.

BROCK,G.C., HARVEY,D.I., KOHLER,R.J., MYSKOWSKI,E.P.: Photographic Considerations for Aerospace. Itek Corp., IL-9026-8, 1961.

BROOKE,R.K., JR.: Spectral/Spatial Resolution Targets for Aerial Imagery. U.S. Amery Engineer Topographic Laboratories Technical Report ETL-TR-74-3, 1974.

BROOKE,R.K., JR.: A Single-Lens, Four-Channel Multiband Camera, U.S. Army Engineer Topographic Laboratories Equipment Test Report. ETL-ETR-74.

BROWN,E.B.: Prediction and Compensation of Linear Image Motions in Aerial Cameras. The Perkin-Elmer Corp, 1961.

BROWN,F.M.: Photographic Systems for Engineers, SPSE. Washington, 1966.

BROWN,W.R.J.: Substractive Color Reproduction: Evaluation of the Actual Color-Reproduction Equations for a Color Process. J. Opt. Soc. Am. **45**, 539–546 (1955).

CHUTKA,G.E., GERGEN,J.B.: SPSE Unconventional Photographic Systems, 1971.

CLARKE,G.L., JAMES,H.R.: Laboratory Analysis of the Selective Absorption of Light by Sea Water. J. Opt. Soc. Am. **29**, 43–53 (1939).

Color Tone Reproduction-Part I: Theory Manual, Itek Corporation Technical Report (AFAL-TR-67-164), AD 829-679, 1968.

CONDIT,H.R.: Spectral Reflectance of Soil and Sand. Seminar on New Horizons in Color Aerial Photo. Eastman Kodak Research Laboratories, ASP and SPSE, 1969.

CONROD,A.C.: Investigation of Visible Region Instrumentation for Oceanographic Satellites. Report RE-31, Vol. 1. Experimental Astronomy Laboratory, M.I.T., 1967.

COOPER,P.W.: The Hyperplane in Pattern Recognition. Cybernetica **5**, No. 4, 215–238 (1962).

COOPER,P.W.: The Hyperplane in Pattern Recognition. Information and Control **5**, 324–346 (1962).

CRANE,R.B.: Preprocessing Techniques to Reduce Atmospheric and Sensor Variability in Multispectral Scanner Data. Proceedings of the Seventh International Symposium on Remote Sensing of Environment, pp. 1345–1350. University of Michigan, 1971.

CRONIN,J.F.: Terrestial Science Laboratories. NASA/CR 84495, 1967.

CRT Recording: Industrial Photography, February 1972.

CURCIO,J.A.: Evaluation of Atmospheric Aerosol Particle Size Distribution from Scattering Measurements in Visible and Infrared. J. Opt. Soc. Am. **51**, 548 (1961).

DAHLQUIST,J.A.: Preprints. Conference SPSE, 16, 1970.

DAY,G.F., JEKSEN,R.L.: Preprints. Conference SPSE, 182 (1970).

DUNTLEY,S.Q.: Light in the Sea. J. Opt. Soc. Am. **53**, 214–233 (1963).

EGBERT, D. D., ULABY, F. T.: Effect of Angular Variation on Terrain and Spectral Reflectivity. Proceedings of XVII Symposium of the AGARD Electromagnetic Wave Propagation Panel on Propogation Limitations in Remote Sensing, 1971.

E. K. Co.: Photointerpretation and Its Uses. Pub. No. M-42, 1969.

E. K. Co.: Kodak Data for Aerial Photography. Pub. No. M 29, 1971.

E. K. Co.: Properties of Kodak Materials for Aerial Photographic Systems 1, 2, Pub. No. M 61, M 62, 1972.

E. K. Co.: Practical Densitometry. Pub. No. E-59, 1972.

E. K. Co.: Applied Infrared Photography 5, 73, Pub. No. M 27, 1973.

E. K. Co.: Specifications and Characteristics of Kodak Aerial Films, Pub. No. M 57, 1973.

ELTERMAN, L.: Vertical Attenuation Model with Eight Surface Meteorological Ranges 2 to 13 km, Air Force Research Laboratories. AFCRL-70-0200, 1970.

EVANS, R. M., HANSON, W. T., BREWER, L.: Principles of Color Photography. New York: John Wiley & Sons 1953.

EYNARD, R. A.: Color Densitometry: Avoiding the Pitfalls. Industrial Photography, February 1974.

FISCHER, W. A.: Color Aerial Photography in Geologic Investigation. Phot. Eng. 28, 133–139 (1962).

FRITZ, N.: Optimum Methods for Using Infrared Sensitive Color Films. Paper presented at 33rd ASP Meeting in Washington D.C., 1967.

FU, K. S., LANDGREBE, D. A., PHILLIPS, T. L.: Information Processing of Remotely Sensed Agricultural Data. Proceedings of the IEEE 57, No. 4, 639–653 (1969).

GELTMACHER, H. E.: Contrast Considerations for Evaluation of Aerial Photographic Images. AFAL-TDR-64-232, AD 452-081, 1964.

Geometrical Atmospheric Effects on Imaging Systems: McDonnell Douglas Reconnaissance Laboratory. MDC A0028, 1969.

GUTTMAN, A.: Line Photometry and the Influence of the Eberhard Effect. J. Opt. Soc. Am. 58, 545 (1968).

HALL, H. J., HOWELL, H. K.: Photographic Considerations for Aerospace. Itek Corp. 1965.

HANSON, C. W., SPANGLER, S. B., NEILSON, J. M.: Influence of High Speed Flight on Photography. North American Aviation, AFAL-TR-328, AD-364-277, AD-364-278, 1965.

HANSON, W. T., HORTON, C. A.: Subtractive Color Reproduction: Interimage Effects. J. Opt. Soc. Am. 42 (1952).

HARALICK, R. M., DINSTEIN, I.: An Iterative Clustering Procedure. IEEE Transaction on Systems, Mass, and Cybernetics SMC-1, No. 3, 1971.

HARALICK, R. M., KELLY, G. L.: Pattern Recognition with Measurement Space and Spatial Clustering for Multiple Image. Proc. IEEE, 57, No. 4, 654–665, 1969.

HARRIMAN, B. R.: SPSE Unconventional Photographic Systems, 1967.

HAYASHI, Y.: Preprints. Conference SPSE, 16, 1969.

HELGESON, G. A.: Water Depth and Distance Penetration. Photogrammetric Eng. 36, No. 2, 164–172 (1970).

HELGESON, G. A., ROSS, D. S.: Remote Sensor Imaging for Oceanography. Oceanology International 5, No. 9, 20–25 (1970).

HENDLEY, J., HECHT, S.: The colors of natural objects and terrains and their relation to visual color deficiancy. J. Opt. Soc. Am. 39 (1949).

HERK, L. F., HAMM, F. A.: SPSE Unconventional Photographic Systems, 9, 1971.

HOFERT, H. J.: XYZ in the Realm of Colors. Zeiss Information No. 24, 1957.

HOFFER, R. M., HOLMES, R. A., SHAY, J. R.: Vegetative Soil and Photographic Factors Affecting Tone in Agricultural Remote Multispectral Sensing. Proceedings of the 4th International Symposium on Remote Sensing of the Environment. University of Michigan, Michigan, 1966.

HOTELLING, H.: Analysis of a Complex of Statistical Variables into Principal Components. J. Educat. Psych. 24, 417–441 (1933).

HULBURT, E. O.: Optics of Distilled and Natural Water. J. Opt. Soc. Am. 35, 698–705 (1945).

JAMES, T. H., HIGGINS, G. C.: Fundamentals of Photographic Theory. New York: Morgan & Morgan Inc. 1960.

JENSEN, N.: Optical and Photographic Reconnaissance Systems. New York: John Wiley & Sons, 1966.

JERLOV, N. G.: Optical Classification of Ocean Water. Physical Aspects of Light in the Sea. University of Hawaii, 1961.

JERLOV, N. G.: Optical Oceanography, Vol. 5. Amsterdam: Elsevier 1968.

JOHNSON, F.: Proceedings of the Twentieth Annual Meeting and Convention, N.M.A. Maryland: Silver Spring 1971.

JONES, L. A.: On the Theory of Tone Reproduction with a Graphic Method for the Solution of Problems. J. Opt. Soc. Am. (1921).

JONES, L. A.: Minimum Useful Gradient as a Criterion of Photographic Speed. Phot. J. (1935).

KEELING, D.: Retinal Resolution, Photographic Applications in Science, Technology, Medicine, March 1974.

KOWALISKI, P.: Applied Photographic Theory. New York: John Wiley & Sons 1972.

KRIEGLER et al.: Preprocessing Transformation and Their Effects on Multispectral Recognition. Proceedings of the Sixth International Symposium on Remote Sensing of the Environment, pp. 97–131. Ann Arbor, Michigan: University of Michigan 1971.

LANKES, L. R.: Optics and the Physical Parameters of the Sea. Opt. Spectra 4, 42–49 (1970).

LEWIS, J. C., WATTS, H. V.: Effect of Nuclear Radiation on the Sensitometric Properties of Reconnaissance Film. AFAL-TR-65-113, AD 436-171, 1963.

LIST, R. J.: Smithsonian Meteorological Tables, Smithsonian Misc. Collections, pp. 425–426. Washington, D.C.: Smithsonian Institute 1958.

LOHMANN, A. W., PARIS, D. P.: Computer generated spatial filters for coherent optical data processing. Appl. Optics 7, 651 (1968).

LOHSE, K.: Investigation of Multiband Photographic Techniques. Final Report, Aeronutronic Division of Philco-Ford Corporation for U.S. Army Corps of Engineers, GIMRADA Contract No. DA-44-009-AMC-1613(X), 1965.

MACADAM, D. L.: Geodesic chromaticity diagram based on variances of color matching by fourteen normal observers. Appl. Optics 10, No. 1, 1–7 (1971).

MACLEISH, K. G.: Transmission densitometer for color films. J. SMPTE 60, 696–708 (1953).

Manual of Photogrammetry (third Ed.): Am. Soc. Phot., Falls Church, Virginia 22044, 1966.

MARBLE, D., THOMAS, E.: Some Observations on the Validity of Multispectral Photography for Urban Research, 4th. Symposium on Remote Sensing of Environment. Ann Arbor: Univesity of Michigan 1966.

MARCHANT, J.: SPSE Seminar on Novel Imaging Systems, 1969.

MARLAR, T., RINKER, J.: A small four-camera system for multi-emulsion studies. Phot. Eng. 43, No. 11, 1252–1257 (1967).

MARSHALL, R. E., KRIEGLER, F. J.: An Operational Multi-Spectral Surveys System. Proceedings of the Seventh International Symposium on Remote Sensing of Environment, pp. 2169–2192. University of Michigan, 1971.

McCAMY, C. S.: Concepts, terminology, and notation for optical modulation. Phot. Sci. Eng. 10, 314 (1966).

MEES, C. E., JAMES, T. H.: The Theory of the Photographic Process. New York: MacMillan Co. 1967.

MEIER, H. K.: Color Correct Color Photography? Translated from paper published in: Bildmessung and Luftbildwesen, No. 5, 1967.

MIDDLETON, W. E. K.: Vision Through the Atmosphere. Toronto: University Press 1952.

MORGAN, D. A.: SPSE Unconventional Photographic System, 1971.

MORRIS, R. H., MORRISSEY, J. H.: An Objective Method for Determination of Equivalent Neutral Densities of Color Film Images-II: Determination of Primary Equivalent Neutral Densities. J. Opt. Soc. Am. 44 (1954).

NAGY, G.: State of the Art in Pattern Recognition. Proceedings of the IEEE 56, No. 5, 836–862 (1968).

NAGY, G., SHELTON, G., TOLABA, J.: Procedural Questions in Signature Analysis. Proceedings of the Seventh International Symposium on Remote Sensing of Environment, pp. 1387–1401. University of Michigan, 1971.

NASA Sp 5099: Photography Equipment and Techniques. A Survey of NASA Developments, Chap. V. Multispectral Photography, 1972.

National Bureau of Standards: Tables of Scattering Functions for Spherical Particles, N.B. St. Appl. Math. Series 4, 1949.

NEBLETT, C. B.: Photography. Its Materials and Processes, 6th Ed. Van Nostrand 1961.

NEBLETTE, C. B.: Fundamentals of Photography. Van Nostrand-Reinhold 1969.

NEILSEN, J. N., GOODWIN, F. K.: Environmental Effects of Supersonic and Hypersonic Speeds on Aerial Photography. Phot. Eng., June 1961.

NISENSON, P. (Ed.): Recent Advan. Evaluation Photogr. Image. SPSE, 1971.

NORMAN, G. G., FRITZ, N. L.: Infrared Photography as an Indicator of Disease and Decline in Citrus Trees. Proc. Florida State Horticultural Soc. **78**, 59–63 (1965).

NORTON, C. L.: Aerial cameras for color. Phot. Eng. **34**, 1968.

ONLEY, J. W.: Analytical densitometry for color print evaluation. J. Opt. Soc. Am. **50**, 177 (1960).

ORR, D. G.: Multiband Color Photography. Manual of Color Aerial Photography (Smith and Anson, eds.), p. 441. Am. Soc. Phot. 1968.

PARIS, D. P.: Influence of Image Motion on the Resolution of a Photographic System. Photogr. Sci. Eng. January 1960.

PARRENT, G. B., THOMPSON, B. J.: On the Fraunhofer (far field) diffraction patterns of opaque and transparent objects with coherent background. Optica Acta **11**, 183 (1969).

PEASE, R. W., BOWDEN, L. W.: Making Color Infrared Film a More Effective High Altitude Sensor. Interagency Report NASA-117 Contract No. R-14-08-0001-10674, 1974.

PERRIN, F. H.: Methods of appraising photographic systems. J. SMPTE **69**, 151–156 (1960).

PETTINGER, L. R.: Analysis of Earth Resources on Sequential High Altitude Multiband Photography. Special Report, Forestry and Conservation. University of California, December 1969.

PHILLIPS, T.: Corn Blight Data Processing Analysis and Interpretation. Fourth Annual Earth Resources Program Review, NASA/MSC, January 17–21, 1972.

PINNEY, J. E., VOGLESONG, W. F.: Analytical densitometry of reflection color print materials. Phot. Sci. Eng. **6**, 367 (1962).

PLASS, G. N., KATTAWAR, G. W.: Radiant intensity of light scattered from clouds. Ap. Optics **7**, 699 (1968).

POLCYN, F. C., ROLLIN, R. A.: Remote Sensing Techniques for the Location and Measurement of Shallow-Water Features. Contract N 62306-67-6-0243, 1969.

PRYOR, P.: The Performance of Imaging Sensors Aloft. Astronautics and Aeronautics, September 1971.

READY, P. J., WINTZ, P. A., WHITSITT, S. J., LANDGREBE, D. A.: Effects of Data Compression and Random Noise on Multispectral Data. Proceedings of the Seventh International Symposium on Remote Sensing of Environment, pp. 1321–1342, 1971.

ROBILLARD, J. J.: New approaches in photography. Phot. Sci. Eng. **8**, 18 (1964).

ROSENFELD, A.: Automatic recognition of basic terrain types from aerial photographs. Phot. Eng. **28**, 115–132.

ROSS, D. S.: Experiments in Oceanographic Aerospace Photography—I, Ben Franklin Spectral Filter Tests. Philco-Ford Corp., U.S. Naval Oceanographic Office, Contract No. N 62306-69-C-0072, 1969.

ROSS, D. S.: Enhanced Oceanographic Imagery. Proceedings of the Sixth International Symposium of Remote Sensing of Environment, pp. 1029–1044, 1969.

ROSS, D. S.: Enhancing aerospace imagery for visual interpretation. Proceedings 15th annual technical symposium. Soc. Photo Opt. Instr. Eng. **3** (1970).

ROSS, D. S.: Remote determination of gross atmospheric effects in multispectral photography. Phot. Eng. **39**, No. 4 (1973).

SEBESTYER, G. S.: Decision-Marking Processes in Pattern Recognition. New York: MacMillan Company 1962.

SELWYN, E. W. H.: Scientists' way of thinking about definition. J. Phot. Sci. **7** (1959).

SIMONDS, J. L.: A quantitative study of the influence of tone-reproduction factors on picture quality. Phot. Sci. Eng. **5**, 270 (1961).

SIMONDS, J. L.: Reproduction of fine structure in photographic printing-I: Mathematical simulation. Phot. Sci. Eng. **8**, 172 (1964).

SIMONDS, J. L.: Analysis of nonlinear photographic systems. Phot. Sci. Eng. **9**, 294 (1965).

SLATER, P. N.: Multiband Camera Monograph. Tech. Report 44, Optical Sciences Center, University of Arizona, 1969.

SMITH, J. T., JR. (Ed.): Manual of Color Aerial Photography. Am. Soc. Photogrammetry 1968.

S.P.S.E.: Photographic Systems for Engineers. Washington D.C., 1969.

S.P.S.E.: Handbook of Photographic Science and Engineering. New York: John Wiley & Sons 1973.

STEINER, D., HAEFNER, H.: Tone distortion for automated interpretation. Phot. Eng. **31**, No. 2, 269–280 (1965).

STULTZ, K. F., ZWEIG, H. J.: Relation between graininess and granularity for black-and-white samples with nonuniform granularity spectra. J. Opt. Soc. Am. **49**, 693 (1959).

SULLIVAN, P. A.: J. Soc. Photo-Opt. Inst. Eng. **9**, 201 (1971).

SWING, R. E., SHIN, M. C. H.: The determination of modulation-transfer characteristics of photographic emulsions in a coherent optical system. Phot. Sci. Eng. **6**, 350 (1963).

TODD, H. N., ZAKIA, R. D.: Tutorial: A review of speed methods. Phot. Sci. Eng. **8**, 249 (1964).

TODD, H. N., ZAKIA, R. D.: Sensitometry. RIT Press 1965.

TODD, H. N., ZAKIA, R. D.: Photographic Sensitometry, The Study of Tone Reproduction. Morgan & Morgan Inc. 1969.

TODD, H. N., ZAKIA, R. D.: 101 Experiments in Photography. New York: Morgan & Morgan Inc. 1969.

TROTT, T.: The Effects of Motion on Resolution. Phot. Eng. December 1960.

TUPPER, J. L., NELSON: The effect of atmospheric haze in aerial photography treated as a problem in tone reproduction. Phot. Eng. **6** (1955).

TWEET, A. G.: SPSE Unconventional Photographic Systems, 16, 1971.

TYLER, J. E., SMITH, R. C.: Measurements of Spectral Irradiance Underwater. New York: Gordon and Breach 1970.

UMBERGER, J. Q.: Color reproduction theory for subtractive color films. Phot. Sci. Eng. **7**, 34 (1963).

U.S.A.F. Avionics Laboratory: Airborne Photographic Equipment (3 Vols.), U.S.A.F. Avionics Laboratory, Research and Technology Division, USAF Systems Command, Wright-Patterson AFB, RC 013200, 1965.

U.S.A.F. Avionics Laboratory: Aerial Camera Lenses, RC 027000, 1967.

WACKER, A. C., LANDGREBE, D. A.: Boundaris in Multi-Spectral Imagery by Clustering. Proceedings of the 1970 IEEE Symposium on Adoptive Processes (0th) Decision and Control. University of Texas, Austin, 1970.

WAYWOOD, D. J.: SPSE Seminar on Recent Advances in the Evaluation of the Photographic Image, 1971.

WENDEROTH, S.: Hydrographic and Oceanographic Applications of Multispectral Color Aerial Photography. Seminar Proceedings New Horizons in Color Aerial Photography, ASP & SPSE, pp. 115–126, 1969.

WENDEROTH, S., YOST, E., KALIA, R., ANDERSON, R.: Multispectral Photography for Earth Resources. West Hills Printing Co. (printed under NASA L. B. Johnson, Space Center, Contract No. NASA-11188), 1974.

WERNICKE, B. K.: Collection of Graphs to be Used as Tables for Determining or Evaluating the Combined Effect of Image Motion and Camera-Lens-Film Resolving Power on the Capabilities of Aerial Cameras. WADC-TN-58-321, AD 204-661, 1958.

WERNICKE, B. K.: Effect of Image Motion on Resolving Power of a High-Acuity Lens-Film System. WADD-TN-60-2, AD 240-993, 1960.

WERNICKE, B. K.: General Discussion of Several Technical Methods for Image Motion Compensation. ASD-TDR-62-497, AD 332-753, 1962.

WHEELER, C., HALL, J.: Design and Fabrication of an Experimental Multiband Camera. Report No. ETL-CR-71-28, USAETL Contract No. DAAK02-70-C-0121. (The reader is reffered to this report for specific details of the optical system), 1971.

WHEELER, C., HALL, J.: Design and Fabrication of Experimental Multiband Camera. Report No. ETL-CR-71-28, pp. 25–30, 1971.

WILLIAMS, F. C.: Objectives and methods of density measurement in sensitometry of color films. J. Opt. Soc. Am. **40** (1950).

WINTER, D. C.: Matching an Image Display to a Human Observer. Electro-Optical Systems Design, August 1971.

YOST, E., WENDEROTH, S.: Multispectral Color Aerial Photography. Photogrammetric Engineering 33, No. 9, p. 1020 (1962).

YOST, E., WENDEROTH, S.: Precision Multispectral Photography for Earth Resources Application, NASA ICR 92360 (1967).

YOST, E., WENDEROTH, S.: Multispectral color aerial photography. Photogrammetric Engineering 33, No. 9 (1967).

YOST, E., WENDEROTH, S.: Coastal Water Penetration Using Multispectral Photographic Techniques. Proceedings of the Fifth Symposium on Remote Sensing of Environment, pp. 571–586 (1968).

YOST, E., WENDEROTH, S.: Additive Color Aerial Photography. Manual of Color Aerial Photography (Smith and Anson, eds.), American Society of Photogrammetry, p. 451, 1968.

ZWEIG, H. J.: Autocorrelation and granularity-part I: Theory. J. Opt. Soc. Am. 46, 805–820 (1956).

ZWEIG, H. J.: Autocorrelation and granularity-part II: Results on flashed black-and-white emulsions. J. Opt. Soc. Am. 46.

ZWEIG, H. J.: Autocorrelation and granularity-part III: Spatial frequency response of the scanning system and granularity correlation effects beyond the aperture. J. Opt. Soc. Am. 49, 238 (1959).

3. Infrared Sensing Methods

P. W. SCHAPER

3.1 Introduction

The ever increasing rate at which the population of the world is growing has intensified concern over the environment which supports life and over the availability of resources needed to sustain it. Realizing that the state of the environment is directly tied to molecular processes occurring in it, one is led to the conclusion that the measurement of the effects of molecular processes may provide an excellent technique for monitoring the condition of the environment. Because electromagnetic radiation at infrared frequencies interacts with molecular processes, it provides a source from which the desired information relating to the state of the environment can be obtained. One is then faced with the problem of detecting this radiation and performing a quantitative analysis of the data collected. This has traditionally been accomplished in two basic ways: by direct sampling or by remote sensing. While direct sampling methods can be extremely accurate, they are limited in spatial coverage because of their reliance on a physical or chemical action on the sample. Remote sensing surveys, on the other hand, offer two significant advantages: they are readily adapted to wide area investigations, and the sample's properties are determined without the danger of contamination or alteration through a physical contact with, or a confinement in a foreign material.

In the 1940 ies sophisticated techniques of remote sensing of infrared radiation were primarily devoted to military applications for such purposes as fire control, missile guidance, night vision, etc. Through extensive efforts by the military, the technology was advanced to a point where it became economically feasible to contemplate a wide range of peaceful uses for infrared sensors. The early space program gave an additional impetus to the development of instrumentation, so that we now stand on the threshold of new fields of applications. We find ourselves in a position to meet the needs of modern society for monitoring its effect on the earth's environment and resources, while increasing our knowledge of the neighboring planets and ultimately the universe.

In this chapter, devoted to the subject of remote sensing of infrared radiation, we have purposely resisted the temptation to go into detail on any or all the aspects of the topic; a task too enormous to be accomplished in the limited space available for a review. In-depth discussion of the various topics associated with this field are available in the literature and the interested reader is directed to the references as a guide for further study. What is attempted here is to present a review—a "perspective" of the subject with the intent to connect in a meaningful manner the physics of the infrared with the instrumentation designed for its measurement, and some practical interpretations which have resulted from the analysis of a few of the data collected. The reader will find the chapter arranged in the following manner:

— We begin with the statement of some of the most important principles governing radiation in the infrared. We will also briefly describe the physical processes which affect the radiation field at infrared frequencies.

— We will then devote some discussion to typical instrumentation which has been developed to measure the phenomena one might observe in each of the categories established.

— This will be followed by a few examples of typical applications of the techniques described.

3.2 The Infrared Radiation Field

In the introduction we postulated that the infrared radiation field is influenced by molecular processes in the environment with which electromagnetic radiation interacts. Let us now examine this statement more closely [1].

Electromagnetic radiation of wavelength longer than that usually denoted as "gamma rays" and "X-rays" interacts with both atoms and molecules by adding or substracting energy from the electrons which surround and are bound to the atom or molecule. The essential difference between an atom and a molecule is that the electrons in an atom are bound by the nucleus, whereas in a molecule which has, by definition, two or more nuclei, the electrons are both bound by the individual nuclei and provide the forces that hold the nuclei together. We shall not consider atomic interactions further because the difference alluded to above causes molecular interactions with electromagnetic radiation to differ both qualitatively and quantitatively from atomic interactions, and because we wish to direct our attention to phenomena associated with molecular processes.

If we consider a simple, mechanistic model of a molecule, consisting of two or more spherical masses (the nuclei) bound together by light springs, we find that the masses may vibrate freely at their resonant frequencies and may also rotate about a common center-of-mass (some very complex molecules are also subject to internal rotation of one part with respect to another, but this is quite uncommon and will not be discussed further). As the molecule rotates, the "spring" can stretch by centrifugal forces thereby changing the "spring constant" and, furthermore, the spring is unclassical in that it can partake of anharmonic vibrations which permit overtones of the resonant vibration frequency to occur and, in polyatomic molecules, for various modes of vibration to be coupled.

The "spring" of course, is provided by the orbital motion of the electrons. When the molecule interacts with a photon the electronic energy, the vibrational

---

[1] The author is indebted to Dr. R. BEER of the California Institute of Technology, Jet Propulsion Laboratory for much of the discussion on the methodology of infrared spectroscopy which is to follow.

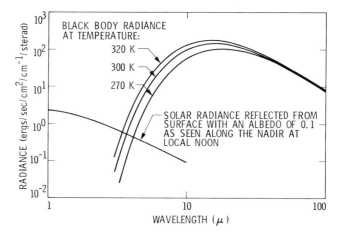

Fig. 1. Graph of the Planck black body function and reflected solar radiation in the infrared

energy or the rotational energy may be changed or any combination of these three phenomena may occur. If it were not for the fact that the energies involved in the three phenomena are markedly different, analysis of the result would be nearly impossible. In general, transitions of the electronic state are a result of interaction with ultraviolet and visible light; vibration and rotation are associated with energies appropriate to near and far-infrared radiations, respectively. That is, if a visible/ultraviolet photon interacts with the molecule, all three states are affected; if the photon is of near infrared energy, both vibration and rotation states are changed, and if the photon is from the far infrared, only the rotational state is changed. Microwave photons may also interact with molecules, either through very low energy rotational transitions or through certain energy changes involved in the detail sub-structure of a molecule.

As this chapter is limited to the discussion of infrared remote sensing methods, let us, from this point on, neglect the effects of radiation in other spectral regions, and let us examine the radiation field of the planet earth in our spectral region of interest. For this purpose we position ourselves as an observer in space, looking at the daylight side of the planet with instruments which are sensitive to infrared radiation. The distribution of radiation intensity we receive varies in the manner shown in Fig. 1. At the shorter wavelengths in the near infrared, between 1 micrometer (μ) and about 3μ, the source of radiation is the reflection of energy incident upon the planet from the sun, which varies in absolute intensity primarily as a function of albedo and angle of incidence. For obvious reasons this region is referred to as the "solar reflection" region. At wavelengths beyond about 3μ the radiation field is predominated by energy originating from the planet itself— emitted by the earth's surface and atmosphere. The energy at these wavelengths is strongly dependent on source temperature, and therefore we refer to this part of the infrared spectrum as the "thermal emission" region.

As we examine the curves plotted in Fig. 1, we can draw some very general conclusions regarding the potential use of the infrared for determining properties of the radiation source. These are:

— In the solar reflection region the daylight side of the earth offers a source infrared radiation which is independent of its temperature.

— In the thermal emission region infrared radiation contains information descriptive of the temperature of the radiating body.

These two factors are fundamental to the discussion to follow and divide remote sensing in the infrared into two categories. Without further explanation at this point, let it be stated that they are:

1. Determination of molecular composition with high spectral resolution instrumentation (fine detail in the near-infrared spectrum), and

2. determination of bulk properties of the radiation source with low spectral resolution instrumentation.

In summary, then, we have seen that the infrared radiation field surrounding the earth interacts with the atmosphere and the surface on a molecular level. It therefore would follow that measurements of the radiation field will yield data which should be interpretable in terms of the properties describing the state of the molecular make up of the surface and the atmosphere, yielding the desired capability to monitor the state of the environment in which we live.

3.3 Fundamentals of Measurement

Energy changes in a molecule are subject to the rules of quantum mechanics. These rules prescribe that only certain discrete changes in energy are permissible. Consequently, if radiation having a broad spectrum of energies traverses a cloud of molecules of a given species, only certain energies are absorbed by the molecules, giving rise to an "absorption spectrum". By the same token, if the cloud is heated by some means, it will be observed to emit radiation at the same energies, giving rise to an "emission spectrum". Both types are observed in nature and can be analyzed by similar, although not identical methods. We shall first concern ourselves with the formation of absorption spectra.

Since an absorption line is formed when a molecule interacts with (i.e. absorbs) a photon, it is clear that as the number of molecules between the source of energy and the receiving/detecting apparatus increases, the number of photons of that energy that are absorbed will increase; that is, the absorption line gets "deeper". Increasing the number of molecules capable of absorption has no effect and what proportionality existed is lost. The proportionality constant is basically a probability, that is, the probability that a "collision" between a molecule and a photon will result in that photon being absorbed. The specific factors which affect this probability are too numerous to mention in this review chapter.

So far it has assumed that the energies of the photons absorbed by the molecule are defined with infinite precision. In fact, it is found that the molecule will absorb not only the specific energy but also a small range around that value. The range is frequently (although not always) smaller than the distance (in energy) from the next absorption so that the technique can retain its basic validity, i.e. the absorptions are separable.

The strength of an absorption is thus to be considered not merely as the ratio of the depth of absorption at a specific energy (or frequency) to the unabsorbed "continuum", but must take into account the energy absorbed away from the specific energy. The specific energy is termed the "line center" because it is usually found that, notwithstanding the foregoing, the strongest absorption occurs at the specific energy and falls away symmetrically on each side. The concept of "equivalent width" is frequently employed. It is simply the width of the rectangle having 100% absorption that has the same area as the area contained under the actual absorption line. The shape of the absorption line is affected by temperature, pressure and electric and magnetic fields. We shall not consider the effects of fields further and only state, without discussion, that the shape of an absorption line produced by a low-pressure, high-temperature gas differs markedly from that produced by a high-pressure, low temperature gas typified by the earth's lower atmosphere (by lower atmosphere we mean the region below about 20 km above the surface). Furthermore, the width of such high-pressure absorptions is strongly dependent upon the nature of the gases producing the pressure. As a specific example, the width of the methane (CH_4) absorptions in the earth's atmosphere is controlled by the dominant molecules oxygen and nitrogen (O_2 and N_2) and is significantly different from that which would pertain were the entire atmosphere CH_4. It may be seen that, for atmosphere of largely unknown composition, such an effect can lead to serious errors in the analysis of spectra.

Even a cursory examination of the spectrum of a molecule will demonstrate that there is a distinct regularity in the arrays of absorption features. Whether this regularity remains on close examination is dependent on the type of molecule involved. Simple diatomic molecules retain considerable regularity even under the most detailed analysis, whereas for others (water vapor [H_2O] being a good example), close inspection suggests a totally random array of individual lines. In all cases, however, the arrays of lines group together into what are called absorption bands. Analysis of such bands shows that they result from transitions between vibrational states of the molecule, and the "fine structure" of the individual absorption lines within the band are a consequence of transitions between rotational states within the vibrational states concerned.

An example of an absorption spectrum is shown in Fig. 2. Depicted here are lines of three of the carbon monoxide (CO) isotopes commonly found as trace elements in the earth's atmosphere. The absorptions occur at wavelengths near 4.75 μ.

If one were to examine the spectrum of a typical molecule starting at a long wavelength near the microwave region and going to shorter wavelengths, one would first of all notice a fairly simple, widely spaced spectrum of relatively few lines. This is the region of pure rotation, wherein the energies involved are too low to cause changes in the vibrational state of the molecule. As the wavelength decreases, vibrational transitions become possible and we are in the region of vibration-rotation bands. The bands are still quite widely spaced with long, clear regions between them. These bands remain as the most obvious source of absorption until the wavelengths of visible light are reached, when the energies become great enough to cause transitions in the electronic state of the molecule. In this region of electronic bands, the character of the spectrum undertakes a dramatic

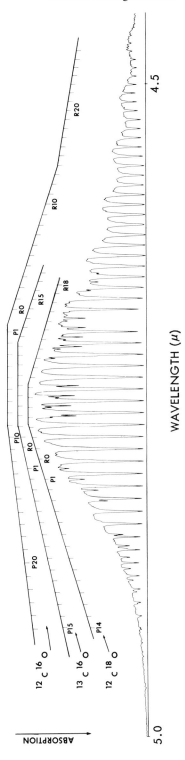

Fig. 2. Laboratory spectrum of carbon monoxide and its naturally occurring isotopes

change. Whereas previously, in the pure rotation and rotation-vibration region, there were large areas free of absorption, we now find it difficult to find any region free of absorption lines. This comes about because of the multiplicative effect of the possibilities of transition: in the pure rotation region, only one kind of transition is occurring; in the vibration-rotation region, there is a rotational band for each vibrational transition; now we have vibration-rotation bands for each electronic transition and it is found that these bands overlap each other in a most confusing manner, even for a single molecular species. If, now, we have many different kinds of molecules, the task of disentangling the spectra becomes an analytical nightmare.

Before showing the reader a spectrum typifying the absorptions manifested by a constituent in our own atmosphere, it is necessary to introduce a little jargon peculiar to spectroscopy. When a molecule undergoes a simultaneous vibrational and rotational transition, giving rise to a component of a vibration-rotation band, the laws of quantum mechanics require that the rotational state either remain unchanged in the new vibrational state or change by one quantum unit (higher or lower). No such rule pertains for changes in vibrational state—any, or no change is equally permissible. It is also true that the "no change" of rotational state in a transition may be forbidden, particularly in very simple molecules. However, if a "no change" transition is permitted, then it is found that the complete set of such transitions all result in absorptions at essentially the same frequency, giving rise to a strong, single, absorption called a Q-branch. Actually, the absorptions are never exactly superposed, so most Q-branches have a noticeably distorted appearance and under special conditions may even be separated into individual lines. Transitions in which the rotational state changes by -1 quantum units give rise to a regular array of lines on the low frequency side of the Q-branch and is called a P-branch. Similarly, the array resulting from a change in $+1$ units results in an R-branch on the high-frequency side. Such a set may be seen in Fig. 2, which shows a position of an actual spectrum of carbon monoxide (CO). The three branches may be clearly seen if the reader follows the schematic presentation shown above the spectrum. The figure also illustrates several other germane points. Overlapping the spectrum of CO are the lines formed by the naturally occurring isotopes of this atmospheric trace compound. These lines are indicated on separate schematic presentations, and are readily identified and separated in the spectrum of a single constituent. In nature, however, such identification and separation is extremely complex when a multiplicity of molecules are present, particularly if their existence is not known. Thus, for example, studies of air pollution utilizing infrared spectroscopy require considerable detail analysis work offering a real challenge to the process of automation.

The discussion of absorption spectroscopy has been considerably simplified in order to make the observation that, given sufficient spectral resolution, it is possible to discern the "signature" of a given atmospheric molecule in the radiation field. A lower resolution presentation of a wider region of the spectrum is shown in Fig. 3. This presentation begins to illustrate some of the complexity with which the spectroscopist is faced in the actual environment. In the lower portion of the figure the individual absorptions of atmospheric constituents have been combined and superposed on an absorption curve for a typical air mass.

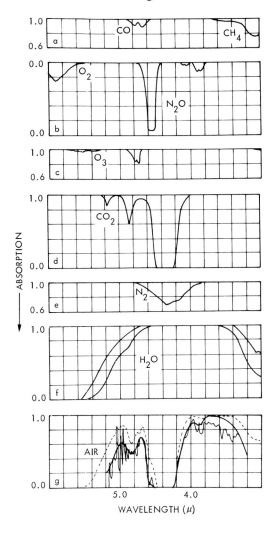

Fig. 3. Absorption characteristics of atmospheric gases. (From J. H. SHAW)

The degree to which we have oversimplified the case for identification of species from infrared spectra might be indicated by the following discussion. The spectrum shown in Fig. 3 might have been obtained by using the sun as a source and observing from the ground, a commonly used technique. Without even considering the fact that the sun has a spectrum of its own (fortunately, in the infrared, relatively sparse in absorption features), the resultant spectra will differ depending on (1) the sun's altitude above the horizon (2) the relative humidity along the line of sight (3) the temperature and pressure profile along the line of sight and (4) any inhomogeneity in the same line of sight. Such variations can result in marked differences in the appearance of spectra, all ostensibly of the same thing. Consequently, spectral analysis is really a pattern-recognition problem.

We have spent considerable time in the vibration-rotation region of molecular transitions. There are obvious analytic simplicities in working in the pure rotation region, but technology and some of the facts of operating in a real world are against us. In order to observe an absorption spectrum, it is necessary to have a source which is very much brighter than the gas being observed. However, Planck's radiation law tells us that at low frequencies, the energy output of a source in a narrow frequency interval (i.e. the monochromatic brightness) is directly proportional to the absolute temperature. In order, therefore, for a source to be, say, 100 times brighter than the gas, it must be 100 times hotter. For a gas near room temperature (300 °K), this would require a source temperature near 30000 °K, a temperature not readily attainable. Furthermore, our detection system for the absorption spectrum is also bathed in energy from the world surrounding it. Although one invariably cools the detector and endeavors to prevent it from "seeing" as much of the outside world as possible, inevitably this background radiation determines the limit to which one can go. Our case is somewhat exaggerated, but not excessively so; some excellent pioneering work has been done in this region of the spectrum, but even its most vehement champions do not understate the difficulties of working there.

Hence, between the Scylla of the analysis of electronic spectra and the Carybdis of far infrared technology we find our safe passage for molecular composition analysis, the near infrared region of the vibration-rotation bands. Not as simple to analyze as pure rotation spectra, but far simpler than electronic spectra; detection systems less sophisticated than the photomultipliers and image tubes of the visible and ultraviolet region but much more manageable than far infrared systems: the near infrared region is albeit a compromise, but one that has paid great dividents.

3.4 Methods of Measurement

Having now arrived at the point where we realize that the information leading to a better understanding of the environment in which we exist is contained in the radiation field which surrounds our planet, we must consider what requirements exist for instrumentation to be devised with which the data are to be collected. In the categorization which we employed in the earlier discussion of the radiation field we distinguished between the two regions of the infrared spectrum: the near infrared solar reflection region which seemed most suitable for compositional measurements, and the thermal emission region which appeared to be most suitable for the determination of bulk properties. It should be emphasized that this generalization regarding the usefulness of each region follows the pattern of all generalizations; i.e., it has numerous exceptions. It does serve our purpose here, however, to continue along the chosen path without expressly noting the exceptions in order to classify the wealth of remote sensing instrumentation which exists to serve the investigator of molecular processes.

In terms of requirements for instruments which measure the data in both categories of radiation, let us first consider the requirements for spectral resolution. For the determination of temperature from data measured at a remote sensor, high resolution in the wavelength region covered is generally not required.

The radiation received at the instrument which describes the temperature of the source varies fairly uniformly over a wide spectral interval. Thus we can gather a larger amount of energy at more wavelengths, making—for this type of experiment—high resolution in fact undesirable, as it will tend to reduce the quality of the measurement through the availability of less signal for the sensor. A similar argument can be made for measurements of radiation which are to be interpreted in terms of surface properties. In a general sense, these variations in reflectivity or emissivity also take place over wavelength intervals which are large in comparison to individual molecular absorption features (i.e. transitions). Requirements for typical spectral resolving powers are on the order of several percent.

If one is interested in the analysis of features in the spectral radiance which are caused by the absorption of atmospheric constituents, a high spectral resolving power is required. We have seen that a gaseous mixture as complex as the earth's atmosphere presents an overabundance of absorption lines due to molecular transitions of the infrared active constituents, including the trace species. If one wishes to distinguish between the effects of the various species (or even between the effects of the various isotopes of a given species, as we saw earlier), we must achieve a resolution in the spectrum better than the separation of individual transitions.

We next consider the requirement of signal to noise ratio. Although in a well designed instrument system the major contribution to the system noise is generated in the detector, a quantity which will be discussed separately, other aspects of this topic will be mentioned at this point. The signal emitted by a source at a given temperature can be determined from Planck's radiation law, assuming the emissivity of the source is known. For reflected solar radiation one calculates the solar flux incident on the reflecting surface at a given angle, assumes a reflectivity as a function of viewing angle, and then calculates the amount being reflected into the solid angle subtended by the instrument. After the energy enters the entrance aperture of the instrument it is modified by the optical components in the instrument head, primarily by the efficiency with which each component reflects or transmits the radiance incident upon it. This efficiency is related to a second effect on the signal to noise ratio: self emission of each component. For example, a mirror which reflects 97% of the infrared energy incident upon it will have an emissivity of 3%. The radiance emitted by the mirror in the field of view of a detector viewing it may be of a magnitude similar to that of the radiance to be measured. Thus, optical components of poor efficiency not only decrease the amount of signal arriving at the detector, but also add to signal uncertainty or "noise" due to high self emission characteristics.

Experience with meteorological experiments has shown that, while a signal to noise ratio in excess of 100:1 is required in any meaningful infrared measurement of emitted radiation, this ratio alone is not always a sufficient functional requirement imposed on the instrument designer. Just as important as the reproduction of a signal free of noise interference is the knowledge that the absolute value of the radiance is established with high precision. High resolution measurements made for the purpose of identifying constituents in the atmosphere, or low resolution measurements made to determine variations of albedo, are less demanding in this respect because the data interpretation requires accuracy only in relative radiance

values; i.e., the comparison of data occurs in generally narrow, adjacent wavelength intervals, determined with high accuracy relative to one-another. However,
all measurements made for ultimate reduction to temperature information, as
exemplified by meteorologically oriented temperature sounding experiments, require that the absolute level of radiance be determined to an accuracy near 1%.
The accuracy requirements vary with wavelength, permissible error in radiance
determination being larger for a given temperature uncertainly as wavelength
decreases. Examination of Fig. 1 shows that there is a significantly greater change
in radiance emitted in the near infrared by a black body for a given change in
temperature than there is at the longer wavelengths. This is a property of the
Planck function and can be viewed from the standpoint that, for a given accuracy
of absolute radiance determination, experiments utilizing data at shorter wavelengths are capable of achieving a higher degree of accuracy in temperature
determination.

This last requirement, pertaining to the necessity of absolute radiance determination, presents a particular problem in the area of instrument calibration,
both pre- and in-flight. It requires careful attention to instrument design to eliminate stray sources of infrared radiance of unknown magnitude.

The common element of all instruments intended to sense electromagnetic
radiation in the infrared is the detector. Its purpose is to convert the power
collected by the optical arrangement preceding it, to a proportional electrical
signal which can then be amplified in a manner suitable for recording. This must
be accomplished without altering the signal or adding extraneous information in
the form of noise. As we noted earlier, most experiments require that the noise is
less than 1% of the information which is to be recorded.

The conversion of infrared radiation (photons) impinging on the detector to
electrical signals can be accomplished in many different ways, all involving processes which create spurious electrons over a wide frequency band. By limiting the
signal detection to a narrow band-width and by integrating the signal and noise
over time intervals which are small with respect to the temporal signal variations,
noise levels can be significantly reduced. Detector manufactures specify the detector's specific detectivity in a parameter, D^*, for a unit area of detector surface
and a unit bandwidth of the system. The units of D^* are cm $H^{1/2}$ watts^{-1}. It is
customary to determine the system signal to noise ratio by first calculating the
apparent power incident upon the detector which is in fact due to the noise inherent
to the detection system, the Noise Equivalent Power. This parameter is then
compared to the irradiance at the detector to obtain the signal to noise ratio.

The process is as follows:

$$NEP = \sqrt{\frac{A_D \times \Delta f}{D^*}}$$

and

$$SNR = \sqrt{\frac{W_\lambda \times \tau}{NEP}}$$

In the above equations

NEP = Noise equivalent power
A_D = Area of detector
Δf = System bandwidth
SNR = Signal to noise ratio
W_λ = Power at the instrument aperture
τ = Optical efficiency of the instrument.

Let us now proceed to review the different ways in which the radiation arriving at the instrument is pre-conditioned, according to the specific experiment which is to be performed, before it is allowed to reach the infrared detector. The examples we have chosen to cite are only those which have achieved a measure of success in space applications. There are many others which are extensively used in the laboratory and on aircraft and balloons. Space does not permit us to present an all-inclusive list.

3.4.1 The Single Element Radiometer

This is probably the most basic of all infrared instruments. The sensor is used to detect radiation in a wide wavelength interval, quite often for the purpose of measuring temperature. A typical optical design is shown in Fig.4. The wavelength region which is to be covered is generally selected out of the infrared by means of a filter which is transparent to the desired radiation, and rejects all other wavelengths. Various types of filters are available, depending on the wavelength coverage desired; in the shorter wavelength intervals, below about 30 μ, the use of interference filters is quite common, and band passes as narrow as one-percent of center-wavelength can be obtained. At longer wavelengths, filters are generally dependent on transmission and/or reflection characteristics of materials and the number of wavelength intervals one can separate from the broad band by filters readily available commercially become more restricted.

In the radiometer the filter is most optimally placed in a portion of the optical train where the energy beam is collimated, as this assures maximum transmission and optimum rejection. However, in order to keep the size of the filter reasonable, it is most expeditious to place the filter as close to the detector as feasible, which generally is a point in the energy beam where the radiation is not collimated, thus requiring design compromises.

In its most fundamental concept the instrument itself is often a very common source of noise or inaccuracies in the determination of absolute radiance levels. As was mentioned before, radiometry generally requires absolute accuracies in radiance levels of a few percent or better; this self emission of components at temperatures near those to be measured is critical and must be either removed by calibration or reduced to negligible amounts. Good radiometer designs achieve this goal by either utilizing primarily reflecting optics—featuring high reflectivity and low emissivity—and low, well controlled component temperatures, or by modulating the radiation entering the instrument aperture and by not modulating radiation emitted by instrument components. An additional aid to the achievement of high quality radiometric data is the "in-flight" calibration source. This

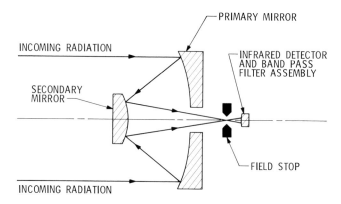

Fig. 4. Typical infrared radiometer optical arrangement

source of radiation of predictable magnitude and wavelength-dependent characteristic in the instrument's band pass generally consists of a black body of emissivity near unity whose temperature is held very uniform at an even distribution over
its active area.

A typical instrument which features the design concepts enumerated above is
shown in Figs. 5 and 6. The Mariner 10 Infrared Radiometer shown in these
figures successfully measured the radiance emitted by the planet Mercury in two
wavelengths intervals, 8–14 μ and 35–55 μ. This instrument design features a
movable plane mirror with which the instrument field of view can be directed at
two separate targets. This "scan mirror" is also used to direct the radiation from
the "in-flight" calibration source to the detectors through the same optical components used to make the measurements over the field of view under investigation.

3.4.2 Imaging Radiometers

The single element radiometer is generally designed so that the target is not
imaged on the detector, but on a field lens preceding it. The lens acts essentially as
a diffuser and thus target inhomogeneities are not projected on the detector active
area, which in itself is quite inhomogeneous in terms of detectivity. Through this
design technique the single element radiometer can integrate the effect of surface
differences in the field of view.

A special case of the simple radiometer is the imaging radiometer. Its optical
design is arranged to image the target on an array of detectors, sensitive to the
same wavelength interval and uniform in detectivity. These detectors are then
scanned by the electronic system, producing a map in the infrared of the changing
features in the field of view.

A different mode of scanning has found application for example in the Earth
Resources Technology Satellite and in surveys performed during the Skylab mission. Here the scan mirror of the basic radiometer sweeps rapidly at right angles
to the spacecraft track and produces a very high spatial resolution image of the

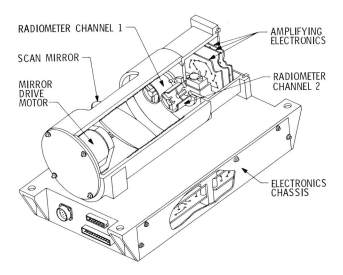

RADIOMETER CHANNEL 1

SCAN MIRROR

MIRROR DRIVE MOTOR

AMPLIFYING ELECTRONICS

RADIOMETER CHANNEL 2

ELECTRONICS CHASSIS

Fig. 5. Infrared radiometer used on Mariner 10

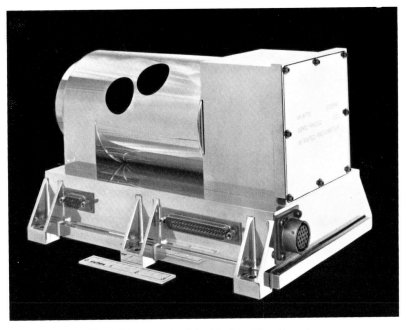

Fig. 6. Photograph of the Mariner 10 radiometer

ground on the detector face. The optical schematic of this scanner is basically identical to the one shown in Fig. 3, save for the addition of a scan mirror preceding the primary (energy collecting) mirror.

We have already noted in the discussion of signal to noise ratio that the scanning instruments require special design considerations because of the rapidly

changing signal on the detector. Furthermore, the mirror motion which is required to complete and repeat a line scan rapidly enough to produce a contiguous
picture, introduces a new source of noise which requires careful design analysis:
the high frequency mechanical vibration inherent in rapidly moving devices. In
addition, this instrument requires an electronic bandwidth which can accomodate
all of the component frequencies of the changing radiation pattern faithfully on
the detectors.

3.4.3 Multi Wavelength Channel Radiometers

A hybrid radiometer, which evolved from the basic radiometer concept, is the
instrument which views the same area of the target in many different wavelength
intervals. Although the instrument we wish to use as an example also bears a
similarity to the spectrometer concept, it is proper to consider it here because of
its application: the determination of the atmospheric temperature profile from the
emission in the CO_2 bands, which establishes the requirement for the instrument
to have the capability of performing radiometrically accurate measurements.

A schematic of this multidetector radiometer is shown in Fig. 7 and a photograph of an instrument designed for use in the 4.3 μ band is shown in Fig. 8.
SCHAPER et al. (1970) describe the design in detail. A diffraction grating is used to
separate the wavelength intervals in the emission band under study. The absolute
value of the radiance emitted in the field of view is measured by a separate
detector for each wavelength increment. As opposed to the conventional spectrometer concept, the grating is fixed in its relation to the incident energy beam.

The optical components of this instrument are arranged so that the grating is
imaged on the detector, thus again effectively integrating over the intensity variations across the target. In the multidetector radiometer the in-flight calibration
source is placed over the entrance slit of the instrument, since no scan mirror is
employed in the design shown.

3.4.4 Spectrometers

The concept of the fixed grating instrument described above utilized the advantage gained by viewing the target in several wavelength intervals simultaneously, to improve the signal to noise ratio obtainable in a short interval of time.
It gained this advantage by sacrificing the ability of determining the exact shape
of an absorption line, or—more correctly—a portion of the absorption band
consisting of a number of lines. In its application to the temperature sounding
problem this is not a serious drawback, since the major CO_2 bands which are
being used have been studied in considerable detail in the laboratory, and the line
shapes are quite well understood.

In the determination of atmospheric composition as well as in other investigations which rely on the measurement of individual lines caused by molecular
transitions of certain gaseous species, it is advantageous to "scan" the spectrum to
describe the exact shape of the lines. This is the task for which the spectrometer
has been used traditionally. A typical instrument design might look very similar
to the optical arrangement shown in Fig. 7, with the essential difference being that

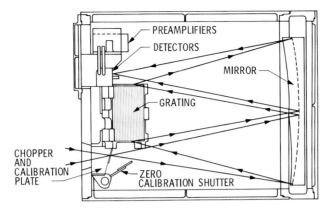

Fig. 7. Infrared multidetector radiometer optical configuration

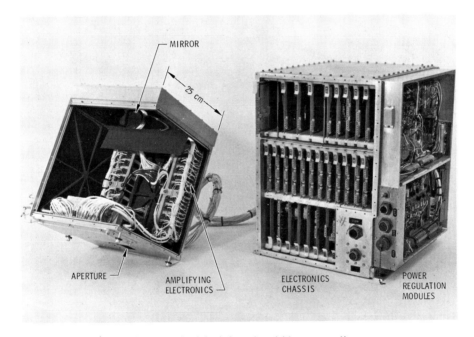

Fig. 8. Photograph of the infrared multidetector radiometer

the grating moves to change the angle between the incident energy beam and the normal to the plane of the grating, and that a single detector is employed. Depending on the scanning time desired per point and the total wavelength interval to be covered, a quasi-continuous spectrum may be reproduced. The instrument shown in Fig. 7 is capable of a 0.40 μ spectral resolution, dictated primarily by the temperature sounding requirement. Flight worthy grating spectrometers can potentially achieve resolution improvement over this by one or two orders of magnitude.

An excellent example of a spectrometer achieving this level of spectral resolution is shown in Fig. 9 (see p. 360). FARMER et al. (1972) describe the instrument performance. Although the primary operating mode of the spectrometer in this instance does not require the grating to move and relies on the output of five separate detectors, the instrument is also capable of scanning the portion of the near infrared spectrum in which the molecular transitions due to trace quantities of water vapor occur. For the detection and mapping of this trace element in the Martian atmosphere, which in comparison to that of earth has a less complex composition and hence fewer absorption features which are overlapping, the throughput obtainable at its spectral resolution is sufficient. For similar measurements on the earth at yet higher spectral resolution and wider wavelength coverage, different instrument concepts are required.

3.4.5 Interferometers

One instrumental concept which promises to bring the conventional high resolution spectroscopy practiced in the laboratory to the satellite applications is the interferometer-spectrometer. Through the development of modern computer technology it has been possible to reconstruct fairly routinely a conventional spectrum in spectral frequency or wavelength space from an interferogram, as generated for the first time by MICHELSON almost a century ago. The familiar interferometer records the constructive and destructive interferences caused between two beams as they recombine after having been separated into two paths, one of which is varied in length either continuously or incrementally be a reflector precisely controlled in alignment to keep the energy beams in a coherent relationship. The amount of interference observed at the detector as a function of path difference is the interferogram, which represents the frequencies present in the spectral interval admitted into the instrument. The spectrum which is used in the data analysis is obtained by performing a standard Fourier transform over the frequency interval sampled by the interferometer.

The difficult and complex data analysis and the engineering problems inherent in a design which requires that optical components be moved in a manner which maintains alignments accurate to dimensions of the order of wavelengths of light, is justified if one considers the large energy throughput advantage gained in an interferometer. These advantages are gained over conventional spectrometers in two ways: the detector can "see" the total energy entering the instrument throughout the recording of the interferogram, while the spectrometer can only observe a narrow wavelength interval at one time; secondly the aperture of the instrument can be made considerably larger than that of a spectrometer, in which case the entrance slit width also defines its ultimate spectral resolution. By overcoming the design difficulties and taking advantage of the increased energy available at the detector, an interferometer-spectrometer was flown successfully on the Nimbus 4 earth orbiting satellite and subsequently performed an important function in the scientific mission of Mariner 9. The Mariner 9 Infrared Interferometer-Spectrometer is shown in Fig. 10.

A more recent development for interferometric measurements at shorter wavelengths, meeting higher spectral resolution requirements for the analysis of

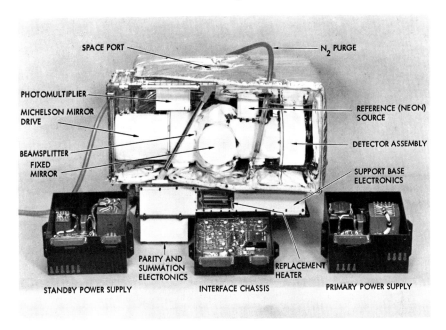

Fig. 10. Infrared interferometer-spectrometer

stratospheric trace gas composition in the earth's atmosphere is described by SCHINDLER (1970). FARMER (1974) discusses some of the observations which were made utilizing this region of the spectrum at the improved resolving power.

3.5 Applications

Our discussion of remote sensing with infrared techniques so far has been devoted to the description of the potential of these methods for making contributions to the increased understanding of our sphere of existence. All scientific discoveries and engineering developments in this area would go for naught, were it not for the fact that meaningful results have been obtained from infrared experimentation on the surface of the earth, in the atmospheric regions near the surface and at stratospheric altitudes, and in more spectacular fashion from earth satellites and interplanetary probes. The number of investigations which have been, and are currently being carried out with the objective of gaining a better understanding of our resources and our environment are too numerous to list. Suffice it to state that interest and activity is world-wide and growing at an ever increasing rate. We are faced, therefore, with a decision which might challenge the wisdom of a Solomon: How to best exemplify what has been accomplished in this field to date without omitting the most powerful examples. The approach which will be used is to cite a few typical examples which resulted in a significant advancement in the understanding of the processes which influence the infrared radiation field; each example being advanced strictly because its existence happens to be known to the author of this chapter.

One of the earliest concepts of truly remote sensing in the infrared from satellites concerned the study of temperature fields on the earth's surface from the first successful satellite in the Nimbus series. This investigation used radiation measured in one of the window channels—a wavelength interval between 3.4 µ and 4.2 µ where the atmosphere was thought to be relatively free of absorption features—to map the variation of surface brightness temperatures along a swath below the path of the Nimbus satellite. In this investigation the variations in the data received could be ascribed either to the variation of the true temperature or to the variation in surface albedo (emissivity). A report on this first infrared experiment from a satellite is given by NORDBERG (1965). We will not attempt to list here the many significant results of this first look at our globe with remote sensors in the infrared, as this is beyond the scope of the chapter. The reader is urged, however, to review this paper, as this quite comprehensive report represents a significant milestone in the history of remote sensing with infrared instruments.

More recently, sensing of thermal radiation emitted in window channels had another quite spectacular demonstration in an experiment flow aboard the Mariner 9 orbiter around the planet Mars. From the data collected by the Infrared Radiometer, in the 8–12 µ region in one channel and the 18–25 µ interval in a second channel, scientists were able to determine for example that the south polar cap consisted primarily of CO_2 ice. The experiment's final report lists these findings as well as other noteworthy results, see NEUGEBAUER et al. (1971).

At the time of the flight of the first Nimbus spacecraft, a much more sophisticated experiment involving the sensing of emitted thermal radiation was being devised by the U.S. Weather Bureau. This experiment and its succeeding versions sponsored by both NOAA and NASA (the National Oceanographic and Atmospheric Administration and the National Aeronautics and Space Administration) uses the radiation emitted by atmospheric CO_2 to derive the temperature of this well known and evenly mixed atmospheric constituent. By carefully analyzing the amount of radiation emitted at various wavelength intervals in the wings of one of the absorption bands of CO_2, emission contributions from different atmospheric layers can be determined and thus an atmospheric temperature profile is obtained. The short wavelength wing of the 4.3 µ band of CO_2, shown in Fig. 3, has been utilized for this purpose by some investigators. The very complex data analysis required to arrive at good temperature profiles is the subject of a considerable amount of research and can be traced by the interested reader from more detailed and complete discussions listed in the references. KAPLAN (1959) was the first to suggest that information on the atmospheric temperature profile was contained in the upwelling radiation. An excellent summary of the development work in the theoretical and experimental fields of temperature sounding which followed the original concept is given by HOUGHTON and TAYLOR (1973) along with some experimental results obtained. Today most meteorological satellites carry some form of temperature sounding instrumentation making infrared measurements in the CO_2 bands.

As a result of extending the measurements of atmospheric temperatures from the very localized balloon launched radio sonde coverage in populated areas of the globe to a global coverage from meteorological satellites, the accuracy of weather prediction—which relies chiefly on accurate knowledge of atmospheric temperature profiles—is steadily improving, and the ultimate goal of meteorologist for long range weather forecasts is moving into the realm of possibilities.

To quantitatively discuss these applications in a few paragraphs is impossible. Since the paper by HOUGHTON and TAYLOR (1973) gives a very complete review of this topic, we shall proceed directly to the applications which relate to surface and atmospheric composition.

Experiments using the data collected on the sunlit side of the earth in the 1–3.5 μ region—the solar reflection region—have recently been introduced into the space program as instrumentation technology has advanced sufficiently to open this region to the interested investigators. Geological features, which had been studied in some detail at wavelengths in the visible part of the spectrum, were subjected to infrared scrutiny by the Multispectral Scanner (MSS) aboard the Earth Resources Satellite. Among the many interpretations of the data collected by this instrument which are available to the investigators are vegetation surveys, mineral mapping and surveys of water resources, all characterized by unique variations in the reflectivity at selected wavelength intervals.

A striking example of this type of investigation is shown in Fig. 11 (see p. 361). Here the marked increase in reflectivity near 0.9 μ of vegetation covering the earth's surface is depicted in the red colored regions of the photograph. By choosing the contrasting red color to represent this wavelength interval (MSS channel 7), vegetation in the Southern California region near Los Angeles is easily discernible to the naked eye.

Three wavelength regions are superimposed on the picture and represented by colors. These are:

channel 4, covering 0.5–0.6 μ and depicted in blue
channel 5, covering 0.6–0.7 μ and depicted in green
channel 7, covering 0.8–1.1 μ and depicted in red.

The overlayed photograph, taken from the satellite on a particularly cloud free day over the Los Angeles basin, demonstrates with striking clarity the potential of this technique of studying the earth's surface. Clearly visible are earthquake prone faults such as the San Andreas, extending across the width of the picture, and the Garlock adjacent to the western edge of the Mojave desert north of the city's mountain range. Urban developments, characterized by regions tinted red from vegetation, contrast with the stark, blue-white city. Water, which is totally non-reflecting in the infrared region of channel 7, shows as totally black areas.

Figure 11 demonstrates the capability of the remote sensing instrument to collect infrared data which enhances information visible to the human eye by accentuating certain features—vegetation in this instance. The applications of this enhancement technique to agriculture and related fields is patently obvious. A less obvious application of the same technique for purposes of mineral detection is exemplified by the MSS frame shown in Fig. 12 (see p. 362). The photograph is a display of the ratios of data points obtained in four channels of the MSS; the additional channel (6) covering the wavelengths between 0.7 and 0.8 μ. The channel 6 to channel 7 ratio is of particular interest for the detection of minerals containing Fe^{+++} in the soil near gold deposits in the area of Goldfield, Nevada which lies in the central portion of the frame. Fe^{+++} shows a very strong absorption in channel 7 and reflects energy quite well in channel 6. A green color was chosen to represent the 6–7 ratio in the photograph. The remarkable result is that the region containing a known gold deposit near Goldfield is indeed ringed by green areas. Other areas in the frame show the same green marking and hold promise of gold or silver deposits, thus providing a valuable map to the mineral resources potentially available. The coloring of the ratios of channels 4–5 and 5–6 were chosen to facilitate visual identification of natural features for quick mapping purposes.

In the photographs shown in the last two figures we tend to lose sight of one of the most important features of the data gathered by the Multispectral Scanner aboard the Earth Resources Satellite: the picture represents a mosaic of data points, each representing a rectangle of about 60 by 80 m on the ground. Since the information is stored on magnetic tape, the researcher is able to retrieve the average reflectivity of each spot in each wavelength interval in 64 steps ranging from totally reflecting to totally absorbing.

The data thus stored on tape were used in a study of water quality of a number of inland lakes in the Wisconsin region of the USA. In this particular instance the infrared channel 7 of the MSS was used to isolate the lakes from the surroundings and depict them on a clear background, as shown in Fig. 13 (see p. 363). The ratio data obtained from the other channels were used to construct a classification of each lake's water quality from the oligotrophic or "clean and low in nutrients", to the eutrophic (high in nutrients). In addition it was possible to identify and map water pollution sources from the reconstructed picture.

The examples cited so far illustrate what can be learned from remotely sensed data obtained in the infrared at low spectral resolutions. In the beginning of the chapter we referred to these determinations as the establishment of "bulk" properties of our environment. We also devoted considerable space to a discussion of molecular processes which are detectable only with instrumentation capable of achieving higher spectral resolution.

From measurements of thermal radiation ermitted at preselected wavelengths in very narrow intervals, sufficiently narrow to encompass isolated emission lines of gaseous molecules, inferences regarding the composition of the atmosphere can be made. Because the absolute value of the emitted radiation is so strongly dependent on the temperature, a quantitative determination must include independently derived information regarding the thermal state of the gas in the entire emission path. The complexity of the required analysis is exemplified in the work of SMITH (1970), reporting on the interpretation of measurements of atmospheric water vapor from instrumentation covering the H_2O rotation band near 20 μ and 40 μ as performed from the Nimbus 3 satellite.

Experimental determinations of the composition of the atmosphere which are less dependent on the exact knowledge of the temperature profile, have been made by utilizing the sun as a source of known, constant radiation intensity and observing an atmospheric column in absorption.

Pioneer work in this field was conducted from balloons by investigators at the University of Denver. MURCRAY et al. (1967) open to the reader this field of absorption spectroscopy from observations of the sun through atmospheric layers. In this particular case of observing the sun directly, the wavelength region over which "source radiation" predominates over self emitted thermal radiation extends out to about 7.5 μ. Results of solar observations through long atmospheric paths have already helped considerably in our understanding of the composition of our atmosphere, particularly in stratospheric regions.

FARMER (1974) describes the techniques of determining atmospheric composition with the aid of high resolution absorption spectroscopy. From this work we have chosen to show one example which led to the detection and abundance determination of Nitric Oxide (NO) in the stratosphere. Figure 14 shows the spectrum obtained with a high resolution interferometer from an aircraft flying at stratospheric altitudes (18 km) and observing the sun at sunrise. The absorptions due to the molecular transitions of some of the atmospheric species present in the path are marked. The transitions due to NO are discernible as small perturbations in the background radiation and, upon analysis, yield an abundance value of less than one part per billion. It is significant that even this seemingly low abundance is thought to have a noticeable effect on the ozone layer in the stratosphere which protects living organisms on the earth's surface from harmful ultraviolet radiation. Thus in pollution studies, where the detection of trace quantities of gases is of paramount importance, the scientist is faced with extremely difficult measurement problems, which—when overcome—are followed by extremely complex analysis problems. It is in this arena that the challenge of effectively utilizing the wealth of information present in the radiation field awaits the investigator bent on protecting the environment and managing the resources of our planet.

3.6 Data Interpretation Error Sources

A discussion of of the methods of remote sensing in the infrared cannot be complete without a few remarks on the subject of the infrared-peculiar difficulties encountered in the interpretation of instrument data. Infrared radiance arriving at the instrument aperture has, as we have seen, imprinted on it a wide variety of information pertaining to the source of the radiation and the intervening medium. Although this is a boon to the experimenter searching for a specific characteristic by means of remote sensing, it requires that he also be acutely aware of all other data present in the information and make appropriate corrections. Thus, for example, the geologist searching for reflectivity changes on the earth's surface must have a good familiarity with the absorptive characteristics of the atmosphere. We list here a few of the most influential overlapping data—sources of error in experiments—which can be expected for the types of radiation we considered earlier.

Since measurements made in emission represent the temperature state of the source, they are quite obviously strongly affected by the temperature state of the intervening medium. For example in the case of an experiment designed to deter-

P. W. SCHAPER

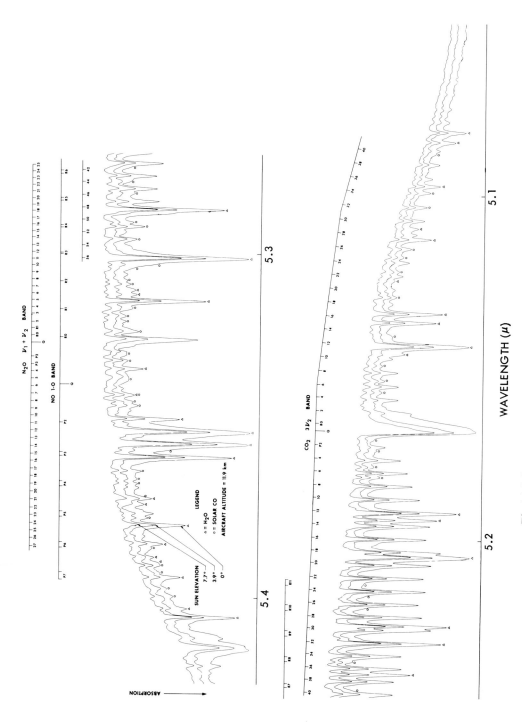

Fig. 14. Infrared solar spectrum taken at stratospheric altitude

mine the composition of the atmosphere by analysis of long wavelength infrared data, it is necessary to ascertain first that there is a measurable difference in source temperature and "unknown constituent" temperature,. lest the absorption of source radiation by the atmospheric constituent is completely nullified by its own self emission. An excellent treatment of this type of error is given by FARMER (1972) for the case of the observations of water vapor on Mars.

Less obvious, but equally important is the effect which the intervening atmospheric layer can exert on the measurement of surface temperature. Although, as we pointed out before, the wavelength interval measured is chosen purposely large enough to minimize this error, it exists and manifests itself in an ultimate accuracy limitation on the measurement of temperature by remote sounding techniques with a single element radiometer. CHAHINE (1970) shows that this error is in the neighborhood of several degrees Kelvin on the earth, and that even the complexity of adding a separate spectrometer to determine the state of the atmosphere in the path does little to improve the best achievable accuracy.

At the shorter wavelengths of the infrared we noted that the existence of high resolution instrumentation has led to the exploitation of this region for the determination of atmospheric composition through the application of absorption spectroscopy to the analysis of data. Temperature effects on the relation of the depth of absorption features to the abundances of gaseous species are generally second order effects. The main source of error found in these types of investigation pertain to the lack of detailed knowledge of the spectroscopic parameters. The infrared region below about $8\,\mu$ is abundant with molecular transitions of common atmospheric constituents, as well as many trace elements of importance to the study of the atmosphere. The numerous overtones of fundamental absorption bands of gaseous species require exact knowledge of the center frequencies of all lines making up the bands, so that overlapping lines can be separated from those which are amenable to quantitative analysis, which in turn requires precise knowledge of line strengths.

Thus the spectroscopist who wishes to use remote sensing infrared data for atmospheric composition studies finds himself limited in the accuracy of his results by the lack of sufficient laboratory data. The higher the spectral resolution employed, the more serious becomes the lack of availability of information. It is also worthy of note that the lack of information on the behavior of absorption lines at low pressures and temperatures is even more pronounced, making stratospheric investigations even more difficult.

The study of reflectivity changes in the near infrared, if carried out in regions relatively free from atmospheric absorptions, poses fewer problems with data interpretation. Its largest common source of error stems from the fact that on the earth reflectivities changes rapidly and frequently, making high spatial resolution mandatory.

Finally we come to the problem of clouds. Although it has been established that, on the average, fifty percent of the globe is covered at any time by clouds (which are opaque in the infrared), information from breaks (or holes) in the cloud layer can be interpreted. In order to identify the source of radiation of a given portion of the field of view containing multilayer, partial cloud cover, several experimental techniques have been proposed. Currently the "clear column radiance" penetrating through a broken cloud layer is being determined by a technique which relies on its knowledge of the surface temperature. This technique was developed by CHAHINE (1970) and its application was demonstrated by SHAW et al. (1970). A more recent technique, which requires no *a priori* information to correct for the presence of clouds, was developed by CHAHINE (1974). This method has not yet been verified experimentally, however. The most successful experiments so far have been conducted in cloud free situations; that these occur at a satisfying frequency is evidenced among many other examples by the success of the Multispectral Scanner aboard the Earth Resources Technology Satellite.

3.7 Expectations of the Near Future

It is not our intent to conclude the discussion of the remote sensing of infrared radiation with speculations of what the future for this experimental technique might be. We have so far attempted to refrain from enumerating techniques which are in the stages of design or preliminary application. Nevertheless these "breadboarding" efforts tend to establish the trend which future developments are likely to take, and thus it appears appropriate to include them at this point.

The field of atmospheric constituent analysis has always demanded high spectral resolution, and there is no reason to expect that this trend will abate in the near future. The most serious difficulty with existing instrumentation for the nadir-viewing composition analysis is brought about by the low signal levels radiated from the earth in the most interesting and promising region of the spectrum: the $3-8\,\mu$ interval. In addition to the fact that the low level signals available require long viewing times (resulting in poor spatial and spectral resolutions), they also prohibit wide spectral coverage with one instrument. The latter inhibits the simultaneous sensing of data which might be used to determine atmospheric temperature profiles which are essential to the data interpretation. The interferometer has been the most successful instrument in answering this demand so far, and more sophisticated instrumentation featuring high resolving powers at continuing high quality signal to noise ratio are in the final stages of development. The most promising technique under development appears to be in the concept of the laser heterodyne spectrometer, which uses a laser as a source of highly concentrated energy of variable wavelength. The return signal reflected from the ground is detected in a heterodyne receiver which produces a spectral output whose resolution approaches the bandwidth of the original laser source. A good, detailed description of the potential of this technique is that of MENZIES et al. (1974).

Along with the higher resolution instrumentation which hopefully will be available for pollution studies, one expects that the longer wavelength regions, especially the $9-15\,\mu$ interval, will become more accessible than it has been so far, primarily through continued improvements in the analysis techniques. The complex molecules involved in the photochemical processes governing urban pollution have their transitions occurring in this region, implying, therefore, that world wide mapping of tropospheric pollutant migrations—from source to potential sink—is dependent on the remote sensing techniques in the thermal emission region.

The much publicized success of the Earth Resources Technology Satellite and the Skylab mission in surveying the earth with scanners operating not just in the visible, but displaying near-infrared-sensed features in photographic format to a wide user community, are an indication of what can be done within today's state of the art. The potential of not only providing data on surface features and cloud cover on an international scale, but displaying air quality information in pictorial form through continual analysis of our atmosphere with remote sensors receiving radiation throughout the infrared region of the spectrum, is indeed a goal worth the considerable effort now being spent on developments in this field. The 1980 era of the Space Shuttle flights is destined to become an early major milestone on the road toward this goal, as one envisions the scientist in the Shuttle-Space Lab developing, through real-time observations of our planet through infrared eyes, the methods of remote sensing in this spectral region to their ultimate application as automated watchdogs over our environment.

References

CHAHINE, M.T.: Inverse problem in radiative transfer: Determination of atmospheric parameters. J. Atmos. Sci. **27**, 960–967 (1970).

CHAHINE, M.T.: Remote sounding of cloudy atmospheres I. The single cloud layer. J. Atmos. Sci. **31**, 233–243 (1974).

FARMER, C.B.: Infrared measurements of stratospheric composition. Canad. J. Chem. **52**, 1544–1559 (1974).

FARMER, C.B., LAPORTE, D.D.: The detection and mapping of water vapor in the Martian atmosphere. Icarus **16**, 34–36 (1972).

HOUGHTON, J.T., TAYLOR, F.W.: Remote sounding from artificial satellites and space probes of the atmospheres of the earth and the planets. Rep. Prog. Phys. **36**, 827–919 (1973).

KAPLAN, L.D.: Inference of atmospheric structure from remote radiation measurements. J. Opt. Soc. **49**, 1004–1007 (1959).

MENZIES, R.T., CHAHINE, M.T.: Remote Atmospheric sensing with an airborne laser absorption spectrometer. Appl. Opt. **13**, 2840–2849 (1974).

MURCRAY, D.G., MURCRAY, F.H., WILLIAMS, W.J.: A balloon-borne grating spectrometer. Appl. Opt. **6**, 191 (1967).

NEUGEBAUER, G., MUNCH, G., KIEFFER, H., CHASE, S.C., MINER, E.: Mariner 1969 Infrared radiometer results: temperatures and thermal properties of the Martian surface. Astron. J. **76**, 719–749 (1971).

NORDBERG, W.: Geophysical observations from Nimbus I. Science **150**, 559–572 (1965).

SCHAPER, P.W., SHAW, J.H.: Performance of a spectrometer for measuring the earth's radiance near 4.3 μ. Appl. Opt. **9**, 924–928 (1970).

SCHINDLER, R.A.: A small, high speed interferometer for aircraft, balloon and spacecraft applications. Appl. Oct. **9**, 301–306 (1970).

SHAW, J.H., CHAHINE, M.T., FARMER, C.B., KAPLAN, L.D., MCCLATCHEY, R.A., SCHAPER, P.W.: Atmospheric and surface properties from spectral radiance observations in the 4.3—micron region. J. Atm. Sci. **27**, 773–380 (1970).

SMITH, W.L.: Iterative solution of the radiative transfer equation for the temperature and absorbing gas profile of an atmosphere. Appl. Opt. **9**, 1993–1999 (1970).

4. Laser Applications in Remote Sensing

R. T. H. Collis and P. B. Russell

4.1 Introduction

The invention of the laser in 1960 was undoubtedly one of the most significant technological advances of a productive century. It marked the beginning of a period of intensive exploration and development as to its use that is still in its infancy. Nevertheless progress has been rapid and whole new areas of capability have already been firmly established. The possibility of applying laser energy for remote sensing was recognized from the very beginning, at least in basic form, and was among the first applications of laser technology to achieve some measure of practical accomplishment. The reasons for this probably lie in the earlier achievements of analogous or similar techniques. As described below, the principal laser application in remote sensing is based upon the radar principle. As it happens some of the very first applications of the radar principle were made with light long before the development of microwaves—and were concerned with remote probing of the atmosphere!

In the 1930ies searchlights were first used for probing the high atmosphere, using what would now be called a bi-static radar configuration with continuous wave (CW) energy. The atmosphere, illuminated along the searchlight beam, scattered light energy (depending upon the gaseous density or the presence of particles), and this energy was detected by a photodetector "receiver" located at some distance from the searchlight. The height from which the scattering was received could be selected by suitable adjustment of the angle of the receiver's beam. Somewhat later, in 1939, a closer analog to radar was developed in France using spark generated pulses of light to measure cloud base in a system in which height was determined by the elapsed time between generation of the pulse and the instant of its reflected return. Although both these concepts have been successfully utilized up to the present time in various forms, their accomplishment has been limited by inherent technical difficulties. The advent of the laser, however, as a source of optical (or near optical) energy of great intensity, particularly in pulsed form, opened the way to much fuller exploitation of these concepts, and also to their extension in more sophisticated techniques that take advantage of the spectral purity and wave nature of the laser energy. There is little doubt that familiarity with microwave technology influenced the conception of many of these techniques. At least one group of innovators (that led by the late Myron G. H. LIGDA) initially considered laser radar (or LIDAR) in basic backscattering applications in terms of weather radar (COLLIS, 1966). Other applications exploiting the wave nature and spectral purity of lasers have, however, been more influenced by classical wave optics experience and spectroscopy.

Whatever the inspiration, we find after over a decade of development that the capabilities of lasers have been widely recognized in a considerable range of atmospheric remote sensing applications, and in applications concerned with remotely observing the earth's land or water surface from aircraft or space vehicles. While in some cases these applications have become firmly established as effective techniques capable of providing unique observational data, it would be misleading at this time to suggest that even these applications are widely used or are of a routine nature. The accomplishment is much more limited—and although they have made and are making very valuable contributions in research applications, for the most part the development of laser applications for remote sensing is still very much in the research stage.

As described in detail below, the achievements to date are, however, by no means inconsiderable and the potential and promise of laser applications is undoubtedly very great indeed.

4.1.1 Scope and Definitions

In this chapter we will explore these potentials after first describing the fundamentals of the various techniques. In doing so we will give major attention to those techniques that have already achieved importance as useful contributors to the study or surveillance of the environment. In view of their potentiality we will discuss other possibilities in some detail, even though they are currently at an embryonic stage. We will be careful, though, to avoid too ready acceptance of ambitious claims that are not uncommon in the heady enthusiasm accompanying the development of a new technology as exciting as the laser.

Following the general context of this book we will cover the concept of remote sensing in respect to the remote observation or measurement of the condition of the earth's surface and atmosphere. This covers a very wide range of viewpoints, from that of orbiting satellites to fixed, upward looking ground stations—and a wide range of targets—from pollutant gases in the atmosphere to concentrations of algae in the ocean. However, we will not treat explicitly those techniques concerned with measurement at a distance of phenomena such as air flow in wind tunnels or rocket exhausts or, say, the regulatory application of pollution sensing devices such as stack monitors—even when these can be employed at a distance.

By "lasers" we mean the various types of generators of electromagnetic energy at optical or near optical wavelengths that employ amplification by stimulated emission, to convert input energy to essentially monochromatic outputs in pulsed or continuous wave form having a high degree of spatial and temporal coherence. (The nature and characteristics of lasers and laser energy are described further below.)

4.1.2 Technical Approaches

In remote sensing application such laser energy sources are used in the following ways:

1. Lidar (or laser radar)—in which information on the distant target is derived from the manner in which the transmitted energy is backscattered, reflected or reradiated by the target material. This technique also provides the capability of ranging and delineating direction, which makes it possible by scanning to determine the distance, shape and structure of extended features of the observed target in addition to its inherent local characteristics.

The target may be a gaseous volume of the atmosphere or the land or water surface, or underwater features.

2. Path absorption techniques—in which information relating to the path along which energy is transmitted or received is derived by comparing radar observations (or point-to-point transmissions) at two or more frequencies. For example, by employing one frequency in the resonant absorption line of a given gaseous pollutant and a second frequency just outside the absorption line, the ratio between radar returns from a distant reflector—e.g., the sea surface, at these frequencies may be measured. This will be indicative of the total amount of the pollutant present along the path.

In a somewhat different manner, the special characteristics of tuneable lasers have been applied to passive radiometric techniques. By employing such lasers as local oscillators in coherent detectors, passive detection of characteristic emissions from distant targets or even pollutants may be achieved with remarkable sensitivity (HINKLEY and KELLY, 1971). This application of lasers in remote sensing is, however, a matter of instrumentation technology rather than a new laser sensing technique.]

4.2 Fundamentals of Physical Processes Involved

4.2.1 Specular Reflection and Elastic (Non-resonant) Scattering

Used in radar fashion, laser devices (lidar) can be used to observe distant targets in various scanning modes, to provide two-dimensional information on the shape or structure of distant solid or liquid surfaces. With such targets, energy is backscattered or reflected by the material surface to a degree depending upon the wavelength of the laser energy and the physical characteristics and spectral reflectivity of the material. In this basic application, the use of laser energy is very similar to microwave radar applications at longer wavelengths of the electromagnetic spectrum. The use of laser radar in this role is obvious in such applications as terrain mapping or investigating the state of the sea surface. In the case of water surfaces, if the water is sufficiently transparent to the laser energy (and if the system is sensitive enough), information may be obtained on the form and depth of the underlying bottom, i.e., bathymetry. As with more conventional forms of light (or near IR) energy, because of the short wavelength, laser energy is also scattered by small particles. Again the degree of scattering is a function of the wavelength and the size, shape and dielectric properties of the particles. At one extreme the molecules of a transparent liquid or the gaseous atmosphere cause scattering. This makes it possible, given adequate performance, to derive information on density (or temperature) of the medium by measuring the magnitude of its backscattering at any given laser frequency. Gaseous backscattering, very small though this is, is readily detectable with current lidars. They thus have a much easier task in observing larger particles—even the sub-micron size particles of the atmospheric aerosol—even in very clear atmospheres. Where such particles are present in considerable concentrations—as in haze layers or smoke plumes—they can easily be mapped and tracked by lidar observations—even when they are not perceptible to the eye. The larger droplets of clouds and fogs or precipitation of course provide very much stronger backscatter signals, and current lidars have no difficulty in observing such targets at considerable ranges. But we are antici-pating, for here our purpose is to identify the basic principles of laser probing techniques, rather than to describe their applications.

We note therefore that the principle upon which such observations of particles and droplets is based is that of elastic scattering i.e., the phenomenon in which the wavelength of the incident energy is unchanged by the scattering process. The interaction of the incident waves of electromagnetic energy with the dielectric discontinuity presented by each particle suspended in the propagating medium modifies the propagating wave energy, so that a certain fraction is scattered. That part which is backscattered in the direction of the lidar receiver is available for detection, and what is scattered out of the forward direction of the transmitted beam is lost and results in attenuation of the transmitted energy. (Significant multiple scattering can and does occur where the assemblage of particles is dense—as in fogs and clouds.) The fraction of energy elastically scattered as a function of angle is, as has been noted, a function of the wavelength, and the size, shape and dielectric properties as described by the refractive index of the material of which the particles consist (for the wavelength in question). Where the refrac-tive index is complex, some energy is absorbed by the particle. At optical wave-lengths an important effect of this absorption is to markedly modify the angular scattering pattern. This further complicates the relationship between the energy backscattered by an assemblage of particles and the incident flux of energy. This relationship is, of course, of prime importance in the interpretation of atmo-spheric lidar observations, since at the very least it is desirable to derive some information on the number or mass concentration of particles detected, even if only qualitatively. The problem is most acute where the particle sizes are compa-rable to the wavelength—for in this range the backscattering cross section of particles is not simply or linearly related to their geometric cross section. A more complete treatment of this complicated subject is available in a number of texts (VAN DE HULST, 1957; DEIRMENDJIAN, 1969; KERKER, 1969), but is mainly limited to a consideration of spherical particles, based upon the classic work of Mie. Where the particles are very small compared with the wavelength, as with gaseous molecules, they act as simple dipoles, and their backscattering properties are readily described by the Rayleigh approximation (in which the backscattering cross section is directly related to the sixth power of the particle diameter). The straightforward relationship between molecular number density and volume backscattering coefficient of a gas makes possible the remote evaluation of atmo-spheric density (temperature). In the so-called Mie scattering range where larger particles are involved (say with radii from 0.1 to 10 times the wavelength), volume backscattering coefficients are less readily relatable to the number or mass concentration of the particles present. Because averaging occurs, however, in a broadly polydisperse assemblage, the amount of backscattered energy may be quite well related to the number (and hence mass) of particles present. The accurate assessment of particle mass or number concentration by remote lidar observations is a matter of some difficulty, to which we will return in more detail later. By the same token, elastic scattering by itself offers no capability for determining the composition of particles, and only a limited, and as yet undemonstrated, capability for determining the particle size distributions of aerosols, if multiple wavelength observations can be made simultaneously. Lidar observations of solid and liquid surfaces offer some possibilities in terms of spectral discrimination for identifying the nature or composition of such surfaces. For a more specific capability of identifying the composition of the material observed we must turn to more sophisticated interactions of the wave energy and the target materials.

4.2.2 Inelastic and Resonant Scattering

Energy emitted by lasers is, to a substantial degree, monochromatic—i.e., it is effectively all concentrated in a narrowly-defined frequency or wavelength interval. This unique capability has opened up the practical exploitation of a number of processes that formerly could only be explored by filtering broadband light sources—an approach that produced only minute residual levels of monochromatic energy. ·

By providing useful amounts of energy at discrete and determinable wavelengths, lasers can make practical use of such processes as Raman scattering, resonance Raman scattering, off-resonance fluorescence, resonance fluorescence, broad fluorescence and resonant scattering. The first five of these scattering processes are inelastic—i.e., the light emitted by a target has a frequency differing from that of the incident light. The sixth, resonant scattering, is elastic, but is closely related to the others as discussed below. The inelastic and resonant elastic scattering processes all occur as a result of the fact that the atoms or molecules comprising a target substance may exist in many different states, each having (in general) different internal energies at discrete values separated by "forbidden" intervals. (The states correspond to different configurations [orbits] of atomic electrons, or to different configurations of atoms within a molecule.) The inelastic and resonant elastic processes are therefore all related to one another, and this relatedness has in the past led to some confusion in the nomenclature describing certain interactions between light and matter. In the present discussion we will follow the nomenclature of Fouche (1974) (see also Chang and Fouche, 1972). Other useful references, which contain more detail than is possible in this short review, are Kildall and Byer (1971) and Schwiesow (1972).

The various inelastic scattering processes are best understood with the aid of an energy level diagram, some examples of which are shown in Fig. 1. In such a diagram, the vertical position of a horizontal line is proportional to the internal energy (relative to that of the minimum, or ground, state) of a given molecular state. The gaps between lines are indicative of the fundamental quantum mechanical result that a molecule may exist only in certain allowed states, each with a

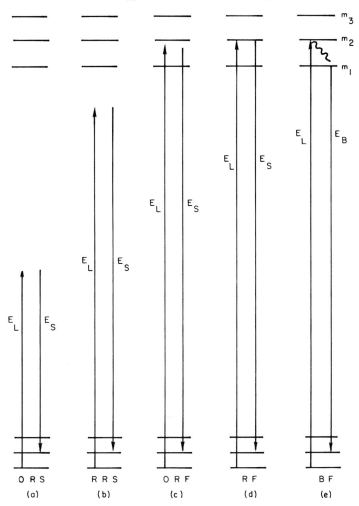

Fig. 1a–e. Inelastic scattering processes. (After FOUCHE, 1974)

corresponding narrowly-defined energy. A band of closely-spaced levels consists of different rotational and vibrational states (i.e., configurations of atoms), all with the same electronic state. The large energy gaps between different bands correspond to a change of the electronic state of one of the molecule's constituent atoms. Each species of molecule has a characteristic energy level scheme or spectrum which, as we shall see, provides a means of species identification.

A light photon of frequency v_l has an energy E_l given by the relation

$$E_l = h v_l,$$

where h is Planck's constant. When such a photon is incident on a target molecule, the photon may be absorbed, causing the internal energy of the molecule to be increased by the amount E_l. The resulting energy of the molecule and the mode of de-excitation or reemission of excess energy, determine the type of inelastic scattering process which takes place. These various alternatives are illustrated by the energy level diagrams of Fig. 1. In each diagram some vibrational levels of the ground electronic state and of an excited electronic state are shown. The initial, final and intermediate molecular states are labeled i, f, and m, respectively. E_l and E_s are the energies of the incident and scattered photons.

Case (a) is ordinary Raman scattering (ORS), where E_l is much less than the energy difference between the ground and excited electronic bands. Here the molecule cannot make a transition to any of the intermediate states in the excited electronic band. However, for a very short period of time ($t \lesssim 10^{-14}$ sec), the molecule may make a virtual transition, in which many intermediate states are very weakly excited. The molecule then returns to a state in the ground electronic band, in the process emitting a photon of energy E_s. If $E_s = E_l$, the molecule returns to its initial state, and the overall scattering process is merely the elastic Rayleigh scattering process described above in connection with the measurement of atmospheric density. If $E_s \neq E_l$, the final molecular state differs from the initial, and the process is called ordinary Raman scattering. The energy difference $E_l - E_s$ $= E_f - E_i$ is characteristic of the species of scattering molecule, so that measurement of this energy (or frequency) difference in a scattering experiment provides a means of species identification. Note also that in such an experiment the incident photon energy (hence laser frequency) is not restricted to specific values determined by the energy level spectrum of the scattering molecule. Thus the Raman scattering process is in principle of rather broad applicability. Its major drawback is the weakness or inefficiency of the scattering process: since the probability of the excited molecule decaying to any given final state with $E_f \neq E_i$ is considerably less than the probability for $E_f = E_i$, the Raman scattering process at any given frequency difference $E_s - E_l$ is much less probable than the Rayleigh scattering process. This makes measurable returns very difficult to obtain for any but the most abundant atmospheric gases (e.g., N_2, O_2, H_2O).

If the incident photon energy is increased, as in case (b) of Fig. 1, the scattering cross section [1] will increase as the excited molecule energy approaches that of the lowest state in the excited electronic band. Such a process, where the excited molecule energy remains below the lowest excited electronic state energy, is called resonance Raman scattering (RRS). It has the advantage over ordinary Raman scattering that the scattering cross section is larger, but a corresponding disadvantage in that incident photon energies (hence laser frequencies) must be chosen to fall near an energy level difference of the molecular species it is desired to detect.

---

[1] The scattering cross section C is a measure of the efficiency or strength of a given scattering process. It is defined as

$$C \equiv \frac{\text{number of photons scattered per unit time}}{\text{number of photons incident per unit area per unit time}}.$$

Alternatively, the cross section may be defined in terms of energies by substituting "energy" for "number of photons" in numerator and denominator.

In case (c) of Fig. 1 the excited molecule energy falls within the excited electronic band, but still outside the width of any vibrational level. Following FOUCHE (1974) we call this process off-resonance fluorescence (ORF), although it has previously often been called resonance Raman scattering (RRS). That is, the remote sensing literature has frequently made no distinction between processes b and c in Fig. 1. For a given molecular band, the cross section for off-resonance fluorescence is slightly larger than that for resonance Raman scattering.

Case (d) of Fig. 1 is resonance fluorescence (RF), where the excited molecule energy falls within the width of a vibrational level. For a given molecular band, the cross section for resonance fluorescence exceeds that of any of the processes described so far.

In case (e) of Fig. 1 the excited molecule energy again falls within the excited electronic band, but, before it can re-emit a photon, the excited molecule collides with another molecule and in the collison process makes a transition to a new intermediate state. The energy of the new intermediate state could actually be between the lines shown on the diagram, because energy may be lost to molecular translation, and a new rotational level may be entered. Therefore E_s can take on a nearly continuous range of values, so that this process gives rise to a broad continuum of energies in the scattered spectrum. Fouche calls this process broad fluorescence (BF), and shows that in general its cross section is greater than that for resonance fluorescence, and hence for any of the other inelastic scattering processes.

The scattering processes shown in Fig. 1 are all drawn with the final molecular state differing from the initial state, so that the processes are truly inelastic. We have already noted, however, that in case a the final state may be identical to the initial state, in which case the process is Rayleigh scattering. In fact, the final state may also be identical to the initial state in any of the other cases shown in Fig. 1. The resulting elastic processes are given the general name resonance scattering, and they have cross sections which greatly exceed the ordinary Rayleigh scattering cross section, thus permitting much greater sensitivity in detection of trace gases. Again we emphasize that resonance scattering is an elastic process, i.e., $v_s = v_l$. It is discussed in this section because its observation requires that the laser output be tuned to a specific frequency, due to its dependence on atomic or molecular absorption spectra. This requirement makes resonance scattering most akin to inelastic scattering in terms of applications, as will be shown in Section 4.

To summarize the sensitivity of the elastic, inelastic and resonant scattering processes, Table 1 lists the sizes of the respective cross sections. Also included are radiative lifetimes of the excited atom or molecule, that is, the mean time between photon absorption and photon emission, provided the excited atom or molecule is not otherwise perturbed. At tropospheric pressures the radiative lifetimes are of practical importance because the time between collisions of gas molecules is of the order of 10^{-9} sec. Collisions tend to de-excite molecules nonradiatively, and thus to greatly reduce (or "quench") the radiation which would otherwise be emitted. This effectively reduces some scattering cross sections and restricts the applicability of otherwise promising techniques to the upper atmosphere, where reduced pressures make collisions less frequent.

Table 1. Typical cross sections and radiative lifetimes of various interactions

Process	Cross section (cm^2 steradian^{-1})	Radiative lifetime (sec)
Rayleigh scattering	10^{-27}	$\lesssim 10^{-14}$
Mie scattering	10^{-27} to 10^{-8}	$\lesssim 10^{-14}$
Ordinary Raman scattering	10^{-30} to 10^{-29}	$\lesssim 10^{-14}$
Resonance Raman scattering	10^{-30} to 10^{-23}	10^{-14} to 10^{-8}
Off-resonance fluorescence	$\sim 10^{-23}$	10^{-10} to 10^{-8}
Resonance fluorescence	10^{-23} to 10^{-16}	10^{-8} to 10^{-1}
Broad fluorescence	$\sim 10^{-16}$	$\sim 10^{-8}$
Resonance (elastic) scattering	10^{-27} to 10^{-20}	10^{-14} to 10^{-6}

Radiative lifetimes are also important because they determine the accuracy with which target ranges may be determined in backscattering applications (see e.g., Kildall and Byer, 1972).

4.2.3 Resonant Absorption

The greatly enhanced scattering cross sections which occur when the frequency of the incident photon is tuned to match a molecular energy level difference correspond to similarly enhanced absorption cross sections. Thus the resonant absorption process may also be used with tuneable lasers to identify atmospheric and marine constituents and to measure their concentrations (through knowledge of their absorption spectra). Especially useful are the differential absorption techniques, where measurements on and slightly removed from an absorption line are made and compared. As shown in Section 4.4, these techniques may be employed in both single-ended and double-ended configurations.

4.2.4 Doppler Effects

Finally we consider special aspects of elastic scattering from which additional information may be derived from remote targets. The first of these is the application of the Doppler principle to targets in motion. The well known shift of frequency of wave energy backscattering from a moving target as a function of radial velocity applies, of course, at light frequencies just as well as at other parts of the electromagnetic spectrum (and sonic energy): $\Delta F = 2 V_R / \lambda$, where F is frequency, V_R radial component of velocity, and λ is wavelength.

The spectral purity and coherence of laser energy makes it possible to apply this principle in practical systems, and techniques have been developed to measure gas flow and the motion of aerosols remotely and also, since molecular motion is a function of the thermal state of gas, to deduce its temperatures at a distance.

4.2.5 Polarization

The second additional physical factor that can be exploited in elastic back-scattering applications is polarization. In most types of laser the emitted energy is largely linearly polarized, and it is readily possible to achieve a higher degree of linear polarization (and —less readily — to obtain various degrees of circular polarization). Since reflected or reradiated energy shows a modification of the incident polarization depending on the surface characteristics of solid surfaces or the shape of suspended particulates, a further source of information is available. In the case of specular reflection from smooth surfaces, the polarization of the incident energy is retained in that reflected. The maintenance of a high degree of linear polarization in the received energy is thus indicative of such a surface. Similarly, linearly polarized energy backscattered from perfect spheres retains its polarization. Thus, in atmospheric observations the degree of depolarization in the backscattered return makes it possible to distinguish between spherical and nonspherical particles.

4.3 Instrumentation

4.3.1 System Concepts

In the principal radar-type or "lidar", applications of laser energy, the following basic elements make up the system:
1. the laser transmitter
2. the collector and detector of backscattered energy
3. the data processing and presentation resources.

The laser transmitter, which is the essence of the active lidar techniques, provides all the energy available for the determination of the distant target's characteristics. Its power is thus a critical factor in the performance of the system, especially for those applications in which the scattering processes are very inefficient.

By far the most effective approach in all forms of active lidar systems is to use laser energy in high intensity pulsed form, although range information can be derived from continuous wave (CW) transmissions by suitable frequency modulation and phase comparison techniques, and CW energy is also used in Doppler applications.

The basic principles of the pulsed lidar approach are indicated by the following general equation for atmospheric scattering relating the received signal P_r, as a function of range, to the transmitted energy:

$$P_r(R) = E\beta(R)\frac{c}{2} \, R^{-2} A \, \exp\{-\int_0^R [\sigma_t(s) + \sigma_r(s)]ds\}$$

where E (joules) is the energy in the transmitted pulse (which may be expressed as $P_t\tau$, in terms of the power transmitted (watts), P_t, in a pulse of duration τ). With light propagating at velocity c, $c\tau/2$ specifies that length of the volume illuminated by the transmitted pulse, from which signals may be simultaneously received at any instant t. R is range, where $R = c(t - t_0)/2$ when t_0 is the time of transmission of the pulse. A is the area of the receiver aperture. β is the volume backscattering coefficient of the atmosphere (per steradian per unit length) and can be applied as appropriate to any of the forms of scattering described above. $\sigma_t(\sigma_r)$ is the extinction coefficient of the atmosphere, per unit length s, at the transmitted (received) wavelength. The term $\exp\{-\int_0^R [\sigma_t(s) + \sigma_r(s)]ds\}$ thus represents the attenuation of energy integrated over the two way path to and from range R, i.e., the two way transmittance. Equations for the various remote techniques described in Section 2 above may readily be derived from this equation. For example, if the observed scattering process is elastic, $\sigma_t = \sigma_r$. Additionally, if energy is reflected from a surface rather than a volume of scatterers, an appropriate coefficient of reflectivity, β_{reflect}, must be used, and P_r becomes almost entirely dependent upon P_t, regardless of the pulse duration τ, since energy is only returned from one range i.e., that of the reflecting surface.

To exemplify the manner in which non-elastic scattering techniques are applied, the following expression shows the return resulting from ordinary Raman scattering (ORS)

$$P_{r_{\lambda_{\text{Raman}}}} = P_{t_\lambda} \frac{c\tau}{2} \beta_{\text{Raman}} A R^{-2} \exp\{-\int_0^R \sigma_\lambda(s)ds - \int_0^R \sigma_{\lambda_{\text{Raman}}}(s)ds\}.$$

Where λ_{Raman} indicates the wavelength shifted from that transmitted, λ, which results from Raman scattering of the gas having a Raman backscattering coefficient β_{Raman} at the transmitted wavelength λ. The wavelength dependence of the extinction coefficient σ can also be used to measure the concentration of gaseous species with resonant absorption techniques by transmitting at two different wavelengths, λ_1 and λ_2, at and removed from a resonance line specific to a given gas. When this approach is used, the received power ratio $P_r(\lambda_1)/P_r(\lambda_2)$ provides an indication of the presence and concentration of the species in question along the path in question.

4.3.2 System Configuration

In practical systems the laser transmitter and the collecting optical system of the receiver are co-located and their beams overlap. A typical arrangement is illustrated in Fig.2 (ALLEN and EVANS, 1972). The high degree of spatial coherence of laser energy makes it possible to achieve great directionality, and very narrow transmitted beams (typically less than a milliradian in angle) are used—to achieve maximum concentration of energy and high resolution. Returned energy collected in the optical system of the receiver is converted to electrical signals by a photo-detector. (Because of the monochromaticity of laser energy, light of other wavelengths can readily be excluded from the detector by filters, thus reducing extraneous noise in the received signal.)

Figure 3 shows a pulsed ruby lidar built on the principles illustrated in Fig.2. The instrument is capable of scanning in elevation and azimuth. The electrical signals derived from the photo-detector (see below) represent the magnitude of the returned laser energy received as a function of time (and hence, range).

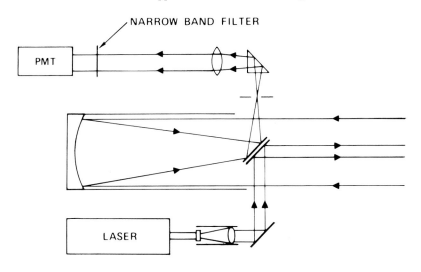

Fig. 2. Typical lidar system configuration. The narrow band filter shown ahead of the photo-multiplier (PMT) is centered on the wavelength of the laser

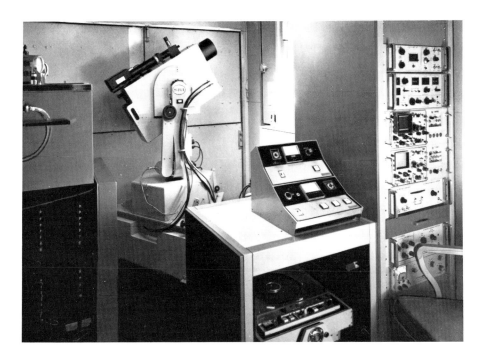

Fig. 3. SRI mark IX ruby lidar. The lidar unit (see Fig. 2 for details) may be automatically scanned in azimuth and/or elevation. Data is recorded magnetically in both analog and digital form. The ruby laser transmits 1 joule pulses at rates up to 1 per second, 15×10^{-9} sec in duration, in a 0.8 milliradian beam

Such signals may be displayed directly on an oscilloscope where the vertical deflection of the trace represents intensity as a function of range i.e., the A-scope presentation of radar practice. Lidar data so displayed is very transitory, however, being acquired pulse by pulse within intervals rarely longer than 200 microseconds—i.e., ranges of 30 km—and frequently less than 1 micro-second—i.e., ranges of 150 m. Photography or magnetic recording is commonly used to record such signals for subsequent display in more convenient form, but recently, digital data processing has been increasingly used to store, display and interpret lidar signals. (See Section 4.4 for illustrations of these techniques.)

4.3.3 Lasers

The first applications of lasers to remote probing were accomplished with the giant-pulsed Q-switched ruby lasers that first became available in the early 1960ies. Such lasers are still the mainstay of pulsed lidars relying on elastic scattering or specular reflection, although other solid-state, crystal lasers, such as neodymium doped glass or Yttrium Aluminum Garnet (YAG) are widely used. The wavelength of the energy transmitted is specific to the laser—0.6943 μm for ruby, 1.06 μm for neodymium or YAG—and is essentially fixed, although temperature changes can be used to vary wavelength to a very limited degree. Ruby and neodymium lasers commonly produce energies of the order 1 joule or so per pulse at intervals of a second or two, and pulse durations of $20–30 \times 10^{-9}$ sec, (equivalent to a range resolution of 3–4.5 m). YAG lasers commonly produce lower power pulses, but at higher pulse repetition rates. Pulsing rates of 1–50-pulses/sec are typical with pulse energy of the order 0.1 joule/pulse. Another solid state laser, the so-called junction laser, has been used in remote sensing applications. This is the Gallium arsenide laser (GAS), generating energy at 0.89 μm. Since the individual output of such lasers is low (e.g., 10^{-6} joule), they must be used in arrays—from which pulses can be transmitted at a rate of as much as 5000 pulse/sec with a combined energy of 10^{-5} to 10^{-4} joules. (The high pulse emission rate compensates for these low energy levels and allows average power of adequate levels to be achieved.)

The fixed wavelength of all such crystal or solid state lasers limits their suitability for certain applications. While it is possible to double (and even quadruple) the frequency of their output by conversion to the second or third harmonic frequencies, energy is lost in the conversion process. In any case only a limited range of fixed frequencies can be achieved by this method. Certain gas lasers are also well suited for various remote sensing applications, particularly where wavelength is a critical factor. Gas lasers are also commonly used to produce continuous wave (CW) energy. The ubiquitous helium neon laser (0.6321 μm) and the helium cadmium laser 0.4416 μm are convenient sources of CW energy that may be amplitude modulated to provide ranging capability or improved signal/noise capabilities in narrow band detection systems. Argon-ion lasers capable of generating energy at a number of wavelengths between 0.460 μm and 0.515 μm, but most intensely at 0.488 and 0.514 μm, are also valuable sources of CW energy and can also be pulsed by a technique known as cavity dumping at repetition frequencies up to 10^7 Hz. Neon lasers (0.5401 μm wavelength) may also be pulsed electrically but at much lower frequencies (0.1–100 Hz typically). Pulsed nitrogen lasers operating at a wavelength of 0.3371 μm, (i.e., in the UV) are also available with pulse rates of as much as 100 pulses/sec. In such pulsed gas lasers, short pulses of the order of 10^{-8} sec are typical, but the energy achievable per pulse varies considerably and does not compare with the more intense emissions of the crystal lasers. Pulse energies measured in millijoules are normal. CW gas lasers also have relatively low outputs, with power outputs typically measured in milliwatts.

For flexibility in the selection of wavelength and, above all, a capability to vary wavelength in a controlled manner, tuneable dye lasers are now available in practical form and have opened the way to many hitherto unattainable techniques. Using various formulations and concentrations of organic dyes in aqueous or other solutions, these lasers provide a whole new range of capability. Although commonly having longer pulses than their solid state counterparts (10^{-7}–10^{-6} sec—i.e., providing 15–150 m range discrimination) and of limited energy output (typically 0.1–1 joules), this type of laser can produce high pulse rates (e.g., > 1/sec) and can concentrate energy in very narrow spectral intervals. This latter characteristic makes possible the sophisticated remote sensing techniques that rely on critical wavelength exploitation—such as resonant scattering, resonant Raman and differential absorption.

The solid state and dye lasers used in most remote sensing applications to date have produced energy at optical or near optical wavelengths—essentially ranging from the near UV at 0.32 μm of doubled ruby emissions to the near IR at 1.54 μm of erbium. In the exploration and development of simple elastic backscattering applications, the wavelength available has generally been a less important factor than power output and pulse rate characteristics in the choice of a laser, and most use has been made of ruby lasers (0.6943 μm) and neodymium lasers (1.06 μm). For the use of lidars in more routine fashion in research studies or in proposed operational roles, the question of eye safety has, however, become increasingly significant. Accordingly, the development of efficient systems either with low peak power at optical wavelengths or lasers operating outside the visual spectrum has been pursued. The question of eye safety is discussed in more detail below, but it may be noted here that since energy at wavelengths longer than about 1.4 μm does not penetrate the eye, lasers operating at such wavelengths are attractive from this standpoint. One such is the erbuim laser which has a wavelength of 1.54 μm.

Certain non-elastic scattering techniques, e.g., ordinary Raman scattering, do not depend critically upon the incident wavelength, and are thus not dependent upon specific laser systems, although better performance is achieved with shorter wavelength lasers since the magnitude of the Raman cross section varies with λ^{-4}. Those which do require very specific wavelengths pose rather greater problems. As noted above, tuneable dye lasers are capable of producing a range of wavelengths extending virtually continuously from the near UV to near IR. As such they meet a number of requirements for remote sensing applications in the atmosphere, but the full exploitation of techniques that rely on such wavelength-critical techniques will only be achieved when tuneable laser sources are available, capable of generating effective energies in the further IR, or UV, than currently available. The reason for this is that with the exception of NO_2 and water vapor

which have resonances in the visible spectrum, for the most part the useful resonances and absorption lines of typical atmospheric constituents or air or water
pollutants occur outside the visible spectrum, mainly in the IR but to a lesser
degree in the UV. [Good references for absorption spectra of atmospheric constituents include DERR (1972), SCHWIESOW (1972), and LAULAINEN (1972). Some
fluorescence spectra for different types of algae and oil, important in hydrographic studies, are given, e.g., by KIM (1973), GROSS (1973), THURSTON and
KNIGHT (1971).]

While CO_2 gas lasers, generating continuous or pulsed energy at 10.6 μm wavelength, have been intensively developed, only limited capabilities currently exist for
generating laser energy throughout the IR spectrum (and even less in the UV).
Semiconductor lead-tin-telluride lasers could, however, be used for these diode
lasers can be tuned to produce useful, although very small outputs in the wavelength range 6.5–32 μm.

The development of lasers (both solid state and dye) capable of effectively
generating energy at a wider range of wavelengths than now available is being
actively pursued, particularly in the IR. The full benefit from progress in these
endeavors will only be realized, however, if suitable detectors are developed, as
discussed below.

4.3.4 Photodetection

In considering the wavelength of laser energy employed in remote sensing
systems, the role of the photodetector must also be considered. In most applications, the received energy is of extremely low intensity. This imposes the need for
great sensitivity in the detection device, as well as freedom from internally generated noise. A detailed discussion of these aspects is beyond our scope here, but the
excellent characteristics of the multistage photomultiplier in these respects makes
this type of detector preeminent in most receiver applications. Here again, this
form of detector is essentially limited to the visual and near visual spectrum.
Specially optimized photocathode surfaces provide maximum performance at
different parts of this spectral range, but the farther one departs from the blue/
green peak in the visual (0.5 μm), the less sensitivity is possible. Thus, at wavelengths longer than say 1.2 μm, other, less efficient forms of detector must be used.

These include both photo-voltaic and photo-conductive devices using such
materials as lead selenide, lead sulfide, indium arsenide, mercury cadmium tellur-
ide, etc. The performance of some of these can be improved by cryogenic cooling.
Thermistor bolometers are also used where energy levels are high.

4.3.5 Eye Safety

In discussing laser wavelengths and performance, the question of eye hazard
must be noted. With the power levels appropriate for remote sensing applications,
the possibility of direct thermal damage to human tissue (such as even the outer
elements of the eye) may be discounted even close to the laser. But laser energy at
optical wavelengths that enters the eye can cause eye damage. At wavelengths in

the visual spectrum, such energy may be focussed on the retina by the eye's lens, forming an image of the distant source in the normal manner of the visual process. If the intensity level of the incident energy is sufficiently high, the resulting concentration of energy on the retinal surface will cause a lesion, with serious consequences if the fundus is affected.

Following the recommendation of the American National Standards Institute, the upper limit of incident intensity may be taken as 5×10^{-7} joules/cm^2. Such intensities are produced by typical ruby lasers having an output of 1 joule, at distances up to 20 km. At longer wavelengths, e.g., 1.06 µm of neodymium lasers, the refractive properties of the lens do not concentrate the energy so sharply and the hazard is somewhat reduced for such lasers (possibly by a factor of 5). It is not until the wavelengths longer than 1.4 µm are used, however, that eye hazards cease to be significant for at this wavelength and above, energy is absorbed by the outer surfaces of the eye and harmlessly dissipated before it reaches the sensitive retina.

While recognizing the eye hazard aspects of lasers used in remote sensing applications, a sense of proportion must, however, be retained. Such risks must not be minimized or neglected in the design or operation of remote sensing systems. On the other hand, the nature of the risks and the degree of harm that may result are such that they may readily be mitigated with reasonable precaution, and will not restrict properly conducted experiments and operations.

4.3.6 Viewpoint of Remote Sensing

The wide range of sensing applications making use of laser/lidar systems leads to considerable variability in the available forms of utilization. For atmospheric observations, a fixed or mobile surface location enables the overlying atmosphere to be observed in a variety of scanning programs. Vertical cross sections or time sequences of changes along a fixed line of sight are common forms of observation. By mounting the lidar in an aircraft, similar vertical or horizontal cross sections can be mapped by making series of observations along the path of flight. Again, by directing the lidar downward, surface observations may be obtained along a line. (It should be noted, however, that the radar principle does not lend itself to developing extensive data in a plane normal to the viewing path. In this geometry, information is acquired only from the single point of intersection of the beam and the surface for each pulse. Extensive coverage of the earth's surface is thus difficult to obtain from an overflying aircraft using lidar sensing systems, and data will normally be acquired in series of traverses, since current pulsing rates are inadequate to permit effective lateral scanning.)

Although lidars have been successfully operated in aircraft in a number of applications, to date no practical attempts have been made to use satellites as vantage points for their use. The advantages and possibilities of such platforms have of course been recognized and explored in a number of studies. The engineering problems of obtaining adequate performance in space vehicle installations are very great though, particularly when the magnitude of the technical demands are considered. At satellite altitudes the range at which observations of the surface or atmosphere must be made is of the order of hundreds of kilometers as a minimum. Since even for extended or volume targets an inverse range squared factor applies, the problem of achieving effective systems becomes very considerable. Again, to provide meaningful observational coverage from a satellite with a high relative motion (i.e., in low orbits), the data rate problem noted in connection with aircraft arises again, in even more severe terms. Nevertheless, for certain applications such as sampling stratospheric dust layers on a global basis, the space satellite vantage point is attractive and is receiving increasing consideration (see, for example, ELTERMAN et al., 1973).

4.4 Applications

4.4.1 General

Laser techniques have been available for barely more than a decade. Enterprising and energetic though development has been, laser technology must still be considered as being in its infancy. Technological progress, although rapid, for the most part has not attained a level that allows fully effective realization of the many remote probing techniques that have been recognized and tentatively explored. In this section, therefore, we describe a number of applications that have potential rather than actual value. In fact, laser techniques of remote probing have only achieved practical utility in various research studies, and at the present time the laser's role in remote sensing is as a research tool of limited performance. The range and scope of the potential, however, is unquestionably very great and the accomplishment of certain laser/lidar techniques to date, even in their early form, has been noteworthy in a number of areas, particularly in atmospheric research. In this section, without attempting to cover all possibilities we will therefore describe a number of examples of such applications in detail. In addition we will note and direct attention toward a number of other approaches that have been demonstrated or suggested. In doing so we hope to stimulate consideration of the various techniques for use in as yet unrecognized areas—particularly in connection with ecological studies. The wider the appreciation of such possibilities, the more rapid is progress likely to be in developing necessary technological capabilities to make such applications available on a routine, operational basis.

A convenient classification for such applications is provided by the pattern established above in describing the fundamental processes involved. The examples given are intended to illustrate the types of application, and the references cited are by no means exhaustive or necessarily the very latest available. For the most part these references are selected as being generally accessible and useful starting points for obtaining more detailed information.

4.4.2 Backscattering Techniques — Non-resonant Elastic Scattering

4.4.2.1 Laser Profilometry

The most direct application of laser probing in remote sensing is the use of the radar principle to derive information on the form of the earth's surface from an overflying aircraft or, ultimately, from an orbiting satellite. Using the radar ranging capability along a fixed angle of view (normally directly below the aircraft), a profile of the surface can be derived and relative heights determined along the flight track. This form of "terrain mapping", as has been noted, is generally limited to series of traverses due to the difficulties of scanning a surface.

Although pulsed laser ranging systems have obvious ability in this role, continuous wave (CW) systems can provide more continuous coverage above the flight path and greater height resolution. By impressing an amplitude modulation on the laser energy transmitted, variations in the relative range of surfaces from which energy is reflected result in changes in the phase of the return signal. Such changes can readily be detected and measured by reference to the transmitted signal, and evaluated in terms of height differences. In more sophisticated approaches, frequency modulation (FM) techniques may be used to avoid range ambiguity and determine absolute range in the manner of radar altimeters.

Such profiling techniques have been successfully used in terrain mapping, as for example in surveying in an area of some 188000 sq miles in Australia from a height of 2000 m using an argon laser system. Accuracies of 0.5 m height and a horizontal resolution of 2 m along the traverses were achieved (LINES, 1972). Similarly, the surface of sea ice masses has been mapped (KETCHUM, 1971).

The high degree of precision in range resolution achievable by this technique has made it possible to observe sea surface waves from aloft. OLSEN and ADAMS (1970) for example describe results obtained with a helium neon system of modest performance obtained from a height of 60 m, which reveal ripples of 2.5 cm or less and waves of any height. SCHULE et al. (1971) describe similar results, addressing in particular the significance of spectral analyses made of such wave height returns.

4.4.2.2 Bathymetry and Turbidity Observations

In other hydrographic studies, the transparency of water to laser energy of suitable wavelength makes possible a number of applications in which the subsurface conditions are probed. Energy in the blue-green (0.540–0.580 μm) is optimum for transmission through sea water and can be generated at sufficient intensity by such lasers as the pulsed argon-ion or neon lasers. Lidars operating with such lasers have been able to penetrate to depths of some 25–30 m in fairly clear water. This makes possible the profiling of the bottom surface in appropriately shallow waters. This is a valuable capability in estuaries and coastal areas. The technique is not without problems, however. For example, the very strong return from the water surface must be discriminated against by suitable filtering (polarization and spatial filters have been successfully employed for this purpose). Again, refractive effects at the air/sea interface, and scattering by the water and suspended particulates cause the beam to diverge rapidly as it penetrates into the water, with a corresponding loss of resolution. Despite these difficulties, however, encouraging results have been obtained in experimental programs under laboratory conditions and in at least one airborne prototype development. This is the "Pulsed Light Airborne Depth Sounder" ("PLADS") being developed by Raytheon, Inc., for the U.S. Naval Oceanographic Office. This system, operating at a wavelength of 0.58 μm from an aircraft, penetrated to depths of some 30 m in relatively clear waters off Panama City in 1970 (BRIGHT, 1973). Other studies reported by HICKMAN (1973) have established the transmission/scattering charac-teristics of a pulsed neon laser as a function of water turbidity ($\lambda = 0.54$ µm), while similar studies are being carried out in Canada (CARSWELL and SIZGORIC, 1973) with an argon-ion lidar ($\lambda = 0.460$–0.515 µm). The latter studies include observa-tions from a ship on Lake Erie which revealed a spatial "structure" in the volume-tric return from the water below the surface. (The manner in which such informa-tion is obtained will be more readily appreciated after reading the section on atmospheric probing, below). The possibility of evaluating turbidity or distin-guishing layers of differing turbidity has important implications—both for the remote sensing of water quality and for investigating variations of ecological significance. The possibility of detecting shoals of fish by laser underwater sound-ing techniques has also been recognized.

4.4.2.3 Atmospheric Probing

Clouds. The use of lidar to determine cloud base or "ceiling" from ground level was one of the most obvious and earliest applications of the technique (COLLIS, 1965). In research applications, the capability of relatively modest lidars to deline-ate the structure and shape of clouds has been much used. A good example is illustrated in Fig. 4, which shows a vertical section through a cirrus cloud layer in

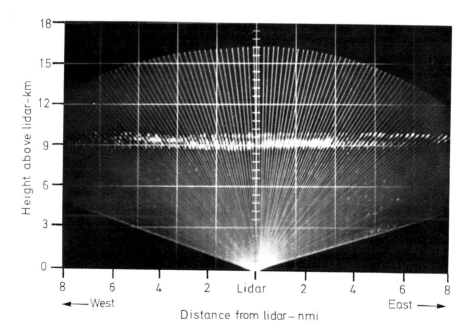

Fig. 4. Cirrus cloud layer. This vertical cross section derived by scanning the lidar through the zenith shows a tenuous cirrus cloud layer in which the effect of a mountain lee wave is apparent

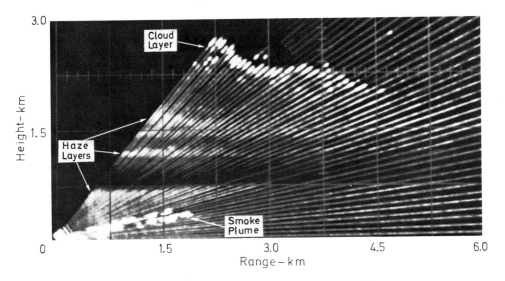

Fig. 5. Haze and cloud layers. This vertical cross section shows (1) the base of visible cloud at approximately 2 km altitude, (2) haze layers, from the surface to approximately 0.75 km and at 1.1 and 1.5 km altitude, (3) a smoke plume passing through the plane of the section, from a power plant located approximately 1.5 km upwind

which wave structure is apparent. This example was obtained in the course of studies of mountain lee waves that occur in airflows over mountain ranges under certain conditions (VIEZEE et al., 1973). The perturbation in flow that results causes standing waves aloft, of the type revealed by the tenuous cirrus cloud delineated in the illustration. Other lidar studies have provided new insights into the way in which clouds form and dissipate showing, for instance, the manner in which they are related to layers of haze. This concept is described further below.

Particulates and the "Clear Air". Although visible haze layers have always been familiar characteristics of the atmosphere (particularly in urban areas), it was not until the advent of lidar observations that the nature of aerosol distribution could be fully realized. At all levels, including the stratosphere, a completely "clear", gaseous atmosphere is the exception rather than the rule. In the atmospheric boundary layer, even in very clear conditions, suspended particulates are a significant component of the atmosphere. The capability of lidar for detecting such particles, even in what to the eye appears to be perfectly clear air, has been one of the most exciting contributions of the technique to atmospheric research. Figure 5 shows the way in which a vertical profile of the atmosphere, made by scanning a ruby lidar in the vertical, reveals structure that could not be perceived by any other method. Because of its ability to map the variability of such aerosol concentrations, lidar has opened up a new understanding of the atmosphere, particularly in the boundary layer. Lidar has also provided a better insight into the way aerosols affect radiative energy transfer. Figure 6 provides an excellent indication of lidar's capability in this regard (UTHE, 1972). It shows a height/time sequence of the atmospheric condition at a location in downtown St. Louis,

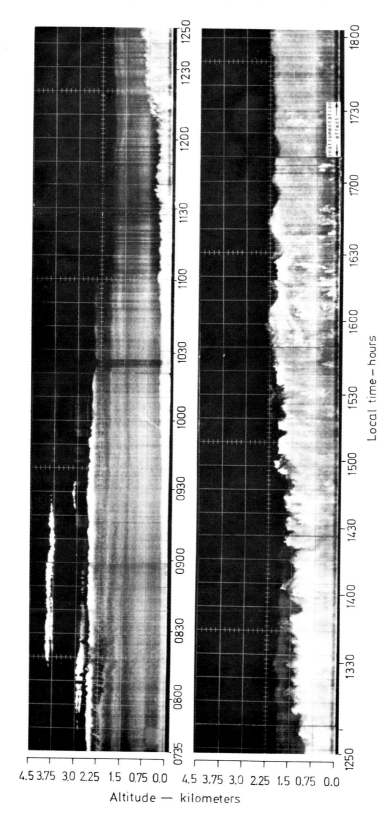

Fig. 6. Height/time cross section of the aerosol structure over St. Louis, Missouri, on 13 August 1971

Missouri, during the course of a typical summer day. In this form of presentation, the cross section shows the variation with time along a vertical line of sight above the lidar. At 0730 local time, the section shows a fairly homogeneous haze layer with some stratification capped by a thin layer of stratus cloud that had formed at its upper surface. As the morning progresses, this cloud layer dissipates under the influence of solar heating (or "burns off" in the jargon of the weather forecaster), but the layered structure of the haze remains (1030 local time). Meanwhile, the denser layer of particulates that has formerly been confined to the immediate surface layer is seen to rise and, around 1200 noon, begins to reveal cumulus shapes indicative of convection. The heads of these thermal "bubbles" are accompanied by deformation of the overlying layers showing the vertical extent of the uprising motion, and by about 1400 local time, the bubbles have penetrated the stable layers and the deep convective layer is the dominant feature. At its upper level, some condensation occurs and visible cloud forms (as evidenced by the more intense echoes—confirmed by visual observations), while at the surface a clearance is apparent as visibility improves. (Some injections of particulates are apparent, probably smoke plumes.) By the end of the observation period (1700 local time), the more uniform appearance of the haze layer suggests a reduction of convective activity and the achievement of a fairly well mixed haze condition. These observations, during which the surface visibility varied in the range 5–10 km in urban haze, were made during a study of the effect of the aerosol on urban climate. They well illustrate the role lidar can play in such studies, even on a semi-qualitative basis, (where the mass or number concentration of the particles is considered only in relative terms).

In this and other studies, however, much effort has been expended on developing methods for interpreting the intensity of the lidar returns on a quantitative basis. As noted in the discussion of Mie scattering above (4.2.1), it is necessary to have independent knowledge of the particulate size distribution before it is possible to evaluate the magnitude of the lidar backscattered signal in absolute terms. COLLIS and UTHE (1972) review such concepts, with special reference to the problems of measuring atmospheric turbidity or the concentration of particulates in air pollution studies. This paper also describes airborne lidar observations of dust concentrations which were detected at a height of some 2 km in atmosphere over the Caribbean sea, in the course of the Barbados Oceanographic and Meteorological Experiment, (BOMEX) in 1969. This layer is interpreted as being the stream of dust carried across the Atlantic by the northeast trade winds from the Sahara desert.

It is suggested that lidar observations of atmospheric stratification on airflow trajectories revealed by variations in dust or haze concentration could be useful in studying bird or insect behavior.

The capability of lidar for tracking airflow was, incidentally, one of the first practical applications of lidar in research studies concerned with ecological problems. COLLIS (1968) describes how lidar observations revealed the dispersal of insecticide sprayed from aircraft in a U.S. Forest Service investigation concerned with the control of the spruce bud-worm. The pesticide was dispensed in low volumes in small droplet form, and drifted over the forest slopes under the influence of natural air motions in the valleys. Lidar observation enabled these motions to be observed and studied.

Finally, the capability of lidar for observing particulate layers in the upper atmosphere should be noted. Many such observations have been made (see for example, FIOCCO and GRAMS, 1964; GRAMS and FIOCCO, 1967; KENT and WRIGHT, 1970; CLEMESHA and NAKAMURA, 1972; FOX et al., 1973), and recent concern with the effects of pollution by fleets of high flying aircraft (SST's) sustain interest in these levels. Such lidar observations of the stratosphere indicate that the particulate layer which is characteristically formed at about 21 km decreased in magnitude over the decade 1964–1974, and in early 1974 caused only about 10–20% increase in backscattering (over that from a purely gaseous atmosphere) from that level. It is believed that this observed reduction in particulate concentration reflects the dispersal of particulates injected into the upper atmosphere by the Mt. Agung volcanic eruption in Bali in 1963. The eruption of Vulcan de Fuego in Guatemala, however, in October 1974 has given rise to a widely observed increase in the scattering from the stratosphere.

In the higher atmosphere (i.e., 30 km and above), particulate layers are only infrequently observed, and lidar returns from these levels can be interpreted in terms of the density (and hence temperature) of the purely gaseous atmosphere. SANDFORD (1967) reported the observation of molecular density variations on a seasonal scale in the 50–80 km region. KENT et al. (1972) show evidence of a diurnal tidal variation in atmospheric density in the layer 70–100 km.

4.4.3 Inelastic and Resonant Backscattering

4.4.3.1 Raman Scattering

The application of Raman scattering processes to the measurement of atmospheric constituents was reviewed by INABA and KOBAYASI (1972). The first published report of lidar observations of Raman scattering in the atmosphere was made by LEONARD (1967), who observed scattering from O_2 and N_2 using a 100 kw pulsed N_2 gas laser and a transmitted wavelength of 337.1 nm. Operating at nighttime, backscattered returns were obtained from a useful range of 1.2 km. Since this initial observation, more powerful lasers and larger receiving telescopes have greatly extended the useful range (see, e.g., COONEY, 1968), permitting measurement of atmospheric density profiles without the contamination by aerosol scattering which would strongly influence on-frequency scattering measurements. Recently GARVEY and KENT (1974) have reported Raman measurements of atmospheric N_2 density profiles extending to 40 km and above, by using a very large and powerful laser radar. The measurements were shown to compare favorably with a concurrent radiosonde density measurement.

Raman scattering from atmospheric water vapor was observed by MELFI et al. (1969) using a frequency-doubled ruby laser ($\lambda = 347.2$ nm). Water vapor profiles obtained (to a height of 2.5 km) by this method have been compared to radiosonde measurements by COONEY (1970, 1971), and also by MELFI (1972). The results of one such comparison are shown in Fig. 7. Raman measurements of

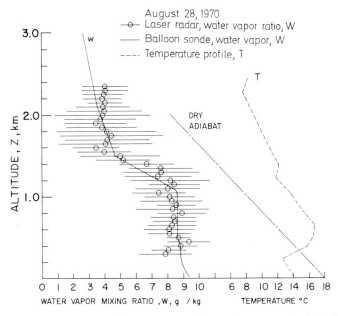

Fig. 7. Laser-Raman measurement of water vapor mixing ratio compared with balloon-sonde data. (After MELFI, 1972)

water vapor density have been made and compared to microwave refractometer measurements by STRAUCH et al. (1972). Because of the small Raman cross section and possible skylight interference, these measurements were conducted at night-time. However, COONEY (1973) has suggested a "differencing" method for extending the measurements to daytime.

The small cross sections of Raman scattering severely limit its capability for detection of pollutant gases at typical atmospheric concentrations (several parts per million or less). However, successful measurements of Raman scattering by SO_2 and CO_2 in smokestack plumes have been reported by KOBAYASI and INABA (1970), NAKAHARA et al. (1972), and MELFI et al. (1973). The SO_2 concentrations were several hundred to 1000 parts per million, and the viewing range was between 20 and 230 meters. Other measurements (INABA and KOBAYASI, 1972) have also detected the presence of CO_2, C_2H_4, H_2CO, NO, CO, H_2S, CH_4, and liquid H_2O in a tenuous oil smoke plume, as well as CO_2, C_2H_4, NO, CO, and liquid H_2O in an auto exhaust plume. These measurements were made using a pulsed N_2 laser delivering 10 mW of average power at 337.1 nm, and 20 kW peak power of 10 ns pulse length with a repetition rate of 50 pps. The resulting spectra, obtained with an f/8.5, 0.5 m single-grating monochromator, are shown in Fig. 8.

The possibility that Raman measurements of pollutant gases may be contaminated by aerosol fluorescence has been noted by GELBWACHS and BIRNBAUM (1973). (See also the section Fluorescence below.) In addition, fluorescence by NO_2 can contribute another source of interference. Techniques for overcoming this difficulty, and successful Raman measurements of both SO_2 and CO_2 in a boiler exhaust plume, have been described by NAKAHARA et al. (1973).

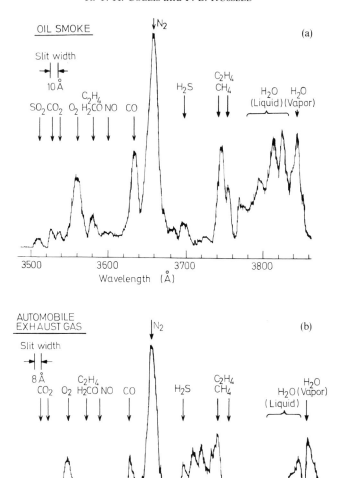

Fig. 8a and b. Spectral distribution of Raman-shifted components in (a) an oil smoke plume, and (b) an automobile exhaust plume. The spectra were analyzed remotely using a pulsed nitrogen laser ($\lambda = 337.1$ nm) and the laser-Raman radar method. (After INABA and KOBAYASI, 1972)

A technique for measurement of atmospheric temperature by using Raman backscatter has been demonstrated by STRAUCH et al. (1972). Basically, the technique deduces temperature from the nitrogen density by application of the universal gas law and the hydrostatic equation. Another technique, which directly employs the temperature dependence of the Raman scattering spectrum, has been described by COONEY (1972), but results have not yet been reported. CHANG and YOUNG (1972) have demonstrated the feasibility of a Raman scattering technique

Table 2. Relative Raman cross sections of selected molecular transitions

Molecule, M (and line)	Wave-number shift, Δv (cm^{-1})	Shifted wave-length, $\lambda_R{}^a$ (nm)	Relative Raman cross section $\left[\dfrac{dC(M)}{d\Omega}\right] \div \left[\dfrac{dC(N_2)}{d\Omega}\right]_Q{}^b$ ($\lambda_0 = 337.1$ nm)	References[c]
CCl$_4$	459	342.4	9.0	MHB
SO$_2$(v_2)	519	343.1	0.11	FHKP
NO$_2$(v_2)	754	345.7	8.6	IK
SF$_6$	775	346.1	4.3	MHB
C$_3$H$_8$	867	347.2	2.1, 2.7	FC, MHB
C$_6$H$_6$(v_2)	991	348.7	15.1, 8.2	MHB, FHKP
C$_2$H$_6$(v_3)	993	348.8	1.5, 3.2	FHKP, MHB
SO$_2$(v_1)	1151	350.8	5.0, 4.9 (Q), 6.1, 4.8	FC, PGL, IK, FHKP
N$_2$O(v_1)	1285	352.4	2.0, 2.5	FHKP, FC
CO$_2$($2v_2$)	1286	352.5	0.83, 0.92, 0.85	FHKP, FC, MHB
NO$_2$(v_1)	1320	352.8	18.2	IK
CO$_2$(v_1)	1388	353.7	1.3, 1.4, 1.2, 3.7, 1.4 (Q)	FHKP, FC, MHB, SCW, PGL
O$_2$	1556	355.9	1.2, 1.1, 1.1, 1.1, 1.1 (Q)	FHKP, FC, MHB, SCW, PGL
C$_2$H$_4$(v_2)	1623	556.6	1.9 (Q)	MHB
NO	1877	360.0	0.26, 0.53, 0.44 (Q)	FHKP, FC, PGL
CO	2145	363.5	0.99, 0.90, 0.93, 0.97 (Q)	FHKP, FC, MHB, PHL
N$_2$O(v_3)	2224	364.4	0.51, 0.53	FHKP, FC
N$_2$	2331	365.8	1.2	MHB
ND$_3$(v_1)	2420	367.0	3.0	FHKP
H$_2$S	2611	369.7	6.8, 6.8	MHB, FC
CH$_3$OH(v_2)	2846	372.8	5.0, 5.0	MHB, IK
C$_3$H$_3$	2890	373.5	6.1	FC
CH$_4$(v_1)	2914	373.9	6.3, 8.5, 7.3, 8.4, 8.1 (Q)	FHKP, FC, MHB, SCW, PGL
CH$_3$OH($2v_6$)	2955	374.4	2.6, 2.7	MHB, IK
CH$_4$(v_3)	3017	375.2	0.83, 1.9	FC, MHB
C$_2$H$_4$(v_1)	3020	375.3	5.7 (Q)	MHB
C$_6$H$_6$(v_1)	3062	376.0	7.4, 10.6	FHKP, MHB
NH$_3$(v)	3334	379.8	5.4, 3.9	FHKP, MHB
H$_2$O	3652	384.4	2.8, 2.8 (Q)	PGL, IK
H$_2$	4161	392.2	2.8, 2.7, 3.1, 3.2	FHKP, FC, MHB, SCW

[a] Assumes incident wavelength $\lambda_0 = 337.1$ nm (pulsed nitrogen laser).

[b] $\left[\dfrac{dC(N_2)}{d\Omega}\right]_Q \equiv$ the absolute differential Raman cross section of the 2331-cm^{-1} Q-branch vibrational transition in N$_2$. In preparing the table, published results have been converted to an incident wavelength of $\lambda_0 = 337.1$ nm, where necessary, using an assumed wavelength dependence of $\lambda_R{}^{-4}$. Values followed by the symbol (Q) are relative cross sections of only the Q-branch of the transition of molecule M.

[c] FC = Fouche and Chang (1972a, b); FHKP = Fenner et al. (1973); HCFP = Hyatt et al. (1973); IK = Inaba and Kobayasi (1972); MHB = Murphy, Holzer, Bernstein (1969); PGL = Penny, Goldman, Lapp (1972); SCW = Stansbury, Crawford, Welsh (1953).

Table 3. Absolute Raman backscattering cross section $[dC(N_2)/d\Omega]_Q$ of the 2331-cm^{-1} Q-branch vibrational transition in N_2. All cross sections in units of $(cm^2 \, sr^{-1}) \times 10^{-31}$

Incident wavelength (nm)			Reference[b]
337.1[a]	488.0	514.5	
28 ± 1	5.4 ± 0.3	4.2 ± 0.2	HCFP
29 ± 1		4.4 ± 0.2	PGL
29 ± 11		4.4 ± 1.7	FC
29 ± 2	5.5 ± 0.4	4.4 ± 0.3	MHB
25	4.8	3.8	SCW·

[a] Values computed from column 2 or 3 assuming λ_R^{-4} wavelength dependence.
[b] See Table 2 for key to reference symbols.

to measure subsurface ocean water temperatures to depths of 30 m. HOUGHTON (1974) has described measurements of Raman backscattering from the sulfate ion ($SO_4^=$) in seawater, and described how such measurements could be used to remotely monitor salinity.

In remote Raman measurements of both temperature and species concentration, it is frequently necessary to know accurately the Raman scattering cross sections of various molecular transitions, or alternatively their cross sections relative to that of N_2. As a result, measurements of these relative cross sections and of the absolute vibrational cross section of the 2331-cm^{-1} transition in N_2 are actively being pursued. A compendium of published results is given in Tables 2 and 3. In addition to the tabulated values, absolute cross sections for various rotational transitions in N_2, O_2, and CO_2 have recently been published by PENNEY et al. (1974).

4.4.3.2 Resonance Raman Scattering

The enhancement of ordinary Raman scattering cross sections that occurs when the incident laser frequency is tuned towards an electronic absorption line has long held promise as a means of remotely monitoring pollutant gases at typical atmospheric concentrations (several ppm or less). Despite this, successful atmospheric observation of resonance Raman scattering has not yet been achieved. One reason for this is that the electronic absorption frequencies of many pollutant molecules lie outside of the visible spectrum, and thus are inaccessible to the present range of laser output frequencies. However, ozone (Huggins band between $\lambda = 0.3$ and 0.35 µm), SO_2 (absorption beyond 0.34 µm), NO_2 (absorption throughout the visible) and halogen gases have absorption bands which are readily attainable by existing laser sources. Laboratory studies have been conducted of the behavior of the scattering cross sections of some of these gases (e.g., NO_2, SO_2, I_2) in the approach to resonance (see, e.g., FOUCHE and CHANG, 1971; 1972a, 1972b; FOUCHE et al., 1972; ST. PETERS et al., 1973). These studies have indicated complicated behavior which is not well understood. The distinction between resonance Raman scattering and the fluorescence processes is difficult to make, and some results are still considered controversial. It presently appears that other inelastic processes, especially fluorescence and differential absorption, will be more useful than resonance Raman scattering in the remote detection of air and water pollutants.

4.4.3.3 Fluorescence

The application of fluorescence processes to the remote detection of atmospheric pollutants has been studied for several years by a number of groups. Successful remote observations have not yet been reported, but GELBWACHS et al. (1972) have successfully demonstrated the *in situ* use of the technique to monitor NO_2 in samples of ambient air. GELBWACHS and BIRNBAUM (1973) have pointed out that both Raman and fluorescence detection of gaseous pollutants could be seriously contaminated by the broad fluorescence excited in typical concentrations of ambient aerosols. They have, however, described how this limitation can be overcome by use of a differencing method in which the aerosol fluorescence excited at two wavelengths is constant while the gaseous signals differ. The method has been demonstrated in the laboratory for fluorescence detection of NO_2 at concentrations of several parts per hundred million (pphm). While use of this method thus maintains the possibility that fluorescence may eventually be successful in remote atmospheric measurements of gaseous pollutants, it decreases the sensitivity (signal-to-noise ratio) of the measurements by a factor of two or more. This makes differential absorption techniques (described below) appear relatively more promising in most circumstances. However, as pointed out by GELBWACHS (1973), because of changing atmospheric conditions the two techniques are actually complementary rather than competitive, and he recommends that future NO_2 lidars be equipped to observe both fluorescence and differential absorption.

The achievement of fluorescence detection of water constitutents is somewhat more advanced than that for air pollutants. Principally, these applications have focused on the detection of oil spills and algae concentrations. KIM (1973) has reported a comparison between measurements of chlorophyll-a concentration as measured in the Chesapeake Bay by laser-induced fluorescence and conventional wet chemistry. Using a helicopter-borne system, he has also obtained a two-dimensional map of chlorophyll-a concentrations in a portion of Lake Ontario, as shown in Fig. 9. The feasibility of fluorescence detection of oil spills has been amply demonstrated, and one actual airborne laser detection has been reported by HICKMAN (1974). Whereas the simple detection of both oil and algae is quite achievable, quantitative measurement of concentration or identification of types is considerably more difficult. This will require more laboratory measurements to determine features of the comparatively broad luminescence spectra (see, e.g., THURSTON and KNIGHT, 1971; VICKERS et al., 1973), and more detailed spectroscopic analysis techniques to compare field-measured signatures with laboratory-measured or computed spectra. Some of the techniques involved in this process, and the prospects for future measurements, have been discussed by GROSS (1973). The possibility of employing both Raman and fluorescence excitation for the remote detection of pesticides in the environment has also been considered (VICK-

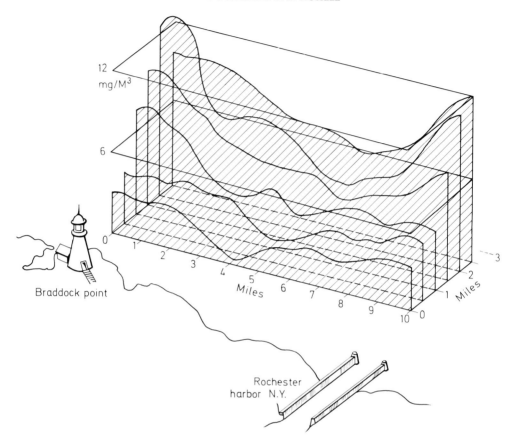

Fig. 9. Chlorophyll-a distribution in a portion of Lake Ontario on 6 October 1972, as measured by laser fluorescence. (After Kim, 1973)

ers et al., 1973). Examples of combined Raman and fluorescence spectra of several pesticides as measured in the laboratory are shown in Fig. 10. The sharp major peaks are due to Raman scattering, while the broader peaks are due to fluorescence. Spectroscopic characteristics of various rocks and minerals have also been measured in the laboratory. These could be employed for the remote identification of rocks and minerals, and there is the prospect of applications to crops, vegetation and other materials. To date, however, no successful field observations involving lasers have been reported.

4.4.3.4 Resonance (Elastic) Scattering

Because it is strongly quenched by molecular collisions in the troposphere, resonance elastic scattering has been applied largely to observations of the upper atmosphere. The most common application has been to ground-based observations of the layer of free atomic sodium near 90 km. Numerous resonance scatter-

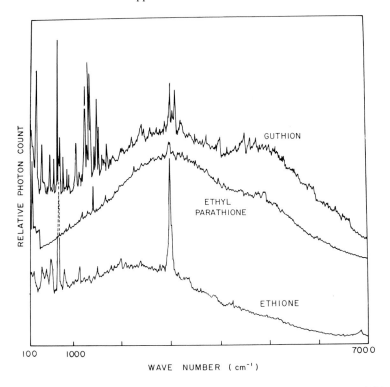

Fig. 10. Laboratory-measured laser-induced spectra of three organophosphate pesticides. The sharp major peaks are due to Raman scattering, while the broader peaks are due to fluorescence. (After VICKERS et al., 1973)

ing observations of the layer, under both nighttime and daytime conditions, have been reported in the literature (e.g., BOWMAN et al., 1969; SANDFORD and GIBSON, 1970; SCHULER et al., 1971; GIBSON and SANDFORD, 1972; BLAMONT et al., 1972). Some of these are summarized by HAKE et al. (1972), who made nighttime measurements with a tuneable dye-laser radar. A typical profile thus obtained is shown in Fig. 11. An interesting result noted by HAKE et al. was the enhancement of resonantly scattered returns at the time of the Geminids meteor shower on the night of December 13–14, 1971. A similar result has been noted in Japan just after the Jacobini meteor shower on October 8, 1972. Together these observations give evidence for meteor production as one source of the 90-km sodium layer.

FELIX et al. (1973) have also reported observations of atomic potassium in the 70–100 km region by using a laser transmitting in the near infrared.

4.4.4 Resonant Absorption

As noted in Section 4.2.3, resonant absorption may be used in two-ended configurations, where a remote retroreflector (e.g., a mirror, a lake surface, a hill) provides a large return signal. This type of geometry provides information on

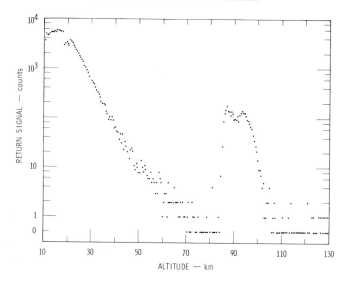

Fig. 11. Return signal from ten-minute observation of the 90-km atomic sodium layer using a
ground-based tuneable dye laser emitting at 0.5890 μm. (Hake et al., 1972)

path-integrated gas concentration, without resolution in the intervening range. A
range-resolved measure of gas concentration may, however, be obtained by measuring resonant absorption of signals elastically scattered by gas molecules and
particles along the laser beam. The method was first demonstrated by Schotland (1964), who has called it Differential Absorption of Scattered Energy (or
DASE). In using the method, a pulsed laser is tuned to the center of an isolated
absorption line of the gas to be measured, and a measurement is made of the
backscattered radiation as a function of range. A similar measurement is then
taken with the laser tuned just off the absorption line, so that the absorption and
scattering processes of the atmosphere can be separated from those of the gas
under investigation. The range-resolved profile of the gas is obtained by differencing (with respect to the lidar range) the logarithm of the ratio of the two measurements of backscattered power.

Schotland first employed the differential absorption technique to measure
profiles of water vapor, by thermally tuning a ruby laser on and off the water
vapor absorption line at 0.69436 μm. The sensitivity of the technique is great
enough, however, so that several investigators (e.g., Measures and Pilon, 1972;
Byer and Garbuny, 1973) have predicted it to be capable of measuring less than
1.0 ppm of NO_2, CO, and SO_2 with a spatial resolution of 15 m and a range
exceeding 1.0 km. Such a capability would clearly make DASE the most promising of the remote spectroscopic techniques for determination of trace gas profiles
(Ahmed, 1973; note however, Gelbwachs, 1973, and the subsection Fluorescence
of Section 4.4.3, above). Moreover, if range resolution can be sacrificed, Byer and
Garbuny (1973) have shown that average concentrations of less than 0.01 ppm
can be measured over a 10-km path by using a topographical reflector (e.g., a hill)
and quite modest laser pulse energy.

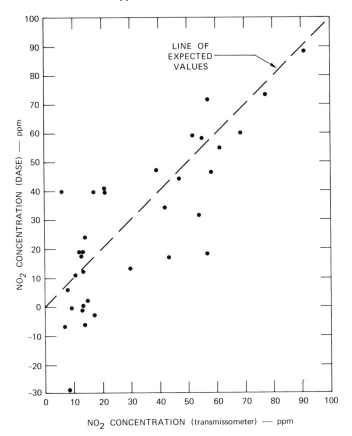

Fig. 12. Values of NO$_2$ concentration in a sample chamber, as measured remotely using differential absorption of scattered energy (DASE), compared with transmissometer measurements. The values shown were obtained using the dye laser wavelength pair 441.8 nm (peak) and 444.8 nm (valley). (After GRANT et al., 1974)

Actual application of DASE to detection of trace gases does not yet match the predicted potential, but some successful atmospheric measurements have been made. ROTHE et al. (1974) reported the first application of the technique to the measurement of ambient atmospheric NO$_2$. GRANT et al. (1974) reported a measurement using elastic backscattering from the atmosphere on both sides of a chamber containing a controlled amount of NO$_2$. Measurements were made at two peak-and-valley wavelength pairs (441.8–444.8 nm and 446.5–448.1), using a tuneable dye laser. The NO$_2$ concentrations measured in this manner were compared to values measured by a conventional transmissometer, giving the results shown in Fig. 12. At a concentration of 20 ppm, the DASE measurements gave a signal-to-noise ratio of 1 in the chamber length of 2.45 m. This is equivalent to an uncertainty of 0.5 ppm in 100 m, or 0.05 km-ppm. This measurement uncertainty includes all significant sources of error as analyzed, for example, by SCHOTLAND

(1974). Grant et al. show how reasonable system improvements could reduce the uncertainty to 0.005 km-ppm, and that this could be achieved in daytime measurements by use of narrow interference filters. The DASE technique also shows much promise for measurements of SO_2 and O_3 in the UV, and many gases in usable portions of the IR.

An example of an IR measurement has been reported by Henningsen et al. (1974), who remotely measured the concentration of CO in plastic bags by monitoring the differential absorption of the signal backscattered from a topographical target at wavelengths near 2.3 µm. The range and sensitivity of the technique were in good agreement with the predictions noted above.

4.4.5 Doppler Techniques

The application of the Doppler principle to determine the relative radial velocity of atmospheric volumes from which laser energy is backscattered offers possibilities for remotely measuring wind velocity, or for detecting turbulent motion in the atmosphere.

Various techniques have been utilized for detecting the frequency shift in the energy backscattered by the particulates of the "clear air", or even by the gaseous atmosphere using CW lasers (argon or CO_2), or pulsed coherent CO_2 lasers.

The approaches used differ in the manner in which the frequency shift is detected and measured, but all rely upon the backscattering of laser energy by the gaseous molecules or suspended particulates. Three main alternatives are available

1. the direct detection of the frequency shift in the received signal compared with the laser output by high resolution interferometric techniques,

2. coherent heterodyne or homodyne detection, where the backscattered signal is compared with the signal from a local oscillator, and

3. the comparison of the signals received from a single volume by two beams originating from the same transmitter, in an arrangement where the geometry of the paths gives rise to a differential in received frequencies due to the Doppler effect. (This geometry restricts its application to close range, however.)

Huffaker (1970) describes the application of the heterodyne and differential approaches to the remote measurement of wind velocity and turbulence from the ground or from aircraft, as well as for observing trailing vortices generated by large aircraft of particular concern. Although this paper describes only observations at short ranges, subsequent reports of this program indicate that wind velocity have been measured at ranges of several kilometers, using a coherent pulsed CO_2 lidar. Benedetti-Michelangeli et al. (1972) using an argon lidar and the interferometric technique, describe their success in measuring wind velocity with good accuracy in the lower troposphere in North Italy.

The same group has also addressed the problem of measuring atmospheric temperature remotely, by observing the spread of the Doppler shifted frequencies backscattered by the gaseous molecules, the motion of which is temperature dependent (Fiocco et al., 1971).

4.4.6 Polarization

The detection of the presence of non-spherical scatterers by measuring the depolarization ratio of lidar returns has immediate value in the observation of clouds, where the presence of such scatterers indicates that ice crystals must be present. This technique is especially suitable for distinguishing between ice crystal clouds and water droplet clouds, particularly at medium levels (SCHOTLAND et al., 1971). The variation of depolarization effects between different clouds and as a function of height which has been observed (PAL and CARSWELL, 1973) could provide unique data for cloud physics research. The correct interpretation of such data requires further research however, in particular to distinguish between the effects of ice crystal scattering and the depolarization caused in varying degrees by multiple scattering by spherical droplets. Similarly, the role of polarization techniques in lidar observations of water quality is being studied (CARSWELL, 1973).

References

ALLEN, R.J., EVANS, W.E.: Laser radar (LIDAR) for mapping aerosol structure. Rev. Sci. Instr. **43**, 1422–1432 (1972).

AHMED, S.A.: Molecular air pollution monitoring by dye laser measurement of differential absorption of atmospheric elastic backscatter. Appl. Optics **12**, 901–903 (1973).

BENEDETTI-MICHELANGELI, G., CONGEDUTI, F., FIOCCO, G.: Measurement of aerosol motion and wind velocity in the lower troposphere by Doppler optical radar. J. Atmos. Sci. **29**, 906–910 (1972).

BLAMONT, J.E., CHANIN, M.L., MEGIE, G.: Vertical distribution and temperature profile of the night time atmospheric sodium layer obtained by laser backscatter. Ann. Geophys. **28**, 833–838 (1972).

BOWMAN, M.R., GIBSON, A.J., SANDFORD, M.C.W.: Atmospheric sodium measured by a tuned laser radar. Nature (Lond.) **221**, 456 (1969).

BRIGHT, D.: "NOAA's Lidar Program". Proc. Symposium on the use of lasers for hydrographic studies, Sept. 12, 1973, NASA, Wallops Island, Va. (KIM, H.H. Ed.). In course of publication).

BYER, R.L., GARBUNY, M.: Pollutant detection by absorption using Mie scattering and topographic targets as retroreflectors. Appl. Optics **12**, 1496–1505 (1973).

CARSWELL, A.I., SIZGORIC, S.: Underwater probing with laser radar. Proc. Symp. on Use of Lasers for Hydrographic Studies, Sept. 12, 1973, NASA, Wallops Island, Va. (KIM, H.H. Ed.). (In course of publication).

CHANG, C.C., YOUNG, L.A.: Remote measurement of water temperature by Raman scattering. Proc. Eighth Int. Symp. Remote Sens. Env., **II**, 1049–1068, Ann Arbor, 2–6 October (1972).

CHANG, R.K., FOUCHE, D.G.: Gains in detecting pollution. Laser Focus **8**, 43–45 (1972).

CLEMESHA, B.R., NAKAMURA, Y.: Dust in the upper atmosphere. Nature (Lond.) **237**, 328–329 (1972).

COLLIS, R.T.H.: Lidar observation of cloud. Science **149**, 978–981 (1965).

COLLIS, R.T.H.: Lidar—a new atmospheric probe. Quart. J. Roy. Met. Soc. **92**, 220–230 (1966).

COLLIS, R.T.H.: Lidar observations of atmospheric motion in forest valleys. Bull. Amer. Meteor. Soc. **49**, 918–922 (1968).

COLLIS, R.T.H., UTHE, E.E.: Mie scattering techniques for air pollution measurements with lasers. Opto-electronics. **4**, 87–99 (1972).

COONEY, J.: Measurement of the Raman scattering of laser atmospheric backscatter. Appl. Phys. Letters **12**, 42–44 (1968).

COONEY, J.: Remote measurement of atmospheric water vapor profiles using the Raman component of laser backscatter. J. Appl. Meteor. **9**, 182–184 (1970).

COONEY, J.: Comparison of water vapor profiles obtained by radiosonde and laser backscatter. J. Appl. Meteor. **10**, 301–308 (1971).

COONEY, J.: Measurement of atmospheric temperature profile by Raman backscatter. J. Appl. Meteor. **11**, 108–112 (1972).

COONEY, J.: A method for extending the use of Raman lidar to daytime. J. Appl. Meteor. **12**, 888–890 (1973).

DEIRMENDJIAN, D.: Electromagnetic scattering on spherical dispersions. New York: American Elsevier 1969.

DERR, V. E.: The spectra of molecules of the earth's troposphere. In: DERR, V.E. (Ed.): Chapter 9 in Remote Sensing of the Troposphere. Washington, D.C.: U.S. Gov't Printing Office 1972.

ELTERMAN, L., TOOLIN, R. B., ESSEX, J. D.: Stratospheric aerosol measurements with implications for global climate. Appl. Optics **12**, 330–337 (1973).

FELIX, F., KEENLISIDE, W., KENT, G.: Laser radar observations of atmospheric potassium. Nature (Lond.) **246**, 345 (1973).

FENNER, W. R., HYATT, H. A., KELLAM, J. M., PORTO, S. P. S.: Raman cross section of some simple gases. J. Opt. Soc. Amer. **63**, 73–77 (1973).

FIOCCO, G., GRAMS, G.: Observations of the aerosol layer at 20 km by optical radar. J. Atmos. Sci. **21**, 323–324 (1964).

FIOCCO, G., BENEDETTI-MICHELANGELI, G., MAISCHBERGER, K., MADONNA, E.: Measurement of temperature and aerosol to molecule ratio in the trophosphere by optical radar. Nature (Lond.) **229**, 78–79 (1971).

FOUCHE, D. G.: Fluorescence, and Raman and resonant Raman scattering. Appendix A of Study od Air Pollution Detection by Active Remote Sensing Techniques, prepared for NASA, Langley Research Center, Contract NAS1-11657 SRI (1974).

FOUCHE, D. G., CHANG, R. K.: Relative Raman cross sections for N_2, O_2, CO, CO_2, SO_2, and H_2S. Appl. Phys. Letters **18**, 579–580 (1971).

FOUCHE, D. G., CHANG, R. K.: Relative Raman cross section for O_3, CH_4, C_3, NO, N_2O, and H_2. Appl. Phys. Letters **20**, 256–257 (1972a).

FOUCHE, D. G., CHANG, R. K.: Observation of resonance Raman scattering below the dissociation limit in I_2 vapor. Phys. Rev. Letters **29**, 536–539 (1972b).

FOUCHE, D. G., HERZENBERG, A., CHANG, R. K.: Inelastic photon scattering by a polyatomic molecule: NO_2. J. Appl. Phys. **43**, 3846–3851 (1972).

FOX, R. J., GRAMS, G. W., SCHUSTER, B. G., WEINMAN, J. A.: Measurements of stratospheric aerosols by airborne laser radar. J. Geophys. Res. **78**, 7789–7801 (1973).

GARVEY, M. J., KENT, G. S.: Raman backscatter of laser radiation from the stratosphere. Nature (Lond.) **248**, 124–125 (1974).

GELBWACHS, J.: NO_2 lidar comparison: fluorescence vs. backscattered differential absorption. Appl. Optics. **12**, 2812–2813 (1973).

GELBWACHS, J., BIRNBAUM, M.: Fluorescence of atmospheric aerosols and lidar implications. Appl. Optics **12**, 2442–2447 (1973).

GELBWACHS, J., BIRNBAUM, M., TUCKER, A. W., FINCHER, C. L.: Fluorescence determination of atmospheric NO_2. Optoelectronics **4**, 155–160 (1972).

GIBSON, A. J., SANDFORD, M. C. W.: Daytime laser radar measurements of the atmospheric sodium layer. Nature (Lond.) **239**, 509–511 (1972).

GRAMS, G., FIOCCO, G.: Stratospheric aerosol layer during 1964 and 1965. J. Geophys. Res. **72**, 3524–3542 (1967).

GRANT, W. B., HAKE, R. D., LISTON, E. M., ROBBINS, R. C., PROCTOR, E. K.: Calibrated remote measurement of NO_2 using the differential-absorption backscatter technique. Appl. Phys. Letters **24**, 550–552 (1974).

GROSS, H. G.: Progress in the application of higher specificity laser induced luminescence of the remote sensing of environment and resources. In: SHAHROKI, F. (Ed.): Remote Sensing of Earth Resources, Vol. 2. Tullahoma: University of Tennessee 1973.

HAKE, R. D., JR., ARNOLD, D. E., JACKSON, D. W., EVANS, W. E., FICKLIN, B. P., LONG, R. A.: Dye-Laser observations of the nighttime atomic sodium layer. J. Geophys. Res. **77**, 6839–6848 (1972).

HENNINGSEN,T., GARBUNY,M., BYER,R.L.: Remote detection of CO by parametric tuneable laser. Appl. Phys. Letters **24**, 242–244 (1974).

HICKMAN,G.D.: Recent advances in the applications of pulsed lasers in the hydrosphere. In: KIM,H.H., RYAN,P.T. (Eds.): Proceedings, Symposium on the Use of Lasers for Hydrographic Studies. Wallops Island, VA: NASA 1974.

HINKLEY,E.D., KELLY,P.L.: Detection of air pollutants with tuneable diode lasers. Science **171**, 635–639 (1971).

HIRONO,M., FUJIWARA,M., UCHINO,O., ITABE,J.: Observations of aerosol layers in the upper atmosphere by laser radar. Rep. Ionospheric Space Res. in Japan **26**, 237–244 (1972).

HOUGHTON,W.M.: Measurement of Raman spectra of H_2O and $SO_4^=$ in seawater. In: KIM,H.H. (Ed.): The Use of Lasers for Hydrographic Studies. Symposium Proceedings. Wallops Island, Va.: NASA 1974.

HUFFAKER,R.M.: Laser Doppler detection systems for gas velocity measurement. Appl. Optics **9**, 1026–1039 (1970).

HYATT,H.A., CHERLOW,J.M., FENNER,W.R., PORTO,S.P.S.: Cross section for the Raman effect in molecular nitrogen gas. J. Opt. Soc. Amer. **63**, 1604–1606 (1973).

INABA,H., KOBAYASI,T.: Laser-Raman radar. Opto-electronics **4**, 101–123 (1972).

KENT,G.S., KEENLISIDE,W., SANDFORD,M.C.W., WRIGHT,R.W.H.: Laser radar observation of atmospheric tides in the 70–100 km height region. J. Atmos. Terrest. Phys. **34**, 373–386 (1972).

KENT,G.S., WRIGHT,R.W.: A review of laser radar measurements of atmospheric properties. J. Atmos. Terrest. Phys. **32**, 917–943 (1970).

KERKER,M.: The scattering of light and other electromagnetic radiation. New York: Academic Press 1969.

KETCHUM,R.D., JR.: Airborne laser profiling of the arctic pack ice. Remote Sens. Environ. **2**, 41–52 (1971).

KILDALL,H., BYER,R.L.: Comparison of laser methods for the remote detection of atmospheric pollutants. Proc. IEEE **59**, 1644–1663 (1971).

KIM,H.H.: New algae mapping technique by the use of an airborne laser fluorosensor. Appl. Optics **12**, 1454–1459 (1973).

KOBAYASI,T., INABA,H.: Spectroscopic detection of SO_2 and CO_2 molecules in polluted atmosphere by laser-Raman radar technique. Appl. Phys. Letters **17**, 139–141 (1970).

LAULAINEN,N.: Minor gases in the earth's atmosphere: A review and bibliography of their spectra. Nat. Sci. Foundation. Available from National Technical Information Service, U.S. Dept. of Congress, Springfield, Va. (1972).

LEONARD,D.: Observation of Raman scattering from the atmosphere using a pulsed nitrogen ultraviolet laser. Nature (Lond.) **216**, 142–143 (1967).

LINES,J.D.: Dept. National Development, Australia, quoted in Laser Focus (1972).

MEASURES,R.M., PILON,G.: A study of tuneable laser techniques for remote mapping of specific gaseous constituents of the atmosphere. Opto-electronics **4**, 141–153 (1972).

MELFI,S.H.: Remote measurements of the atmosphere using Raman scattering. Appl. Optics **11**, 1605–1610 (1972).

MELFI,S.H., BRUMFIELD,M.L., STORY,R.W.: Observation of Raman scattering by SO_2 in a generating plant stack plume. Appl. Phys. Letters **22**, 402–403 (1973).

MELFI,S., LAWRENCE,J., McCORMICK,M.: Observations of Raman scattering by water vapor in the atmosphere. Appl. Phys. Letters **15**, 295–297 (1969).

MURPHY,W.F., HOLZER,W., BERNSTEIN,H.J.: Gas phase Raman intensities: a review of "pre-laser" data. Appl. Spectroscopy **23**, 211–218 (1969).

NAKAHARA,S., ITO,K., ITO,S.: Detection of SO_2 and NO_2 in stack plume by Raman scattering. Fifth Conference on Laser Radar Studies of the Atmosphere. Williamsburg, VA., 4–6 June (1973).

NAKAHARA,S., ITO,K., ITO,S., FUKE,A., KOMATSU,S., INABA,H., KOBAYASI,T.: Detection of sulfur dioxide in stack plume by laser Raman radar. J. Opt. Electr. **4**, 169–177 (1972).

OLSEN,W.S., ADAMS,G.W.: A laser profilometer. J. Geophys. Rev. **75**, 2185–2187 (1970).

PAL,S.R., CARSWELL,A.I.: Polarization properties of lidar backscattering from clouds. Appl. Optics **12**, 1530–1535 (1973).

PENNEY,C.M., GOLDMAN,L.M., LAPP,M.: Raman scattering cross sections. Nat. Phys. Sci. **235**, 110–111 (1972).

PENNEY, C. M., ST. PETERS, R. L., LAPP, M.: Absolute rotational Raman cross sections for N_2, O_2, and CO_2. J. Opt. Soc. Amer.

ROTHE, K. W., BRINKMAN, U., WALTHER, H.: Applications of tuenable dye lasers to air pollution detection: measurement of the atmospheric NO_2 concentrations by differential absorption technique. Appl. Phys. (Germany) **3**, 115 (1974).

SANDFORD, M. C. W.: Laser scatter measurements in the mesosphere and above. J. Atmos. Terrest. Phys. **29**, 1657–1659 (1967).

SANDFORD, M. C. W., GIBSON, A. J.: Laser radar measurements of the atmospheric sodium layer. J. Atmos. Terr. Phys. **32**, 1423 (1970).

SCHOTLAND, R. M.: The determination of the vertical profile of atmospheric gases by means of a ground based optical radar. In: Proceedings of the Third Symposium on Remote Sensing of the Environmental. Ann Arbor, Univ. of Michigan, 215–224 (1964).

SCHOTLAND, R. M.: Errors in the lidar measurement of atmospheric gases by differential absorption. J. Appl. Meteor. **13**, 71–77 (1974).

SCHOTLAND, R. M., SASSEN, K., STONE, R.: Observations by lidar of linear depolarization ratios by hydrometeors. J. Appl. Meteor. **10**, 1011–1017 (1971).

SCHULE, J. J., SIMPSON, L. S., DELEONIBUS, P. S.: A study of fetch-limited wave spectra with an airborne laser. J. Geophys. Res. **76**, 4160–4170 (1971).

SCHULER, C. J., PIKE, C. T., MIRANDA, H. A.: Dye laser probing of the atmosphere using resonant scattering. Appl. Optics **10**, 1689 (1971).

SCHWIESOW, R. L.: Atomic, molecular, particulate, and collective generalized scattering; and Remote spectral sensing of pollutants. In: DERR, V. E. (Ed.): Chapters 10 and 23 of Remote Sensing of the Troposphere. Washington, D.C.: U.S. Gov't Printing Office 1972.

STANSBURY, E. J., CRAWFORD, M. F., WELSH, H. L.: Determination of rates of change of polarization from Raman and Rayleigh intensities. Canad. J. Phys. **31**, 954–961 (1953).

ST. PETERS, R. L., SILVERSTEIN, S. D., LAPP, M., PENNEY, C. M.: Resonant Raman scattering or resonance fluorescence in I_2 vapor. Phys. Rev. Letters **30**, 191–192 (1973).

STRAUCH, R. G., CUPP, R., DERR, V. E.: Atmospheric temperature measurement using Raman backscatter. Appl. Optics **10**, 2665–2669 (1972).

STRAUCH, R. G., DERR, V. E., CUPP, R. E.: Atmospheric water vapor measurements by Raman lidar. Remote Sens. of Env. **2** (1972).

THURSTON, A. D., KNIGHT, R. W.: Characterization of crude and residual-type oils by fluorescence spectroscopy. Env. Sci. Tech. **5**, 64–69 (1971).

UTHE, E. E.: Lidar observations of the urban aerosol structure. Bull. Amer. Meteor. Soc. **53**, 358–360 (1972).

VAN DE HULST, J. C.: Light scattering by small particles. New York: John Wiley 1957.

VICKERS, R. S., CHAN, P. W., JOHNSEN, R. E.: Laser excited Raman and fluorescence spectra of some important pesticides. Spectroscopy Letters **6**, 131–137 (1973).

VIEZEE, W., COLLIS, R. T. H., LAWRENCE, J. D.: An investigation of mountain waves with lidar observations. J. Appl. Meteor. **12**, 140–148 (1973).

5. Radar Methods

G. P. DE LOOR

5.1 General Aspects

5.1.1 Introduction

Shortly before World War II the principles of radar were investigated at a number of places in the world, all working separately. One of these places was the Physics Laboratory, where development had gone so far that production of a system was considered. The outbreak of the war for the Netherlands on May 10, 1940, prevented the execution of these plans (VAN SOEST, 1949). Similar histories can be told about the work done at other places.

World War II saw the coming of age of radar. After the war development took place at an accelerated rate. The major advantage of radar over other observation systems is its independency of the weather. Although the absorption of radar waves by the atmosphere increases when going to the higher microwave frequencies (shorter wavelengths), it remains small as compared with the visible and thermal infrared part of the spectrum. Figure 1 gives a survey of the microwave frequency band with band designations in common use.

The ordinary radar with rotating aerial and imaging on a PPI screen (plan—position—indicator) has found wide application, also in the air, especially for navigational purposes. For mapping applications, however, this type of radar has its limitations. With the development of the side looking airborne radar or SLAR, many of these limitations have been overcome.

Radar is an active system. It transmits a short pulse of electromagnetic energy and records the echoes received back in order of arrival and puts them on a line on an image tube e.g. in the form of small light-dots (see Fig. 2). After the reception

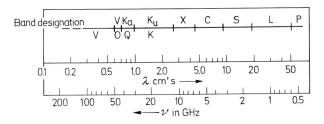

Fig. 1. The microwaves. Band designations in common use

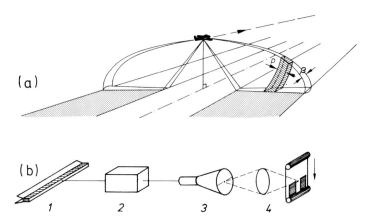

Fig. 2a and b. Side looking radar. (a) Scan configuration; (b) radar system: *1* antenna;
2 transmitter/receiver; *3* image display tube on which one intensity modulated line; picture
is built up to a continuous image on moving film in camera system *4*

of the last echo—determined by the distance over which observation is wanted—a
new pulse is emitted, etc. Since an electromagnetic wave propagates with the
speed of light, the turn around time is short and many pulses can so be transmitted per second. By this technique radar in fact measures distances to the reflecting
objects (the time between the transmission of a pulse and the reception of an
echo). This has its consequences on image formation. In SLAR two antennas can
be used which concentrate the emitted energy, one on each side of the airborne
platform. Figure 2 gives the system build-up. Only one line is given on the
image tube at a time. By imaging this line on a film which moves at a speed
proportional to the speed of the aircraft, a continuous image is obtained. Scale
accuracy in the flight direction is determined by the accuracy of maintaining the
right proportionality between the speed of the moving platform carrying the
radar and that of the film.

Each radar works at one single frequency (wavelength). This is in contradis-
tinction to most other remote sensing systems as e.g. the scanners where, even in
monochromatic operation, a wavelength band is always sensed. The frequencies
used most for SLAR lie between 6 and 40 GHz ($\lambda = 5 - \lambda = 0.8$ cm). Lower frequencies are possible and even the use of the *P*-band ($\lambda = 50$ cm) is reported. These last
items of equipment, however, require very intricate systems to obtain sufficient
resolution (coherent radar with synthetic aperture). The wavelength bands most
commonly used at this stage are the *X*- ($\lambda = 3$ cm) and K_a- or *Q*-bands ($\lambda = 0.8$ cm).
Table 1 gives the parameters of two representative systems, one for each band.
For a more complete survey, vide the Easams study for Esro (Easams, 1972).

The considerations given above apply to imaging radars. There are other
systems based on the radar principle such as the radar altimeter for the measurement of aircraft altitude, and the scatterometer for the airborne measurement of
the radar backscatter coefficient (Moore, 1966; Lundien, 1967). It would go too
far to treat these systems also, and we shall restrict ourselves in the following to
the imaging system: the side looking airborne radar or SLAR.

Table 1. Average parameters of an X- and a K_a-band radar

	X-band	K_a-band
Operating frequency (GHz)	9.5	35.0
Aerial length (m)	4.6	4.0
Transmitter peak power (kW)	100	50
Pulse repetition frequency (sec^{-1})	2500	3800
Pulse length (μsec)	0.2	0.1
Resolution:		
across track (m)	30	20
along track (m/km)	8	2.7
Size (without aerials) (m^3)	0.35	0.4
Weight (without aerials) (kg)	155	156

5.1.2 Resolution

5.1.2.1 Real Aperture SLAR

The parameters determining resolution are: antenna aperture in azimuth (along track) and pulse length in range (across track). See Figs. 2 and 3. The antenna concentrates the electromagnetic energy emitted by the radar in a narrow beam with an opening in azimuth of β radians, as Fig. 2 shows.

Along track resolution is then given by the well-known formula: antenna opening:

$$\beta = 1.2 \frac{\lambda}{D} \text{ radians} \tag{1}$$

with λ the radar wavelength and D the antenna aperture, both in the same units. Absolute resolution along track thus deminishes with range, vide Fig. 4. Resolution in meters will never be better than the size of the aperture D. Across track resolution—the ability of the radar to discriminate two targets situated behind each other in range—is determined by pulse length. Since the radar pulse travels with the speed of light (300 m/μsec) and the radar records the time for a pulse to travel to and from a target, each μsec corresponds with 150 m on the screen. Current systems use pulse lengths between 0.05 and 0.3 μsec—which corresponds with a resolution between 8 and 50 m—dependent on the wavelength used and the maximum range required.

Because of both effects described above, a reflecting object is imaged longer than it really is and a non-reflective object (as e.g. a lake, water reflects the electromagnetie energy away from the radar like a mirror so that no echo is received back at the radar) smaller because of the fact that the reflective surroundings appear to be larger.

Care must be taken that the transfer function of the whole radar system is such that resolution is not further reduced by the other components (image tube, camera, etc.).

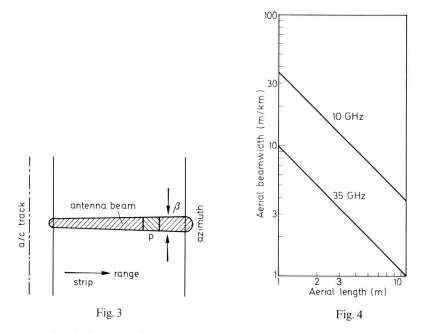

Fig. 3   Fig. 4

Fig. 3. Resolution in azimuth (along track) diminishes with range

Fig. 4. Antenna beamwidth vs. antenna length

5.1.2.2 Synthetic Aperture SLAR; Improvement of the Along Track Resolution

Along track resolution decreases with increasing range (see Figs. 3 and 4). Since the size of the aerial is determined by the size of the aircraft, this means that at a certain range along track resolution will become lower than that obtained across track. Formula (1) gives resolution in azimuth. As Fig. 5 demonstrates, objects that lie farther away from the radar are illuminated longer than nearby objects. Where object 1 in Fig. 5 is illuminated only when the aircraft is in position II, object 2 is already illuminated in position I. This phenomenon can be utilized. With a coherent radar system all observations of object 2 are integrated,

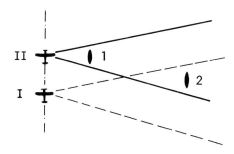

Fig. 5. Longer illumination of distant objects

going from position I to position II. The forward movement of the aircraft carrying the SLAR is used by treating the succesive antenna positions as if they were the individual elements of a very long linear antenna array.

The further away from the radar an object is, the longer it is illuminated and the longer the synthetic antenna array becomes. In this way it is possible to make the along track resolution b independent of range:

$$b = 1.2 \, D/2 \, .$$

For a further treatment of the principles of synthetic aperture SLAR, the reader is referred to the literature (BROWN and PORCELLO, 1969; HARGER, 1969; EASAMS study, 1972). It will be clear to the reader, however, that the procedure is a sophisticated one with the necessary consequences in technique and prize of the system (GOODYEAR, 1970). For a good reconstruction of the long antenna array, the positions of the individual elements of this synthetic array have to be known up to $\frac{1}{4}\lambda$. For an X-band system this means an accuracy of 8 mm. This is done by using an inertial platform as a reference and antenna stabilisation. The recording made in the air is a sort of hologram made on film because of the high information content. This hologram is later treated in an optical processor to a readable image.

For the longer wavelengths (say longer than $\lambda = 5$ cm) this system of aperture synthesizing is a necessity, since the resolution of a real aperture system will become prohibitively low. For these longer wavelengths only experimental systems are yet available, such as the L-band ($\lambda = 25$ cm) system of the Willow Run Laboratories (Ann Arbor, USA) and the P- ($\lambda = 50$ cm) and L-band system of the Naval Research Laboratory (Washington), (GUINARD, 1971; DALEY et al., 1971).

5.1.3 Image Build-up; Flight Procedure

As remarked before, radar records the distance to a reflecting object. This introduces typical distortions in the imagery as compared with a map of the terrain surveyed. See Fig. 6. The radar records the distances OA, OB, OC, etc. and not NA, NB, NC, etc. It is possible to correct this in the imaging system, but it may be preferred to do this later on the ground together with the elimination of other errors; a correction mechanism may be incomplete in its working in the air, and any information lost in the air is irretrievable.

It will be clear that the height of the radar above the reference plane must be accurately known. For low altitudes or very large distances [large values of range (R) over height (H): R/H], corrections become small and a nearly orthographic image is obtained.

Figure 6 also shows that it is not very useful to look right down under the aircraft due to an increasing distortion. In SLAR, registrations are therefore only made for angles larger than 45° away from the vertical. This means that an area with a width of about twice the height of the radar is not recorded. A special flight pattern is therefore required to overcome this. Figure 7 indicates such a procedure. Although this figure suggests that it would be sufficient to lie the following

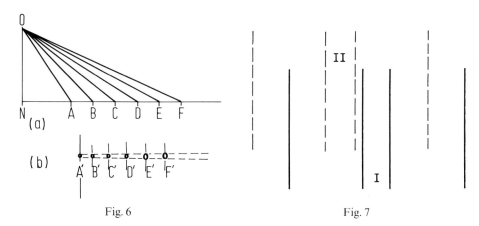

Fig. 6   Fig. 7

Fig. 6a and b. Imaging by SLAR of distances: (a) real situation, (b) image. Since imaging starts at A (the distance to A is suppressed), the distances $(OB-OA)$, $(OC-OA)$ etc. are recorded

Fig. 7. Flying pattern for SLAR

two flight lines besides the two indicated, it can be useful to shift only one stripwidth per flight line. In this way a large part of the area is flown twice, which especially in mountaineous terrain can give a larger freedom in the choice of flying height.

5.1.4 Parallax

In radar imagery two types of parallax are met: the echo parallax and the shadow parallax (see Fig. 8).

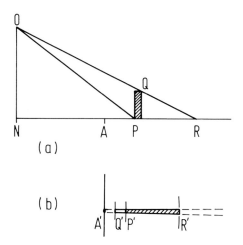

Fig. 8a and b. Radar parallax; (a) real situation; (b) image: A: start of imaging, $Q'P'$: echo, $P'R'$: radar shadow

5.1.4.1 The Echo Parallax

Because of the fact that radar measures distances, the top of a high object will be recorded first, before its bottom, when flying at a certain height. Since (Fig. 8) $OQ < OP$, Q will be recorded before P. This "inverted" echo rapidly decreases in size when the distance to the radar increases (larger values for R/H).

This effect has been used. LA PRADE (1963) made a theoretical study and considered several possibilities, choosing flight lines in such a way that the imagery can eventually be evaluated in ordinary stereo comparators. One can distinguish the same side stereo configuration and the opposite side stereo configuration. In the first case the object is illuminated in both flights on the same side, in the second case on opposite sides. The last case can give trouble to the observer in obtaining stereopsis because of the fact that objects can scatter radar waves very differently at different sides, which can lead to severe disparities between the images. The same side stereo configuration therefore seems a better solution. Figure 9 gives the result for this configuration given by LA PRADE (1963) as an optimum.

Another configuration is suggested in Fig. 10 (DE LOOR, 1960, MOORE, 1971; CARLSON, 1973). The radar now radiates 2 beams, one slanting forward, the other slanting backward. Here also, parallax decreases with distance.

In all these cases, it must be kept in mind that the effect is usable only over a limited range and also that the minimum height difference observable is always larger than one resolution element of the radar system. The best resolution obtainable (military classification) is somewhere between 10 and 20 m. Many radar systems, especially those covering long ranges, have a lower resolution.

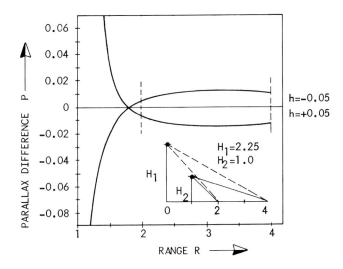

Fig. 9. Optimum configuration for flying "same-side-stereo" as proposed by LA PRADE (1963). Range R, flying heights H and altitude to be measured h all in the same units

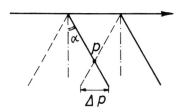

Fig. 10. Stereo by using two squint angles

5.1.4.2 The Shadow Parallax

For a radar in O (Fig. 8), the object PQ screens the terrain between P and R from observation. PR, therefore, cannot be recorded (no signal) and shows up in the image as an area without echoes, thus as a black area or "shadow".

This radarshadow can eventually also be used for a height measurement when it is clearly painted and lying on the reference plain. The radarshadow increases with distance (larger values of R/H). Higher objects in the foreground can screen other objects lying behind them (foreground screening).

The shadows can also be used to increase contrasts in an image. Small altitude differences can so be enhanced by flying low (e.g. old river banks in a river plain, see e.g. Fig. 28).

5.1.5 Image Correction

SLAR is a line-scan system, imaging a continuous strip of the earth's surface line by line. This means that for obtaining a geometrically correct depiction of the earth's surface, a line by line correction of the image is necessary taking into

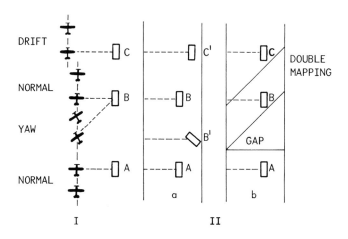

Fig. 11. Influence of yaw and drift on line-scan imagery. *I* Flight configuration. *II* Image: *a* uncorrected, *b* corrected

account all aircraft movements as: tilt (pitch), tip (roll), swing (yaw), crab (drift), velocity and altitude changes. Figure 11 shows the influence of yaw and drift errors. This necessitates the introduction of altitude and attitude sensors in the aircraft, of which the signals are recorded simultaneously with the image signal. In Section 5.3.2.2 we shall derive the requirements to be made upon such equipment.

5.1.6 Registration of the Radar Signals

Figure 2b shows the general set-up of SLAR. An image is made by imaging the line obtained on the CRT (cathode-ray-tube, here: image display tube) on a film which moves with a speed proportional to the speed of the aircraft.

The dynamic range ("contrast range") of the signals received and accepted by the system, however, usually far exceeds the possibilities of today's image tube/film combinations. Since the radar system can handle them, solutions to this registration problem are being sought. ARNOLD (1968) describes a procedure by which it is possible to extend the dynamic range of the recording on film from 20 to 30 dB.

Another possibility is direct recording on magnetic tape. Proposals have been made (EASAMS, 1972) to use good TV-recorders for this purpose. In this way one tries to record the full dynamic range received by the radar and to form the final image on the ground to the user's wishes. In this way it becomes possible to lose less information in the air.

A third possibility is compression. An object is seen by the radar for many sweeps. A SLAR for example, using an antenna system with an opening of 3.5 mrad and mounted in an aircraft flying with a speed of 150 m/sec and using a pulse recurrency frequency (P.R.F.) of 2000 (emitting 2000 pulses per second), keeps an object in the beam:

at   5 km for 0.117 sec or 232 sweeps,
at 10 km for 0.234 sec or 467 sweeps,
at 15 km for 0.35   sec or 700 sweeps.

The integration over so many sweeps is thus done normally on film. If this could be done electrically it would mean a possible bandwidth reduction for the recording of at least a factor 100. The bandwidth required for direct recording is from 5 to 20 MHz. A reduction of a factor 100 would bring bandwidth down to a better manageable value. Several solutions are possible:

a) Using a scan converter (see for example EICHEL, 1973).

b) Using capacitor banks in which the signal for each range element is integrated and read (see HARRIS, 1970).

c) Both solutions are problematic, and a more recent approach goes digital since fast A/D converters up to 15 MHz are now becoming available.

As yet there is no commercially available solution based on the above principles, and all imagery is made on film with the limitations mentioned. This requires an accurate planning of the missions. The next chapter will, among others, indicate the variations in echo strength met when flying SLAR.

5.2 Backgrounds

5.2.1 Introduction

In the first chapter we discussed the general aspects of radar and how an image comes into being: the more metric properties of radar. The difference with other sensors, however, goes deeper than that.

In this chapter we shall treat what is known about the interaction of radar-waves with the object(s) under observation. Two properties in which radar deviates from other remote sensing systems become obvious immediately. First of all, radar transmits electromagnetic waves at a single frequency instead of a frequency band. Even in monochromatic operation, most other sensors work in a frequency band. Furthermore, radar carries its own light source. In this way illumination is better in hand than in many other remote sensing systems which depend on sources outside the system (as e.g. the sun in aerial photography).

Still another point of interest is the fact that radar uses centimetric wavelengths. As electromagnetic radiation radar waves are similar in behavior to light. The fact, however, that the wavelength becomes comparable with the dimensions of the radar system has its consequences. For example, antennas are always in the order of $100\lambda–200\lambda$. In optics, the relation between the diameter of a lens and the wavelength is more than a factor 10^4. Diffraction and interference effects thus play an important role in radar. Roughness or smoothness of a surface is another example which is expressed in wavelengths. A surface is called smooth when height variations remain below $\frac{1}{4}\lambda$. For radar, these effects take place on another scale than in optics. This means, for example, that a surface (e.g. a plowed field) is rough for a radar system working with short radar waves (K_a- or X-band), but can be smooth for another system working at the long wavelength end of the microwaves (e.g. L- or P-band).

The above just mentions a few physical characteristics which have to be taken into account in describing the interaction of the radar waves with an object. The reader must always bear in mind that when he looks at a radar image, he looks at a transformation. A radar image is a transformation from the microwave part of the electromagnetic spectrum, where the radar "sees" to the visible part where our eye can see.

It must be clear that for such an image other criteria apply than used in the visual part of the spectrum.

5.2.2 The Radar Equation

The amount of energy received by a radar, P_{rec}, can be expressed as (Skolnik, 1970):

$$P_{rec} = \frac{P_{trans}G_t}{4\pi R^2} \times \frac{\sigma}{4\pi R^2} \times A_r .$$

With P_{trans} : energy transmitted by the radar
$\quad\quad G_t$: antenna gain (energy gain with respect to an antenna radiating over a complete sphere)
$\quad\quad R$: range to target.

The first factor describes the energy transmitted to the target. The second factor gives the energy reradiated by this target with a radar-cross-section σ. Of this reradiated energy, the antenna of the receiver with effective aperture A_r intercepts a portion given by the product of all three factors. Using the same aerial for transmission and reception G_t and effective receiving aperture A_r are related by:

$$G_t = \frac{4\pi A_r}{\lambda^2} \quad \text{or} \quad A_r = \frac{\lambda^2}{4\pi} G_t$$

which finally leads to the common way of expressing the "radar equation":

$$P_{rec} = P_{trans} \cdot \frac{G_t^2 \lambda^2 \sigma}{(4\pi)^3 R^4}. \tag{2}$$

This equation can also be written shortly as:

$$P_{rec} = Q \cdot \frac{\sigma}{R^4}$$

where in $Q = G^2 \lambda^2/(4\pi)^3 \; P_{trans}$ we have put together all parameters determined by the radar used.

In fact σ is defined through this "radar equation". Expressing λ and R in m's gives σ in m^2. There is rarely a relation between the physical size of a target and its radar-cross-section. Only for a limited amount of targets is calculation of the radar-cross-section possible. An example is the sphere. When the size of a sphere is much larger than the wavelength of the illuminating radar, its cross-section is equal to its optical cross-section ($1/4\pi D^2$, with D: diameter of the sphere; $D \gg \lambda$). Spheres are therefore often used as reference targets. Other references in common use are the corner reflector and the rectangular plate.

Radar targets are either point targets or distributed targets: Point targets are objects smaller than the illuminating beam, where distributed targets are built up of more scatterers and can be larger than the illuminating antenna beam.

5.2.3 Ground Returns

5.2.3.1 General

In earth resources we deal with ground returns. They can be considered as distributed targets larger than the illuminating radar beam with eventually point targets on them. When dealing with ground returns only (vegetation, the sea), the definition of radar cross-section must be expanded and tied to a unit surface (1 m^2) to become independent of illuminated area (determined by antenna opening). Two definitions are in common use: σ_0 and γ (Fig. 12)

$$\sigma_0 = \sigma/A \tag{3}$$

where: σ defined by the radar Eq. (2).
         A surface illuminated.

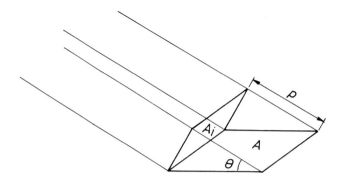

Fig. 12. Definition of γ and σ_0: $\gamma = \sigma/A_i$; $\sigma_0 = \sigma/A$

The second definition divides σ by the cross section A_i of the illuminating antenna beam at the place of the target or

$$\gamma = \sigma/A_i \tag{4}$$

the relation between γ and σ_0 is:

$$\sigma_0 = \gamma \sin\theta \quad \text{or} \quad \gamma = \sigma_0/\sin\theta \tag{5}$$

with θ the grazing angle. γ the "return parameter" was introduced by COSGRIFF et al. (1960) as a kind of reflection coefficient, and was preferred by them because it remains a constant as a function of angle of incidence for rough surfaces in their measurements. For this same reason we shall also use γ instead of σ_0.

When an electromagnetic wave hits a surface, it is reflected and refracted into the underlying medium (penetration depth). The amount of energy actually reflected and refracted depends on the electromagnetic properties of the medium involved (complex permittivity) and its surface properties (surface roughness).

For an electromagnetic wave (radar wave) in air (vacuum, but the permittivity of air is practically 1) reflected from a flat surface (e.g. water), STRATTON (1941) gives for the reflection coefficient the following equations:

$$R_{VV} = \frac{\varepsilon \sin\varphi - \sqrt{\varepsilon - \cos^2\varphi}}{\varepsilon \sin\varphi + \sqrt{\varepsilon - \cos^2\varphi}}$$

for vertical polarisation and

$$R_{HH} = \frac{\sin\varphi - \sqrt{\varepsilon - \cos^2\varphi}}{\sin\varphi + \sqrt{\varepsilon - \cos^2\varphi}}$$

for horizontal polarization,

with ε: the complex permittivity (or dielectric "constant" $\varepsilon = \varepsilon' - j\varepsilon''$),
$\qquad\quad \varphi$ = angle of incidence, measured away from the vertical.

Here R is the specular reflection coefficient (bistatic). In radar observation we are interested in the backscatter. At normal incidence ($\varphi = 0$), specular- and back-scatter reflection are the same and we find the well known relation:

$$R = \frac{\sqrt{\varepsilon} - 1}{\sqrt{\varepsilon} + 1} \quad \text{or} \quad R = \frac{n - 1}{n + 1}$$

with $n = \sqrt{\varepsilon}$ the refractive index. Dependent on the refractive index, energy will be transmitted into the medium observed.

With the aid of the formulae:

$$P = P_0 e^{-2\alpha D}$$

with P = power, P_0: power at the surface (just beneath it) D: depth and

$$\alpha = \frac{2\pi}{\lambda} \cdot \sqrt{|\varepsilon|} \sin \tfrac{1}{2}\delta$$

with λ the wavelength used and δ: the loss angle, the attenuation of power can be calculated as a function of depth D when $|\varepsilon| = \sqrt{\varepsilon'^2 + \varepsilon''^2}$ and $\delta(\tan\delta = \varepsilon''/\varepsilon')$ are known. The quantity D_{skin}, the depth where the power is attenuated to $1/e$ (37%), is called the skin depth. This quantity can thus be given by:

$$D_{skin} = \frac{1}{2\alpha} = \frac{\lambda}{4\pi\sqrt{|\varepsilon|} \sin \tfrac{1}{2}\delta}.$$

Since even for the driest materials (e.g. dry sand) $\sqrt{|\varepsilon|} > 1.5.$, $\sin\tfrac{1}{2}\delta$ must be smaller than $1/6\pi = 0.053$ and thus $\tan\delta < 0.106$ to make D_{skin} larger than λ, the wavelength used. When $|\varepsilon|$ increases, this value becomes even smaller. For an average soil in Western Europe SMITH-ROSE (1933, 1935) found a conductivity of $1.1 \cdot 10^{-4}$ mho/m. Taking $|\varepsilon| = 10$ at $\lambda = 30$ cm (PAQUET, 1964) and accounting for the relaxation of water, we find at $\lambda = 30$ cm: $D_{skin} = 10$ cm and at $\lambda = 10$ cm: $D_{skin} = 2$ cm, values much smaller than the wavelength used.

Only for very dry materials can the skin depth become equal and possibly larger than the wavelength used. The fact that a fair amount of energy will be lost at the surface by reflection and the fact that in most SLAR observations incidence angles are far from normal incidence, ensure that the contribution of the subsurface to the backscatter can only be small in most cases.

The refractive index, and thus the complex permittivity of the materials under observation is thus an important quantity. Most of these materials are heterogeneous systems containing water (e.g. soils but also plant material etc.). The dielectric properties of such heterogeneous systems are very complex. They were summarized by DE LOOR (1956, 1964, 1968). For the frequencies now under consideration ($v > 1000$ MHz), the dielectric properties of such systems are mainly governed by the free water they contain, where some forms of bound water also play a role at the lower frequency end. For more information the reader is referred to the literature cited above.

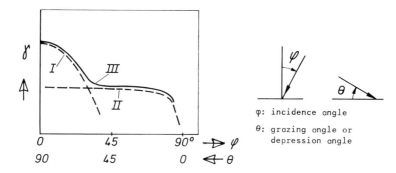

Fig. 13. Dependency of γ on angle of incidence φ (grazing angle θ). *I* Specular part; *II* diffuse part; *III* total

The second important parameter is surface roughness, where roughness must be considered in relation with the wavelength used. As remarked, Cosgriff et al. (1960) introduced the radar return parameter because of the fact that this quantity remained reasonably a constant in their measurements through all angles of incidence. This "semi-lambertian" behaviour, however, is not complete.

As a function of angle of incidence, the backscatter coefficient can be described as consisting of two components: a specular and a diffuse part (Moore, 1970; Ruck et al., 1970; Renau and Collinson, 1965). See Fig. 13. The specular part will be important around the vertical, where the diffuse component will govern the behavior of γ at larger incidence angles φ (smaller depression- or grazing angles θ). Measurements have shown that for the angles of interest when observing the earth's surface with SLAR equipment (grazing angles $2° < \theta < 45°$), many surfaces can be considered as rough and behave as diffuse scatterers (Grant and Yaplee, 1957; Goodyear, 1959; Cosgriff et al., 1959; Oliver and Peak, 1969). γ then becomes a constant between these angles. This also applies for the sea. Therefore we shall use $\gamma \ (= \sigma_0 \sin \theta)$.

Measurements of the backscatter are reported in the literature in several places. It is difficult, however, to obtain a consistent picture. See also Skolnik (1970).

5.2.3.2 Sea Echoes

In principle, a water surface mirrors the radar waves away from the radar. A watersurface, therefore, usually appears as an area of "no-show" and is so easily distinguished in a radar image. Waves on the water surface, however, may have faces or facelets so placed that they return some of the radar energy falling on them back to the radar. The capillary and short gravity waves brought about by the wind and riding on the sea- and swell waves are the cause of the radar echo.

The sea, eternally in motion, always gives a sea return. This return can be a hindrance when we want to see targets on the sea's surface as ships, buoys, etc. In this case one usually speaks about "sea clutter". It can also be a help, e.g. when one wants to know something about the surface for instance the direction of sea

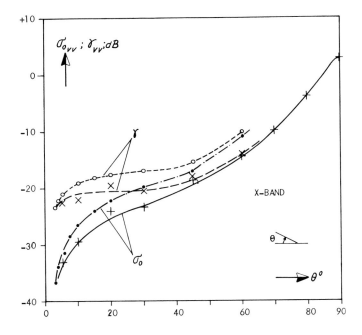

Fig. 14. Sea clutter. $\gamma(\times)$ and $\sigma_0(+)$ vs. grazing angle θ for windspeed $v=15$ knots; JOSS I after DALEY et al. (1971) and clutter of wind generated waves in a tank after MYERS and FULLER (1969): $\gamma(\circ)$, $\sigma_0(\cdot)$

waves and swell waves; oil films dampen the surface structure of the water and this change can be detected, etc.

It is difficult to obtain exact figures on sea echoes, and when given this is usually done in σ_0. Many measurements are reported by the Naval Research Laboratory (Washington DC), taken with their experimental aircraft equipped with a four frequency radar system (P-, L-, C-, and X-band). Figure 14 gives a measurement of $\gamma(\sigma_0)$ vs. grazing angle (JOSS I, DALEY et al., 1971) for a windspeed $v=15$ knots. These measurements are compared with laboratory measurements of clutter of wind generated waves in a tank (MYERS and FULLER, 1969). The same general behavior as sketched in Fig. 13 is found. Measurements for vertical polarization are reported. The data for horizontal polarization lie below those given here. For very small grazing angles this behavior changes, as SITTROP'S (1969) measurements given in Fig. 15 demonstrate.

The NRL data taken near Bermuda in 1970 (JOSS I) and Puerto Rico (1965) reported by DALEY et al. (1971) were further studied by us. We recalculated their σ_0 values to γ and averaged over the grazing angles: $5°$–$45°$. This gives figures as Fig. 16 where γ measured upwind is plotted vs wind speed v for different polarization conditions, at X-band. For downwind and crosswind, similar plots are found with γ_{VV} downwind a 1 dB lower than for upwind. All other values for γ at X-band are at least 9 dB lower than γ_{VV} for the upwind condition.

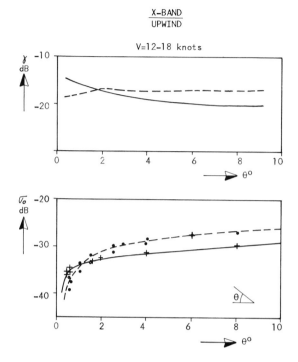

Fig. 15. Sea clutter. σ_0 and γ vs. grazing angle θ for very small angles. (After SITTROP, 1969)

Similar lines are found at C-band, but with another inclination. These data so thus suggest the following dependency of γ on wind speed v:

$$\gamma \sim v^c$$

where c is a constant that diminishes with frequency. When v is expressed in knots, c becomes about 1.7 at X-band and 1.4 at C-band and varies somewhat around these values for these two bands, dependent on polarization used and the wind direction. The NRL data as used here show practically no dependency on wind (thus $c=0$) for the P-band and a very small one, if any, for the L-band. This may suggest a change in reflection mechanism. A very crude average over all upwind, crosswind and downwind data cited gives for L-band: $\gamma_{VV} \approx -26$ dB and $\gamma_{cross} \approx -42$ dB. For P-band these values are $\gamma_{VV} \approx -25$ dB and γ_{HH} and $\gamma_{cross} \approx -40$ dB.

In Fig. 17 we compare these NRL data taken at X-band (downwind) with others found in the literature. The direction of measurement with regard to wind direction is not known except for SITTROP's data (crosswind). It must be remarked that the averages reported by RUCK et al. (1970) do not show the dependency of c on frequency as suggested by the NRL data.

Further work is needed here. It can be questioned, for example, whether γ is the same for oceans (e.g. the Atlantic) and seas (e.g. the North Sea, the Mediterra-

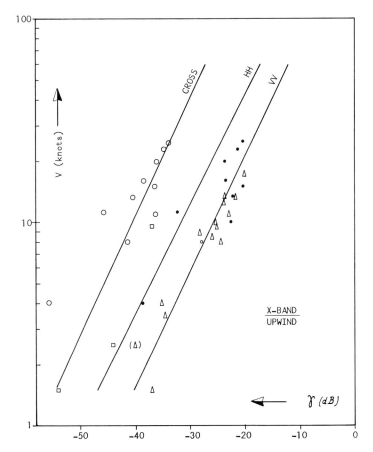

Fig. 16. NRL data: JOSS I (1970) and Puerto Rico (1965) after DALEY et al. (1971). · γ_{vv}, ○ γ_{cross}: JOSS I; △ γ_{vv}, □ γ_{cross}: Puerto Rico. Average for γ_{HH} is included without measuring points

nean). The NRL-data were taken over open ocean, SITTROP's data over the atlantic from a 400 m high cliff on the Norwegian coast, whereas GRANT and YAPLEE (1957) measured above a river. Another problem is the contrast: water-oil. It is known that on oil layer dampens the gravity and capillary waves and thus shows up as a area with a lower sea return (GUINARD, 1971). But an exact knowledge of the magnitude of that contrast is not yet available.

5.2.3.3 Land Echoes

For the land echoes the situation is not much better than with the sea measurements. There are many measurements available. Again, however, it is difficult to extract from them a homogeneous and consistent picture. MOORE (1970) gives a good survey.

Measurements are done from aircraft and from ground based reflectometers (GRANT and YAPLEE, 1957; COSGRIFF et al., 1960; ULABY, 1973). The aircraft

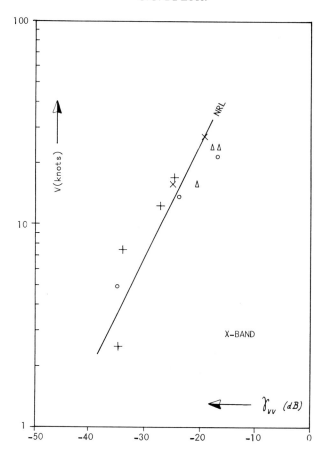

Fig. 17. Literature data compared with NRL data (downwind) at X-band. + Grant and Yaplee (1957) (10°—45°); × Wiltse et al. (1957) (15°—45°); △ Sittrop (1969) (2°—10°, crosswind); ○ Ruck et al. (1970) (1°—30°)

measurements (Goodyear, 1959; Daley et al., 1968) often have trouble with a good definition of the target observed, where ground based measurements can have problems in obtaining a measurement from a sample of sufficient size.

Figure 18 gives a K_a-band ($\lambda = 8.6$ mm) radar image for the same agricultural area taken at two different times: one in July with most crops in a mature stage, the other in November with most of the fields bare. The variation in image tone in the first image is due to the differences in the radar return parameter γ for the different crops. It has therefore been suggested to use the radar return parameter γ for vegetation inventory. γ is dependent on polarization and wavelength. By using different polarizations in one radar system it is possible to make cluster plots by comparing γ_{HH} (the return parameter for horizontal polarization) with γ_{vv} (vertical polarization) and with γ_{HV} or γ_{VH} (cross polarization). This is illustrated in Fig. 19 (de Loor and Jurriens, 1971) for an observation at K_a-band of an agricultural area in the first week of August (mature crops, the cereals are only two weeks

Fig. 18a and b. K_a-band image, polarization HH. Different radar signatures for different crops. Dependency on season: (a) November, (b) July. (British Crown Copyright)

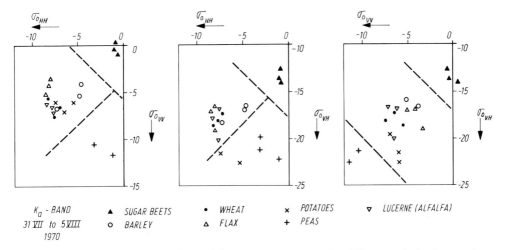

Fig. 19. Cluster plots. Comparison of the return parameters for different polarizations and using them for inventory

before harvesting). A fair discrimination between several species can be made for this single frequency. Using another frequency in combination would probably have improved discrimination further.

Another aspect to take into account is the time of the year, as Fig. 20 illustrates (see also Fig. 18). There are places of crossover for different crops, where in this

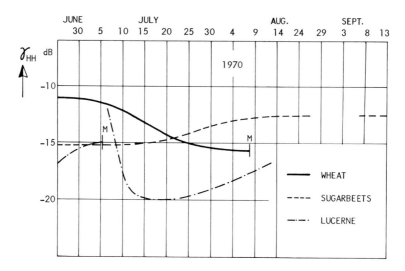

Fig. 20. Variation of γ for some agricultural crops through the growing season. X-band, polarization HH. (After De Loor et al., 1974)

example for the polarization used *(HH)* no discrimination would have been possible (de Loor et al., 1974).

An aspect seen in the last example is that the total variation between different crop species is not large: 8–10 dB. Including bare soils and other vegetation species as e.g. woods, the total possible variation on the return parameter γ for vegetation discrimination increases to a 15 dB. With a SLAR, recording images on film total dynamic range is not much larger: up to 20 dB (Arnold, 1968).

Theory predicts the influence of moisture on the return parameter γ (Moore, 1970; Waite and MacDonald, 1971). In fact we can explain Fig. 20 with such an effect. When for example we examine the curve for lucerne (alfalfa) somewhat closer, we see the following happen: immediately after mowing (*M* in Fig. 20), the echo increases: the echo from short stalks on a field with ample water supply. When the plant grows up again and begins to screen the ground the echo decreases; screening is complete after about 10 days. Then the echo starts rising again: the number of leaves in the top increases and so also the amount of free water above the soil. A similar reasoning is possible for the other two crops shown: in the period covered the wheat ripens, so the amount of free water in the top layers decreases and the echo decreases. For the sugar beets the opposite occurs: more and larger leaves lead to more free water above the soil and thus a higher return. Further research is needed here because such an effect can have an important application. Biomass in situ determinations in a crop can be done through a determination of the free water in that crop. By a calibration procedure, this amount of free water can then be related to the biomass of that crop.

It seems that the radar return parameter is the only possible classifier for vegetation inventory (De Loor et al., 1974) (taking the time of the year into account). The fluctuation or fading spectra (signal fluctuations due to movement of the scatterers) are all very similar for different vegetation species and are only dependent on wind speed. The decorrelation times involved vary between 0.025 and 0.50 sec dependent on wind speed (and for X-band). It can also be shown that vegetation behaves as a Rayleigh scatterer.

Exact figures for the value of the return parameter γ for different vegetation species are difficult to give. General ranges can be given, however, as the following summing up for X-band:

— Sea echoes: γ between -30 and -13 dB dependent on wind speed.
— Echoes of bare soils: γ between -32 and -16 dB dependent on stubble content, treatment (ploughing, harrowing) and wetness.
— Vegetation echoes: γ between -18 and -8 dB, with woods at the high end.
— Echoes of forest borders, single rows of trees, hedges, dams. etc.: γ between -10 and 0 dB.
— Built up areas, and man-made objects: $\gamma > 0$ dB.

5.2.3.4 Actual Value of the Radar Cross-section as Seen by a Real Aperture SLAR

When γ as defined above can be considered as a constant for grazing angles between $2°$ and $45°$, it can be shown that the total radar cross-section of ground returns σ as seen by a SLAR is practically a constant as a function of range.

The amount of energy returned by a piece of ground or sea depends on the area illuminated:

$$\sigma = \sigma_0\, A = \gamma \sin\theta \cdot A . \qquad\qquad\qquad (3) \text{ and } (5)$$

p the length of the pulse transmitted (see Fig. 12) by the radar projected on the ground is:

$$p = c\tau/\cos\theta$$

where c is the speed of light (speed of the radar waves) and τ the pulsetime. When the antenna beamwidth is β radians and $\beta < 1°$:

$$A = pR\beta .$$

$$\begin{aligned} \text{So: } \sigma &= \gamma \sin\theta\, A = \gamma \cdot c\tau\, tg\,\theta\, R\beta \\ &= \gamma \cdot c\tau\, R\beta\, H/R_g \\ &= \gamma \cdot c\tau\, \beta H \cdot R/R_g \end{aligned}$$

where R_g is the groundrange: the distance between object and the projection of the flight line, and H the height of the SLAR above the ground. The variation in $R/R_g = \sec\theta$ between $\theta = 45°$ and $2°$ is only from: 1.41 to 1.001 or from 1.5 to 0 dB. Considering the accuracy with which we know γ, we can thus take σ as a constant $(= \sigma_{\text{clutter}})$.

This means that for each echo on the ground, the signal/clutter ratio $(\sigma/\sigma_{\text{clutter}})$ can be considered as independent of range.

5.3 Applications

5.3.1 Introduction

Much radar imagery has been flown in the course of the years, and its reading has led to the suggestion of a number of applications. Actual operational use of SLAR, however, is still limited. When we look at the crude sketch of the remote sensing system given in Fig.21, we can indicate a reason for that situation. Looking at images is dealing with the output end of the system only. Knowledge of the input end, however, is only scantily available. This input end is the interaction of the radar waves with the ground. Only an improved knowledge of this interaction—thus of the input end of the system—can lead to a better use of the output (the imagery) and thus to a better use of the system as a whole. Chapter 5.2 treated a series of fundamental aspects of this input.

Let us now look at this problem from yet another angle. Radar imagery is sometimes referred to as being "unsharp" and "crude". Indeed it is true that obtainable resolution is in the order of 15 m and more. This may improve somewhat in the future, but it is inherent to this system that one must count with a resolution in the order of meters. When we look to the scale of the imagery finally obtained, however, we see that resolution is always in fair agreement with that scale and the area overseen in one single flight. With SLAR systems available nowadays, resolution is such that mapping is possible to scales 1:100000. In the future this may improve to 1:50000 (with a diminution of the area scanned in one flight in accordance with this change of scale). Photography at the same scale (and comparable resolution) requires aircraft flying at very high altitudes or even spacecraft with all inherent limitations as weather, clouds and daylight, where radar can do the same thing in all weather, day and night, with aircraft flying between 100 and 5000 m.

We must thus live with the resolution belonging to the scales worked at. But resolution is not the only criterion with which one can describe, analyse, identify or map objects. Other properties than the spatial (as shape and geometry where resolution is indeed important) can be used such as:
— the spectral properties ("color"),
— polarization effects,
— temporal effects (variations in time, place).
As we have seen in Section 5.2 radar can use them all:
— the spatial properties: resolution is in accordance with the scales used,
— the spectral properties: the radar return parameter γ is dependent on frequency,

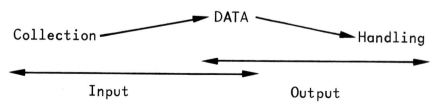

Fig. 21. General survey of the remote sensing system

— polarization effects: the radar return parameter depends on the polarization used (we have γ_{HH}, γ_{vv}, and γ_{cross}) and

— the temporal effects: by using MTI (Moving Target Indication) radar can—by comparison of successive sweeps—identify and display objects moving with speeds of several km/hour and faster.

Basic ground measurements, as e.g. the determination of the reflective properties of the sea, soils, vegetations, rocks (which means more than just taking "groundtruth" data) combined with (in the beginning a limited amount of) SLAR flights in a well balanced program are thus first necessary in most of the fields where applications are foreseen. This means that in the following we can only indicate possibilities for the application of radar. Further research must show in how far these and other applications will finally justify a more operational attack of the problem.

The list given can therefore never be a complete one. We are still at a stage where each flight can give unexpected results that can lead to new applications. The applications are listed crudely to disciplines. Overlaps will be there and possibly placings under a wrong heading.

The author also fails to offer views on the economic or social impact of possible applications either in Europe or elsewhere. He only tries to offer material which may stimulate the reader to investigate the possibilities offered in his own discipline.

All radar imagery used to illustrate this chapter on radar methods was flown by the Royal Radar Establishment (Great Malvern, UK). It is British Crown Copyright and is published by permission of the Controller H.B.M.S.O.

5.3.2 Geography

5.3.2.1 General

Dependent on the SLAR system, all weather mapping is possible at scales between 1:100000 and 1:1000000. A series of applications is then foreseen:

a) The production of base maps: in areas where maps are not available or incomplete, this as a start e.g. for exploration work or further topographic mapping by photography or SLAR. A good example is the mapping of the Darien Province in Panama.

b) Rapid updating of maps. In the range of scales mentioned, turn around times can be short and regular surveys can be planned even with a fairly tight time schedule because of the independence of the weather (except heavy rainfall) and time of day.

c) Town and country planning. This independence of time and weather and its immediate availability make mapping with SLAR also attractive in town and country planning for the determination and regular control of e.g.:

— rural and urban developments

— recreational areas.

d) Traffic control.

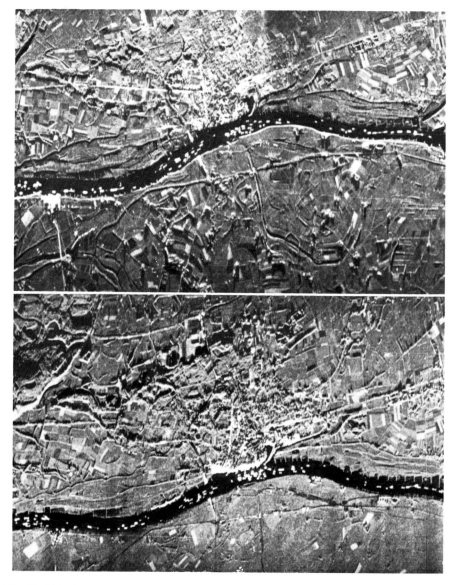

Fig. 22. Shipping on the river Rhine in summer. Time interval between images 4 min.
K_a-band; range 5 km. (British Crown Copyright)

— On land. By the use of MTI an impression of traffic density on (open) highways is possible. Crowding of vehicles in traffic jams is visible. The use of MTI for this purpose has not yet been investigated.

— On inland waterways and in estuaries. Figure 22 gives an example of the density of shipping on the river Rhine near the Dutch border. The two images were made 4 min apart and form part of a whole series. Although traffic density is

Fig. 23. Shipping on the North Sea for the Dutch Coast and the entrance to Rotterdam Harbour, fall 1962. Range 60 km, radar looking North; X-band, negative image. (British Crown Copyright)

easily determined, the study of traffic flow is not easy since the direction of movement of a ship is difficult to determine.

— At sea. The last remark also applies for the detection of ships and other objects at sea. Large areas can be scanned. Danger zones can be found—and eventually controlled—in seas and straits with heavy traffic. A good example is the Channel and the North Sea. Figure 23 gives an example of traffic on the North Sea for the Dutch coast and near Rotterdam harbour in 1962. An exercise with the same X-band SLAR in 1969, where the whole North Sea was mapped during a night with bad weather and limited visibility at the sea's surface, was very illuminating. It emphasized the need of developing traffic regulations, procedures and controls for seas with such dense shipping similar to those common for air traffic control. A surveyance aircraft equipped with a long range SLAR could be of great use in such a system.

e) Special cases. A good example is here damage assessment immediately after disasters. In particular, damages due to flooding are good to determine due to the good land/water contrast. Often the weather will be bad in such cases preventing the use of aerial photography, thus making the radar an efficient aid in planning immediate and well directed rescue actions.

Another application is damage assessment (e.g. in areas where regular flooding occurs) to assess damage claims (insurance).

5.3.2.2 Metric Accuracy

In many of the applications mentioned above, metric accuracy is required. To what extent depends on the actual application. In the following survey this accuracy will be treated in relation to final mapping scale. For some af the applications mentioned in the foregoing, the mapping scale finally used may be larger than corresponds with the accuracy required. This is of no consequence, however, for the following considerations.

SLAR is a line scan system. This means that the image is formed line after line and not over a plane as in aerial photography. This also means that the image has to be reconstructed line after line to make it metrically correct. All motions of the platform carrying the radar have then to be taken into account. These are pitch, roll, yaw, crab, velocity and altitude changes (see Fig. 24). This necessitates the introduction in the aircraft of sensors for altitude, attitude and place of which the signals are recorded simultaneously with the image signal e.g. a good radar or laser altimeter, an inertial platform eventually combined with a hyperbolic location system.

Let us see what accuracies would be required of the system for mapping at scales 1:100000, 1:250000 and 1:500000 (see also LEVINE, 1960). The accuracy of a map may be stated as in the US National Standards of Map Accuracy (SWANSON, 1949). For maps published at the scales mentioned, no more than 10% of the points tested shall be in error by more than $1/50''$ (0.5 mm). This error (error in distance) can be described as being composed of errors along the x- and y-coordinates. When the distribution of errors in both these directions is assumed as being normal with the same standard deviation s, the radial probability function will be a Rayleigh distribution:

$$p(r) = \frac{1}{s^2} \cdot e^{-\frac{1}{2}\left(\frac{r}{s}\right)^2}$$

where $p(r)$ is a maximum for $r=s$. The cumulative distribution function $F(r)$ is:

$$F(r) = \int_0^r \frac{r}{s^2} e^{-\frac{1}{2}\left(\frac{r}{s}\right)^2} dr = 1 - e^{-\frac{1}{2}\left(\frac{r}{s}\right)^2}.$$

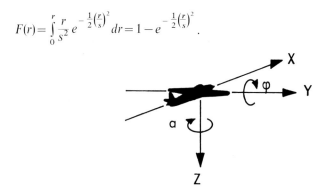

Fig. 24. Aircraft motions which require compensation in mapping

Table 2. Errors allowed for the indicated scales

Scale 1:	r_{10} (m)	s (m)	s_e (m)
100000	51	23.8	16.8
250000	127	59.2	42.0
500000	255	119	84.0

Table 3

SLAR-band	Height H (km)	Slant range min (km)	Resolution Cell (m²)	Slant range max. (km)	Resolution Cell (m²)	For scale 1:
X	8.5	12	50×200	70	30×1000	500000
X	8.5	12	50×200	40	30×600	250000
K_a	0.5	0.7	20×4	10	15×30	100000
K_a	1	1.4	20×4	10	15×30	100000
K_a	2	2.8	20×6	10	15×30	100000

The 10% error limit (r_{10}) is found by putting $F(r)=0.9$. So $s/r_{10}=0.47$ giving for s—the standard deviation of the error distributions in x- and y-directions allowed—$s=0.47\,r_{10}$. This error is caused by:

a) errors in the radar system ("interior orientation"),
b) errors in the coordinates of the platform ("exterior orientation").

Assuming in first approximation that both errors are equal in weight, then s_e (due to the errors in platform coordinates) shall not be larger than $0.47\,r_{10}/\sqrt{2}=0.33\,r_{10}$.

Table 2 then summarizes the values for r_{10}, s, and s_e for the scales under consideration.

For the SLAR we take the two examples given in Table 1. The next table (Table 3) gives resolution for the situations indicated.

HEMPENIUS (1969) showed that the "pointing error" is about 5% of the resolution in the imagery used, which means that its contribution to s can be neglected when we look to the last table. In how far this pointing error will be similar for an automatic system needs investigation when such equipment becomes available.

According to LEBERL (1970), the shifts Δx and Δy in the image points due to shifts in platform coordinates can be written as:

$$k \cdot \Delta y = \frac{G_0}{S}\,\Delta y_f + \frac{H}{S}\,\Delta z_f + \frac{G_0^2}{2S}(\Delta \alpha)^2 \,,$$

$$k \cdot \Delta x = G_0 \Delta \alpha - \Delta x_f + H \Delta \varphi$$

Table 4. Platform motions allowed

SLAR-band	For scale 1:	Height H (km)	s_x (m)	s_y (m)	s_z (m)	s_α mrad	s_φ mrad
X	250000	8.5	25	30	40	0.6	2.7
X	500000	8.5	50	60	80	0.7	5.5
K_a	100000	0.5	10	12	16	1.0	20
K_a	100000	1	10	12	16	1.0	10
K_a	100000	2	10	12	16	0.6	5

with (see Fig. 24):

k scale factor

G_0 $= \sqrt{S^2 - H^2} = $ ground range

H flying height

S slant range

Δx_f variation in x-coordinate (along track)

Δy_f variation in y-coordinate (across track)

Δz_f variation in height

α yaw angle

φ pitch angle.

Assuming that:

    a) the contribution of the term $G_0^2/2S(\Delta x)^2$ can be neglected,

    b) the uncertainties (indicated by Δ) are normally distributed with an average equal to 0 (no systematic errors), statistically independent and of equal weight,

we may put for the respective standard deviations:

$$s_y = \frac{S}{2G_0} \cdot s \qquad\qquad s_\alpha = \frac{1}{G_0\sqrt{6}} \cdot s$$

$$s_x = \frac{1}{\sqrt{6}} \cdot s \qquad\qquad s_\varphi = \frac{1}{H\sqrt{6}} \cdot s$$

$$s_z = \frac{1}{2H} \cdot s$$

which for the radars under consideration leads to the results of Table 4.

    With the assumptions given under b above: the average error then becomes $\sqrt{2/\pi} \cdot s = \Delta_{\mathrm{gem}}$; the 10% error is: $1.65\, s = \Delta_{10}$.

    Although the considerations given above are approximate, they give a good impression of the accuracy required of a sensor system giving the position of the platform carrying the radar. We so learn that the coordinates of the flying platform have to be known with an accuracy of the order of 10 – 50 m—dependent on the scale required—and yaw angle must be known within 1 mrad. These accuracies are fairly high. Relatively, they are possible over short distances say to 6 km, and unless reference points can be found on the ground special measures will be necessary.

5.3.3 Geology

5.3.3.1 General

Not only by rapidly providing base maps is SLAR useful, but also because of its extensive aerial coverage. It now becomes possible to oversee and follow topographic differences over large areas. It thus facilitates physiographic differentiation and reconnaissance mapping on a regional scale. Dependent on the mapping scale used, a fair amount of detail can also be extracted from such imagery. So radar detected faults and lineaments over large areas which had been undetected in aerial photography due to its larger scale (DELLWIG et al., 1966; MOORE and SIMONETT, 1967; MACDONALD et al., 1971).

A good survey of the use of radar for geology is given by MACDONALD (1969) in his evaluation of radar imagery from the Darien Province in Panama. He describes the keys used by him in the interpretation of SLAR imagery:

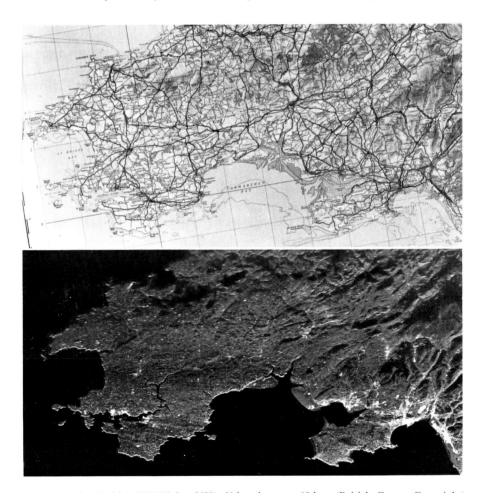

Fig. 25. Pembrokeshire (SW Wales, UK). *X*-band, range 60 km. (British Crown Copyright)

Fig. 26. Menai Street (NW Wales, UK). X-band, range 17 km. See linear feature. $H = 3300$ m.
(British Crown Copyright)

a) Tone. Tonal variations in the imagery are due to variations in the reflectivity of the terrain. Under circumstances it is possible to correlate tonal contrasts with rock type. Often, however, it is determined by vegetation, but information on vegetation can contribute indirectly to the geological interpretation.

b) Texture. Radar terrain texture can be a help in interpretation, and MAC-DONALD (1969) gives a group of keys for the imagery evaluated by him.

c) Pattern and shape. In the same way as in aerial photography pattern and shape aid in the interpretation of radar imagery. Care must be taken, however, to keep in mind the specific geometry of the radar image.

d) Slope and look angle. The radar shadow is an important feature to express relief. A low relief can be enhanced in the image by flying low and using the shadow then formed behind that relief. Radar shadows caused by a high relief, on the other hand, can screen parts of the area. Then more flights are required in such an area illuminating it from different sides. Slope measurements, which form an integral part of any geomorphic or hydrologic analysis, then become possible by a method given by McCoy (1967). The reader is referred to his extensive treatment of the problem, or the abridged version with nomogram given in MAC-DONALD's paper. See also the paper by LEWIS and WAITE (1971).

Using these tools, which in fact are keys to read a SLAR image (in the same way as there exist keys in photogeology to read aerial photographs), it becomes possible to distinguish various structural and stratigraphic features. Linear features as: faults, joint sets, shear zones, fractures, fold structures.

5.3.3.2 Hydrology

a) Drainage Basin Analysis

McCoy (1967) investigated the possible use of radar in drainage basin analysis using imagery of three different types of SLAR equipment. His observations gave a direct relationship between radar data and map derived data. Drainage area, basin perimeter, bifurcation ratio, average length ratio and circularity ratio could be measured directly from the radar data at his disposal. The correlation between radar and map data of stream length and stream number had to be determined for each radar seperately before that radar could be used.

b) Mapping of River Basin Morphology

Due to the good land/water contrast, mapping of rivers and near surroundings and of floodings caused by rivers, etc. is an obvious application. Figure 27 thus gives an example of a marshy area (sweet water) still under the influence of the tide through the river with which it is connected (at the time the image was made, 1964; the open connection with the sea has now been blocked).

The other figure (Fig. 28) shows that low flying makes it possible to enhance small altitude differences in the terrain by the shadow effect. Through this effect old riverbeds become visible in the image.

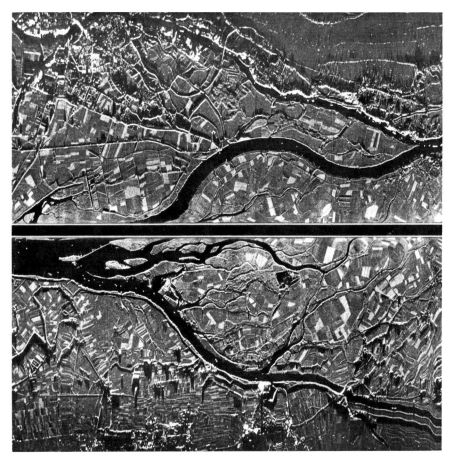

Fig. 27. "De Biesbos", summer 1964; sweet water marsh under influence of the tide. K_a-band, range 2×8 km, $H = 300$ m. (British Crown Copyright)

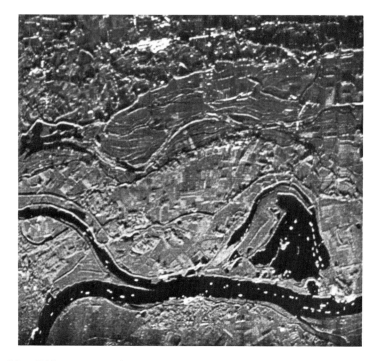

Fig. 28. River Rhine, near Waal, with surroundings, fall. Old river beds become visible by flying low: $H = 350$ m. K_a-band, range 8 km. (British Crown Copyright)

c) Observation of Soil Moisture

It has been suggested that soil moisture would easily be detectable because of the fact that the dielectric constant of most materials varies considerably with the moisture content of that material. As remarked already in Section 5.2, however, surface roughness is just as important a parameter for bare surfaces and for surfaces where vegetation screens the ground. Yet there are ample observations with SLAR where variation in moisture content was the most likely explanation for a variation seen in the image. This may then well have been a secondary phenomenon brought about by a moisture variation. The actual physical properties of such a phenomenon have then to be investigated.

5.3.4 Vegetation Studies

5.3.4.1 General

In vegetation studies SLAR can be useful for:
— reconnaissance and general mapping,
— delineation of boundaries,
— classification of species,
— determination of acreage of different plant communities.

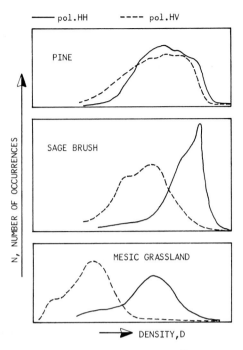

Fig. 29. Probability density curves as a means to describe texture. (After MORAIN and SIMONETT, 1967)

In Section 5.2.3.3 it was shown that the radar return parameter γ is a useful quantity to aid in vegetation inventory. In a radar image, variations in γ become manifest as density variations (see Fig. 18). The film densities found in images taken with different polarizations are then a clue for vegetation inventory (see Fig. 19).

Another key is texture. Many vegetation species show texture in SLAR imagery. MORAIN and SIMONETT (1967) tried to use this effect and made probability density curves (occurrence of a certain film density vs. that density) for some species found on SLAR imagery taken with a high resolution K_a-band radar. Figure 29 gives an example of some distributions found by them. It is known, however, that vegetation is a Rayleigh scatterer (DE LOOR et al., 1974). Curves as given in Fig. 29 should therefore be Rayleigh distributions. Vegetation canopies, however, are living systems and variations are to be expected in them. When such variations are similar in size or larger than a resolution cell, they can become significant as texture. MORAIN et al. (1970) thus showed that texture together with density in SLAR imagery can be used for vegetation inventory, and they developed keys to aid the interpreter in identifying vegetation species. To give texture in a more mathematical form, the curves of Fig. 29 can better be replaced by curves giving the number of density fluctuations (around the average image density determined by γ) in an area covered by one single species vs. number of resolution cells. Such a method is similar as that proposed by LESCHACK (1970) for forest discrimination from aerial photography.

5.3.4.2 Forestry

General mapping of forests with radar has been done at several places in the world. The stratification of forest vegetation is more difficult, but the possibilities of that application have been clearly demonstrated by Morain and Simonett (1967) in mapping at Horsefly Mountain (Oregon, USA) and by Hardy et al. (1971) in a project in the Yellowstone Park.

5.3.4.3 Agriculture

Crop delineation using SLAR imagery with the aid of the keys mentioned in Section 5.3.4.1 has been done (Haralick et al., 1970; Morain and Coiner, 1970). Delineation of boundaries and inventory thus seems possible. Figure 30 so covers two different areas: one agricultural where the different fields with different crops are mapped easily, the other with grasslands used for cattle breeding.

More data on number of frequency bands (and their spacing) and polarization combinations are necessary to become better informed on the effort required to attain a predetermined accuracy. This is a necessary requirement for operational

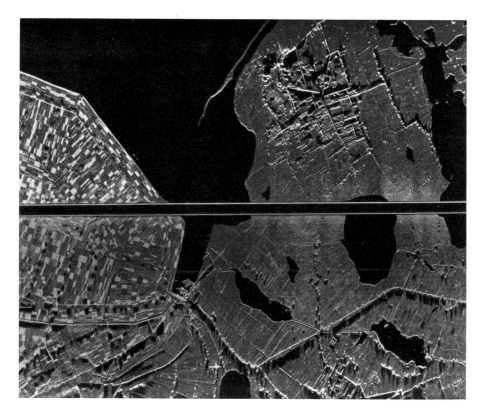

Fig. 30. Agricultural area and grass land area (cattle). K_a-band, $H = 150$ m, range 2×8 km. (British Crown Copyright)

use. Is it possible, for example, to use more than one frequency in one frequency band with the same antenna array? Work done at Kansas University (CRES) (WAITE, 1970; ULABY, 1973; ULABY and MOORE, 1973) showed that γ can vary considerably in one band. Other problems to be investigated are: are irrigation influences visible, and how far can the effect of the amount of free water in the leaves above the soil on the radar return parameter γ be used for in situ biomass determinations and thus for large scale and timely determinations of crop yield?

Well balanced programs where ground based measurements with measuring radars are combined with SLAR flights can give an answer here.

5.3.5 Sea, Coastal and Oceanographic Studies

5.3.5.1 General

Although in general water reflects the radar waves away from the radar, the capillary and short gravity waves caused by the wind and riding on the larger sea- and swell waves cause the water surface to give an echo. With vertical polarization and looking upwind, the highest echoes are obtained for all grazing angles of interest and all frequencies, as was shown in Section 5.2.3.2. In this way it is possible to observe all sorts of variations on a water surface.

Most SLAR equipments in use nowadays are equipped with antennas for horizontal polarization, being designed for working over land. When working above water surfaces becomes more important, it may be worthwile to design equipment specifically for this purpose. The difference in signal strength obtained from a water surface using horizontal polarization as compared to vertical polarization diminishes going to the higher frequencies.

The echoes obtained from a water surface will in most cases be lower than those obtained from land. So a higher gain setting must be used with an imaging radar used over water, and the land echoes eventually visible will saturate and show no details, for today's image tube/film combinations. On the other hand, when gain setting of the radar is right for use over land, no details will be seen in water surfaces which then become areas of "no-show".

5.3.5.2 Mapping of Estuaries

The mapping of rivers and estuaries with a clear indication of water courses, standing water, etc. is a conspicuous application. Setting now the gain for working over land, the land/water contrast becomes optimal for delineating all water bodies and details in the land returns remain visible (see also Fig. 27).

It is also possible, however, to identify stream patterns in tidal flats. Now the gain of the radar receiver must be put higher to look for detail in the water. Figure 31 gives an example. It was made at low tide with a strong wind blowing which caused the clutter in the deeper water, where the wet sand banks became areas of "no-show". In this way it could be shown that watercourses had not changed place due to the making of the dam, which is visible on the radar image but is not yet indicated on the map. The individual spots in the water are wet wooden poles with a diameter comparable to the wavelength of the radar (X-

Fig. 31. Tidal marsh, low tide. Deeper water clutters up and streams become visible. X-band, negative; $H = 150$ m; range 2×3 km. (British Crown Copyright)

band), which become so conspicuous due to a resonance reflection (they are practically invisible on the K_a-band images). They are put there in the water to demarcate oysterbanks.

5.3.5.3 Mapping of Sea- and Swell Waves

Since the reflection mechanism for radar is due to facelets of wind-driven capillary and short gravity waves riding on the sea- and swell waves, these waves can become visible. An example is the K_a-band imagery of the Gulf Stream flown by NASA and reported by MOORE and SIMONETT (1967). The author was also involved in an exercise to detect swell waves on the North Sea. Flights made by the Royal Radar Establishment (UK) with their K_a-band SLAR for the Dutch coast showed a swell wave pattern (wavelength 250 m, estimated height 20 cm) coming from the NorthWest over which a sea wave pattern (wavelength 50 m) coming from the NE was superimposed. Weather was fair that day and sea state was 1.

Thus SLAR can be a useful instrument in the study of sea- and swell waves.

5.3.5.4 Mapping of Ice

The ability of radar to map ice has been the subject of a number of studies which all show that radar has great potential for this purpose (LARROWE et al., 1971; JOHNSON and FARMER, 1971a, 1971b; BIACHE et al., 1971). A practical application was the use of SLAR with the MANHATTAN tanker test in September 1969 in Alaska, where the routing of the tanker was assisted by SLAR imagery (JOHNSON and FARMER, 1971b).

5.3.5.5 Detection and Control of Oil Pollution

Since the oil layer dampens the action of the capillary and small gravity waves which cause the sea clutter, this clutter is reduced in the polluted area. The oil field thus becomes visible as an area of "no-show" in the surrounding sea clutter. Examples of imagery so obtained can be found in the publications of GUINARD (1971a, b).

5.3.5.6 Other Possibilities

In the same exercise described in Section 5.3.5.3, two other phenomena became visible due to changes in the wave pattern on the sea's surface. They are included here as examples of as yet incompletely known possibilities of using the variations in contrast and texture in a SLAR image due to such small changes in wave patterns.

In Fig. 32 the division between the salter water of the North Sea and the sweeter water of the Nieuwe Waterweg (Hook of Holland) is visible. Observation from the air showed that the wave patterns for these two types of water differ markedly.

The second phenomenon was the occurrence of streaks in the radar image due to varying patterns of surface waves created by the tidal current passing along submarine ridges (a dune pattern on the bottom of the sea). The strength of this current increases when going to low tide and the phenomenon so becomes better visible when low tide comes nearer.

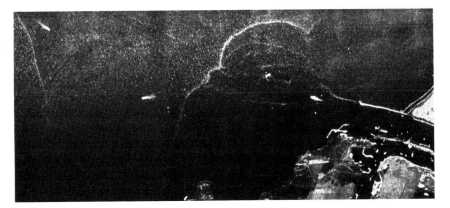

Fig. 32. Outflow of Nieuwe Waterweg into the North Sea before Hook of Holland. Division between sweeter water of the N. Waterweg and the salter water of the North Sea. K_a-band. (British Crown Copyright)

5.4 Conclusion

The foregoing clearly demonstrated that—when properly used—radar is a very versatile tool for remote sensing. Its application requires good use of its real potentialities. Two very important and rather obvious ones are its all weather capability and its ability to oversee large areas. Other potentialities, however, are yet inadequately known, with the result that the introduction of SLAR as a remote sensing tool is slow.

As in other remote sensing systems, a great deal of emphasis has been put until now on the output end: the use of SLAR imagery (see Fig. 21). This is understandable, since with this kind of information it is hoped to reduce costly fieldwork in the end and also, by automation, to reduce the costs of interpretation and turn around times. The images offered, however, are only transformations from the microwave part of the electromagnetic spectrum (where the radar observes) to the visual (where our eye can see). The interpretation techniques which are standard in aerial photography can therefore only be applied with the utmost caution, since different physical properties (of the object under observation) underlie image formation. It is the unsufficient knowledge of this input end of the system which has prevented us from obtaining the full benefits of SLAR. In the foregoing, therefore, it has also been attempted to bring together the material already available as a means to suggest ways for further investigations, since a more systematic approach to investigate SLAR's real potentialities is urgently needed.

SLAR will certainly have many more possibilities than yet foreseen. However, on its own it certainly will not be a panacea to all problems. This, however, applies to other remote sensing systems as well. Only a well directed research program will bring it to its rightful place in the total orchestra of remote sensing methods.

References

ARNOLD, E. M.: Receiving and recording the wide dynamic range of signals of a side-looking radar system. In: Agard Conf. Procs no 29: Advanced techniques for aerospace surveillance, pp. 187–196 (1968).

BIACHE, A., BAY, C. A., BRADIE, R.: Remote sensing of the arctic ice environment. Procs VIIth Symposium on Remote Sensing of Environment, Ann. Arbor, USA, pp. 523–561 (1971).

BROWN, W. M., PORCELLO, L. J.: An introduction to synthetic-aperture radar. IEEE Spectrum **6**, 52–62 (Sept. 1969).

CARLSON, G. E.: An improved single flight technique for radar stereo. IEEE Trans. on Geoscience Electronics **GE-11**, 199–207 (1973).

COSGRIFF, R. L., PEAKE, W. H., TAYLOR, R. C.: Terrain scattering properties for sensor system design (Terrain Handbook II). Engineering Expt. Station Bull. **181**, Ohio State University, 1960.

DALEY, J. C., DAVIS, W. T., DUNCAN, J. R., LAING, M. B.: NRL terrain clutter study. Phase II. NRL Report 6749, Naval Research Laboratory, Washington, Oct. 21, 1968.

DALEY, J. C., RANSON, J. T., JR., BURKETT, J. A.: Radar sea return—JOSS I. NRL Report 7268, Naval Research Laboratory, Washington, May 11, 1971.

DELLWIG, L. F., KIRK, J. N., WALTERS, R. L.: The potential of low-resolution radar imagery in regional geologic studies. J. Geophys. Res. **71**, 4995–4998 (1966).

DE LOOR,G.P.: Dielectric properties of heterogeneous mixtures. Leiden, Thesis, 1956.

DE LOOR,G.P.: Radar and stereoscopy. Report Ph.L. 1960-18. The Hague: Physics Laboratory TNO 1960.

DE LOOR,G.P.: Dielectric properties of heterogeneous mixtures with a polar constituent. Appl. scient. Res. **B11**, 310–320 (1964).

DE LOOR,G.P.: Dielectric properties of heterogeneous mixtures containing water. J. Microwave Power **3**, 67–73 (1968).

DE LOOR,G.P., JURRIENS,A.A.: The radar backscatter of vegetation. Paper 12. In: AGARD Conf. Procs no.90: Propagation limitations in remote sensing, 1971.

DE LOOR,G.P., JURRIENS,A.A., GRAVESTEIJN,H.: The radar backscatter from selected agricultural crops. IEEE Trans. Geosc. Electronics **GE-12**, 70–77 (1974).

EASAMS: Side-looking radar systems and their potential application to earth resource surveys. 7 volumes; report prepared for ESRO under ESTEC Contract 1537/71/EL; August 1972.

EICHEL,L.A.: Design of the display unit for Ku side-looking airborne radar system. CRES Technical Report 2201-1. Kansas University, 1973.

GOODYEAR AIRCRAFT CORP.: Radar terrain return study; final report. Report GERA—463, 30 Sept. 1959.

GOODYEAR AEROSPACE CORP.: Earth resources radar for a remote sensing system. Goodyear Aerospace Corp. Report GAP-4947; Rev. A; 28 Oct. 1970.

GRANT,C.R., YAPLEE,B.S.: Backscattering from water and land at centimeter and millimeter wavelengths. Proc. IRE **45**, 976–982 (1957).

GUINARD,N.W.: The remote sensing of oil slicks. Procs VIIth Symposium on Remote Sensing of Environment, Ann Arbor, USA, pp.1005–1026 (1971a).

GUINARD,N.W.: Remote sensing of ocean effects with radar. Paper 15. In: AGARD Conf. Procs no.90: Propagation limitations in remote sensing (1971b).

HARALICK,R.M., CASPALL,F., SIMONETT,D.S.: Using radar imagery for crop discrimination: a statistical and conditional probability study. Remote Sensing of Environment **1**, 131–142 (1970).

HARDY,N.E., COINER,J.C., LOCKMAN,W.O.: Vegetation mapping with side-looking airborne radar: Yellowstone National Park. Paper 11. In: AGARD Conf. Procs. no.90: Propagation limitations in remote sensing, 1971.

HARGER,R.O.: Synthetic aperture radar systems: theory and design. New York: Academic Press 1969.

HARRIS,D.S.: Switched capacitor storage arrays for AMTI and bandwidth compression. Paper 25 in: AGARD Conf. Procs. no.66: Advanced Radar Systems 1970.

HEMPENIUS,S.A.: Image formation techniques for remote sensing from a moving platform. ITC—publ. series A, no.46; Enschede, International Institute for Aerial Surveys and Earth Sciences (ITC) 1969.

JOHNSON,D.J., FARMER,L.D.: Use of side-looking airborne radar for sea ice identification. J. Geophys. Res. **76**, 2138–2155 (1971a).

JOHNSON,D.J., FARMER,L.D.: Determination of sea ice drift using side-looking airborne radar. Procs. VIIth Symposium on Remote Sensing of Environment, Ann Arbor, USA, pp.2155–2168 (1971b).

LAPRADE,G.L.: An analytical and experimental study of stereo for radar. Photogrammetric Eng. **92**, 294 (1963).

LARROWE,B.T., INNES,R.B., RENDLEMAN,R.A., PORCELLO,L.J.: Lake ice surveillance via airborne radar: some experimental results. Procs. VIIth Symposium on Remote Sensing of Environment, Ann Arbor, USA, pp.511–521 (1971).

LEBERL,F.: Metric properties of imaging produced by SLAR and IRLS systems. In: Procs. Symposium of Commision IV. Delft: International Society for Photogrammetry, ITC 1970.

LEVINE,D.: Radargrammetry. New York, Toronto, London: McGraw-Hill 1960.

LESCHACK,L.A.: ADP of aerial imagery for forest discrimination. Procs. of the annual meeting of the American Society of Photogrammetry, Paper 70–110, pp.187–218 (1970).

LEWIS,A.J., WAITE,W.: Cumulative frequency curves of the Darien Province Panama. Paper 10 in: AGARD Conf. Procs. no.90: Propagation limitations in remote sensing, 1971.

LUNDIEN, J. R.: Analysis of scatterometry data from Pisgah Crater. CRES Technical Report 118-2, University of Kansas 1967.

MACDONALD, H. C.: Geologic evaluation of radar imagery from Darien Province, Panama. Modern Geology **1**, 1–63 (1969).

MACDONALD, H. C., LEWIS, A. J., WING, R. S.: Mapping and landform analysis of coastal regions with radar. Geol. Soc. Am. Bull. **82**, 345–358 (1971).

McCoy, R. M.: An evaluation of radar imagery as a tool for drainage basin analysis. CRES Technical Report 61–31, University of Kansas, Aug. 1967.

MOORE, R. K.: Radar scatterometry—an active remote sensing tool. CRES report no. 61–11 (1966); Procs. IVth Symposium on Remote Sensing of Environment, Ann Arbor, USA, pp. 339–373 (1966).

MOORE, R. K.: Ground Echo. Chapter 25 in: SKOLNIK, M. I. (Ed.): Radar Handbook. New York: McGraw Hill Book Comp. 1970.

MOORE, R. K.: Imaging radars for geoscience use. IEEE Trans. Geosc. Electronics **GE-9**, 155–164 (1971).

MOORE, R. K., SIMONETT, D. S.: Potential research and earth resource studies with orbiting radars: results of recent studies. Paper AIAA no. 67–767 in: Procs. of the AIAA 4th Annual Meeting and Technical Display; 1967.

MORAIN, S. A., COINER, J.: An evaluation of fine resolution radar imagery for making agricultural determinations. CRES Technical Report 177-7, Kansas University, August 1970.

MORAIN, S. A., HOLTZMAN, J., HENDERSON, F.: Radar sensing in agriculture, a socio-economic viewpoint. EASCON Conv. Record, pp. 280–287 (1970).

MORAIN, S. A., SIMONETT, D. S.: K-band radar in vegetation mapping. Photogramm. Eng. **32**, 730–740 (1967).

MYERS, G. F., FULLER, I. W.: Nanosecond radar observations of sea clutter cross section vs. grazing angle. Report NRL 6933, Washington, Naval Research Laboratory, Oct. 8, 1969.

OLIVER, T. L., PEAKE, W. H.: Radar backscatter data for agricultural surfaces. Techn. Rept. 1903-9, Electro Science Laboratory, Ohio State University, 13 Febr. 1969.

PAQUET, J.: Etude diélectrique des matériaux humides. l'Onde Electrique **44**, 940–950 (1964).

RENAU, J., COLLINSON, J. A.: Measurements of electromagnetic backscattering from known rough surfaces. Bell System Techn. J. **44**, 2203–2226 (1965).

RUCK, G. T., BARRICK, D. E., STUART, W. D., KIRCHBAUM, C. K.: Radar cross section handbook; In particular chapter 9: Rough surfaces. New York: Plenum Press 1970.

SITTROP, H.: Unpublished measurements. Physics Laboratory TNO, The Hague, 1969.

SKOLNIK, M. I.: Radar handbook. New York: McGraw Hill Book Comp. 1970.

SMITH-ROSE, R. L.: The electrical properties of soil for alternating currents at radio frequencies. Proc. Royal Soc. London **140**, no. 841 A, 359–377 (1933).

SMITH-ROSE, R. L.: The electrical properties of soil at frequencies up to 100 Mc/sec; with a note on the resistivity of ground in the United Kingdom. Proc. Phys. Soc. **47**, part 5, no. 262, 923–931 (1935).

STRATTON, J. A.: Electromagnetic Theory. New York-London: McGraw Hill Book Comp. 1941.

SWANSON, L. W.: Topographic manual, part II: Photogrammetry. US Coast and Geodetic Survey Spec. Publ. **249**, p. 2 (1949).

ULABY, F. T.: 4–8 GHz microwave active and passive spectrometer (MAPS); Vol. I: Radar section. CRES Technical Report 177-34, Kansas University, April 1973.

ULABY, F. T., MOORE, R. K.: Radar spectral measurements of vegetation. Preprint of a CRES publication; Kansas University 1973.

VAN SOEST, J. L.: Speurwerk, van luistertoestel tot radar. Voordrachten K.I.v.I. (Procs. Neth. Royal Inst. of Eng.), Vol. 1, No. 1, pp. 1–6 (1949).

WAITE, W. P.: Broad spectrum electromagnetic backscatter. CRES Techn. Report 133-17, Kansas University 1970.

WAITE, W. P., MACDONALD, H. C.: Vegetation penetration with K-band imaging radars. IEEE Trans. Geosc. Electronics **GE-9**, 147–155 (1971).

WILTSE, I. C., SCHLESINGER, S. P., JOHNSON, C. M.: Backscattering characteristics of the sea in the region from 10 to 50 kMc/s. Proc. IRE **45**, 220–228 (1957).

6. Passive Microwave Sensing

E. SCHANDA

6.1 Principles of Passive Microwave Remote Sensing

All matter at temperatures different from absolute zero is radiating electromagnetic energy. Most natural solid and fluid media met in the terrestrial environment obey approximately to the spectral behavior of thermal radiation within limited bandwidths as it is described by the Planck radiation law for the so-called black body. At a temperature of 300 K the spectral maximum of the radiative intensity is at a wavelength of about 10 microns i.e. well in the infrared range. In the microwave range (wavelength between one millimeter and several decimeters), the intensity is several orders of magnitude less than this maximum value. In reality no material behaves like a black body, and regarding only a narrow spectral range the various media appear in different gradations of gray. Therefore in general the radiative intensity received from a given real material is composed of a part emitted by thermal radiation and a part of reflected radiation originating from the surroundings. If the ratio of these two parts varies strongly with wavelength, the material appears "colored" i.e. with a characteristic spectral signature. The ratio of emitted and reflected intensities of an object in the natural environment depends on a variety of factors as: The complex permittivity of the medium, its homogeneity, or the mixing ratio with other media (air, water), the surface roughness, the temperature, the look angle, the polarization and the wavelength for which the radiometer is sensitive.

These properties combined produce a great variation of very weak radiative intensities (brightness temperatures) to be received by a microwave radiometer and offer the possibility of remotely sensing differences e.g. between dry and humid soil or between multi year and young ice including temporal changes etc., but the unambiguous interpretation of the sensed intensities is a difficult task.

The radiation received from objects in a natural environment is incoherent i.e. the "signal" has the character of noise and has to be discriminated against the noise originating from within the radiometer (mainly of the preamplifier stages of the microwave receiver). Properly designed radiometers, however, are able to discriminate noise-intensity changes of less than a promille.

The angular resulution of passive microwave sensing systems (as utilized for imagery) is considerably poorer than that of infrared sensors because of the much smaller ratio of the diameter of any practical antenna aperture and the wavelength. The angular resolution is also poorer than in synthetic aperture ("side looking") radar because no coherent signals are present to become treated in a holographic procedure.

The most important advantage of the use of the microwave spectrum (particularly wavelengths longer than 2 cm) for remote sensing purposes is the insensitivity against attenuation by meteorological phenomena, thus offering an all weather sensing tool. On the other hand, the range of wavelengths shorter than about 2 cm contains spectral lines due to molecular transitions of many important atmospheric constituents (oxygen, water vapor, ozon and others). This line radiation can be used for the determination of density distributions and temperature profiles which are essential for the balance between the various constituents.

The physical principles and the technical realizations of microwave radiometry to be discussed briefly in the first two sections have been developed originally for radio astronomical purposes and are presented in more detail in textbooks on Radio Astronomy (e.g. KRAUS, 1966).

6.1.1 Physical Fundamentals of Microwave Radiometry

One can show that for wavelengths longer than one millimeter and at all temperatures occurring in terrestrial environments, the calculation of the spectral radiance S (radiated power per square-meter, per unit of solid angle and per unit bandwidth of frequency) of a black body can be performed by an approximation (Rayleigh-Jeans) of the Planck formula

$$S_b \approx \frac{2kT}{\lambda^2} \quad [W\,m^{-2}\,Hz^{-1}\,rad^{-2}].$$ (1)

The temperature of the radiating black body is T, the wavelength considered is λ and $k = 1.38 \cdot 10^{-23}$ Ws/K is the Boltzmann constant. The condition for the vality of (1) is

$$\frac{v}{T} \ll 2 \cdot 10^{10} \quad [Hz/K]$$

which means that at e.g. $\lambda = 3$ mm ($v = 10^{11}$ Hz) and temperatures sufficiently higher than 5 K, the use of (1) instead of the complete Planck formula is allowed. Figure 3 of the introductory chapter shows the dependence of the spectral radiance of an ideal black body on the frequency (radiation law of Planck) at various temperatures. The most important feature at the long wavelength side of this diagram (range of the Rayleigh-Jeans approximation) is the proportionality between radiance and temperature. Thus a change of the temperature of the radiating medium by one degree say from 10 K to 11 K changes the radiance by the same amount as a heating from 1000 K to 1001 K would do, if measured within a narrow band at a given frequency.

This is different at the short wavelength side, where the radiation obeys the exponential Wien-approximation.

For real media the abilities to absorb and to emit radiation are related by the Kirchhoff radiation law: We suppose a simple experimental set-up consisting of two non-contacted halves of a spherical cavity, the inner wall of one half being an ideal "black" medium (absorption coefficient $a_b = 1$, emission coefficient e_b) while the inner surface of the opposite half is made of real (gray) material ($a < 1$, e). The thermodynamic temperatures of both halves are assumed to be equal and there is no possibility of escape of radiation from, nor of penetration into the cavity. According to the second law of thermodynamics valid for each surface separately i.e. emitted energy equals absorbed energy, in thermodynamic equilibrium we may equate the emitted (left hand side of the equations) and the absorbed powers on the "gray" surface

$$e = ae_b$$

and also for the "black" surface

$$e_b = e + e_b(1 - a)$$

where $(1 - a)$ is the ratio of the reflected to the impinging power at the gray surface. From both equations separately it follows

$$e/a = e_b \tag{2}$$

the Kirchhoff radiation law. The ratio of the emission and absorption coefficient of any medium equals the emission coefficient of the "black" medium at a given temperature. Disregarding the specific radiation mechanism and using the emission coefficient just as a factor in all following considerations, we normalize arbitrarily by taking $e_b = 1$, thus achieving

$$e = a. \tag{2'}$$

Therefore the spectral radiance S of a natural material is related to the radiance S_b of a black body by

$$S = eS_b \tag{3}$$

if the consideration is restricted to small bandwidths and the emission coefficient is frequency independent within these bandwidths.

According to the above reasoning leading to Kirchhoff's law the reflection coefficient of any non-transparent ("optically thick") body, which determines the ratio of the reflected to the impinging intensity of radiation, is defined by

$$r = 1 - e. \tag{4}$$

The apparent brightness (radiance) of an object consists of a part due to the emission eS_b and another part due to the reflection of an averaged brightness surrounding the object rS_s resulting in

$$S = eS_b + (1 - e)S_s. \tag{5}$$

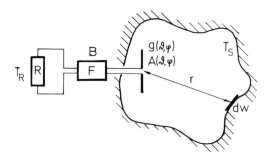

Fig. 1. Schematic arrangement for the definition of the antenna temperature

The contributions to the surrounding brightness S_s are resulting from neighboring objects, the atmosphere and the sky, yielding an integrated brightness over the solid angle which is relevant for the reflection on the considered object. The radiation from neighboring objects is present at all wavelengths, the atmospheric contribution is important at wavelength shorter than say 2 cm while the cosmic radio intensity has measurable effect at wavelengths longer than 10 cm.

By the use of formula (1) the definition of a so-called antenna temperature or radiation temperature of the object is possible. Using Fig. 1 we suppose that the impedances of the resistor R, the bandfilter F with bandwidth B and the antenna A are matched to the cavity with a perfectly absorbing inner surface S. The gain function of the antenna $g(\vartheta, \varphi)$ is normalized by the integration over all space angles $\int g(\vartheta, \varphi)d\Omega = 1$. The surface element of the cavity wall dw receives a fraction of the noise power $P_R = kT_R B$ generated by the resistor and radiated through the ideal (lossless) filter and antenna

$$dP_R = g^2(\vartheta, \varphi) \cdot kT_R B \frac{dw}{r^2}.$$

But the surface element dw itself is emitting radiation due to its temperature according to (1). The solid angle by which the effective antenna area is viewed from dw is $A(\vartheta, \varphi)/r^2$. The amount of radiative power received by the antenna and absorbed in the resistor is

$$dP_W = \frac{kT_W}{\lambda^2} B \cdot \frac{A(\vartheta, \varphi)}{r^2} dw,$$

where a factor $1/2$ takes into account that the antenna is sensitive to only one direction of polarization.

As soon as the thermodynamic equilibrium is reached, $dP_R = dP_W$ has to be fulfilled. From antenna theory it is known that $g^2(\vartheta, \varphi) = A(\vartheta, \varphi)/\lambda^2$.

Thus the equality of $dP_R = dP_W$ yields

$$T_R = T_W . \tag{6}$$

The antenna pointing towards a surface at a temperature T_W is receiving the same amount of power as a resistor is able to generate at the same temperature. Thus a resistor connected to the input of a radiometer instead of the antenna is an equivalent substitute if its temperature is equal to the cavity wall temperature which causes the power received by the antenna (antenna temperature). If the antenna receives only the radiation of one object at a constant temperature (as supposed in the above set-up), the antenna temperature is equal to the brightness (or radiation) temperature of this object.

6.1.2 Semitransparent Media

The discussion of the thermal radiation in Section 6.1.1 was based on the assumption of the opacity of the radiating media. Certainly for air (including clouds and precipitation) but also in cases of layered media or of vegetation, one has to deal with the semi-transparent properties and with the combined effects of the background and the intervening medium on the total radiation.

While propagating through an absorbing medium the flux density $F = S\Omega_s$ of the background source, where Ω_s is the solid angle subtended by the source, becomes attenuated along an elementary path length dz by

$$dF = -F\alpha dz\,,$$

with a constant α characteristic for the considered medium and termed as volume absorption coefficient (unit: m^{-1}). Integrating over a geometric path z yields

$$F = F_0 \exp(-\alpha z)\,,$$

where F_0 is the flux density at $z = 0$. It is widely used practice to characterize the media by their mass absorption coefficient K (unit: m^2/kg) which is useful if the density of the medium is non-uniform; it is connected to the volume absorption coefficient by $\alpha = K\varrho$ where ϱ is the density (kg/m^3). The relation of α to the "penetration depth" δ: see section 6.3.1.

The optical depth (opacity) τ is given by

$$\tau(z) = \int_0^z \alpha(z')dz' = \int_0^z K \varrho(z')dz' \tag{7}$$

it is a measure to which extent the radiative intensity is attenuated on the path from $z' = 0$ to z. Table 1 gives the ratio F/F_0 for a few values of τ

Table 1

τ	0.01	0.5	1	2	4.6
F/F_0	0:99	0.607	0.368	0.135	0.01

Because of the reciprocity of absorption and emission, the semitransparent medium also contributes an emitted power per unit bandwidth $dp = j\varrho dV$ in a

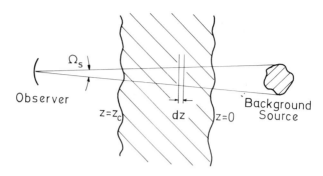

Fig. 2. Radiative transfer in a semi-transparent medium (cloud) of thickness z_c

volume element dV with the emission coefficient j (unit: $\mathrm{W\ kg^{-1}\ Hz^{-1}}$) which adds to the flux density an amount $dF = j\varrho\Omega_s/4\pi dz$, where $dV = z^2\Omega_s dz$ has been used. Combining the effects of absorbing the background flux and of emitting additional radiation yields the net change of the flux density within the elementary path dz (Fig. 2), when scattering effects are neglected.

$$dF = -F(z)K\varrho(z)dz + j(z)\varrho(z)\frac{\Omega_s}{4\pi}dz .$$ (8)

The equation of radiative transfer (8) can be integrated over the thickness of the intervening medium which is assumed to be in local thermodynamic equilibrium, yielding the resulting flux density from a cone of solid angle Ω_s

$$F = F_0 e^{-\tau_0} + \frac{\Omega_s}{4\pi}e^{-\tau_0}\int_0^{z_c}j(z)\varrho(z)\cdot e^{\int_0^z K\varrho(z')dz'}dz .$$ (9)

If j and ϱ are independent of the location, the integration yields

$$F = F_0 e^{-\tau_0} + \frac{\Omega_s}{4\pi}\frac{j}{K}\ (1 - e^{-\tau_0})$$ (9a)

with $j/4\pi K$ the so-called source function or intrinsic spectral radiance (with the same units as the spectral radiance). For the microwave range the validity of the Rayleigh-Jeans approximation allows the simplification of the relation (9a) to

$$T = T_s e^{-\tau_0} + T_c\ (1 - e^{-\tau_0}) .$$ (9b)

The apparent brightness temperature depends on the temperatures of the background source T_s and of the intervening medium T_c as well as from the optical thickness of this medium. Increasing opacity of the intervening medium increases its contribution to the resulting brightness temperature and hides the background source. Figure 3 is a graphic representation of the relationship among these four quantities.

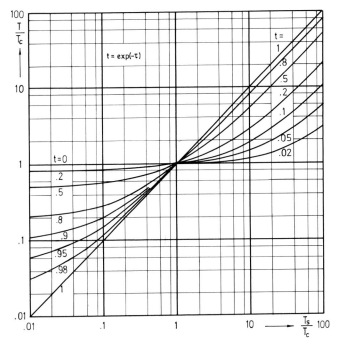

Fig. 3. Graph of the integrated equation of radiative transfer (SCHANDA, 1968). The transmission coefficient t equals $\exp(-\tau)$

The concept of transfer is most important for the study of atmospheric effects, in particular for the sensing of constituents. For the case of a thin (few wave lengths) semitransparent layer on a substrate, the above consideration has to be completed regarding the interference effect of the boundaries between largely different permittivities which play a role as long as the layer is equal or smaller than the coherence length of the wavelengthband.

6.1.3 The Effects of the Atmosphere

The effect of the atmospheric radiation on the antenna temperature in a remote sensing application can best be illustrated by an air- or spaceborne radiometer viewing the ground (Fig. 4). With an emission coefficient of the soil e into the viewing angle Θ, an optical thickness of the atmosphere τ_0 in zenith direction and temperatures T_S and T_c of the soil and the atmosphere respectively, the received total radiation temperature is

$$T_R = e \cdot t [T_S - T_c(1-t)] + T_c(1-t^2) \tag{10}$$

where the abbreviation $t = \exp[-\tau_0/\cos\Theta]$ has been used, the assumption of a horizontally stratified atmosphere has been made and the contributions of the cosmic radio radiation have been neglected. In scanning over a soil of constant

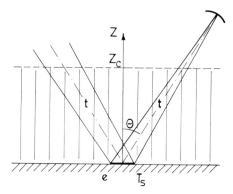

Fig. 4. On the contributions to the radiation temperature sensed by a space borne radiometer

temperature but of changing emission coefficients Δe, the resulting changes of the sensed radiation temperature are

$$\Delta T_R = \Delta e \cdot t [T_S - T_c(1-t)] \tag{11}$$

as long as the path length through the atmosphere does not change noticeably during the scan. In cases of considerable atmospheric attenuation (small values of t as compared to $t = 1$, which latter value represents the complete transparency), not only the effect of the change in emission coefficient Δe is reduced by t but also the temperature of the soil is masked by the atmospheric temperature with the factor $(1-t)$.

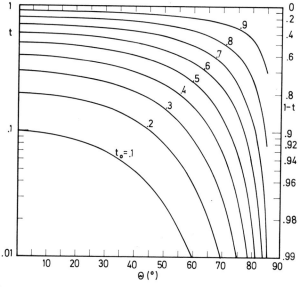

Fig. 5. The single path transmission coefficient of a horizontally assumed layer as a function of the zenith angle Θ for various values of the (zenith) transmissivity t_0

The assumption of a horizontally stratified atmosphere is justified for zenith angles up to about $80°$ or—dependent on the application with strongly increasing errors—even $85°$. Figure 5 is a representation of the $t = \exp[-\tau_0/\cos\Theta]$ relation as a function of the zenith angle for various zenith opacities, which can help in estimations of the atmospheric contributions in scanning.

As mentioned earlier the effect of the atmosphere is small or even negligible for wavelengths increasing from 2 cm on, but the influence is considerable at shorter wavelengths and can even be decisive in the case of precipitation. There are two causes of atmospheric wave attenuation

— the absorption due to transitions between molecular states of the gaseous constituents as oxygen, water vapor and a few minor constituents,
— the scattering and absorption by droplets, hailstones and snowflakes in clouds and during precipitations.

The absorption due to various gaseous molecules and its application to the sounding of the atmosphere will be treated in more detail in later sections. As far as the interference of the "clear" atmosphere is concerned, a number of recent publications (KISLYAKOV, 1972; PENZIAS and BURRUS, 1973; BLUM, 1974; WATERS, 1975) give reviews on the results of experimental and theoretical investigations. Figure 6 shows the zenith attenuation for the microwave part of the spectrum. An idealized dry atmosphere is compared with two real situations, one of an extremely dry climate, the other with about 50% relative humidity at a temperature of $20°$ C. The attenuation is given in decibels (ten times the logarithm of the ratios of the intensities before and after propagating through the atmosphere: $A[db] = 10 \log 1/t$ e.g. 1 db equals about 20% of attenuation). Measurements in the vicinity of the oxygen and water vapor lines are affected very much by these constituents of the atmosphere.

But there are atmospheric spectral "windows" at about 90 GHz and in dry climate even at about 140 GHz and about 230 GHz in which the wave propagation is affected in a tolerable degree.

The effect of precipitation on the propagation of short wavelengths is different in its spectral distribution and much stronger. Quantitative investigations were initiated for Telecommunications and Radar purposes and have been used as basic information for weather Radar interpretations. Discrepancies between the results of various investigators originate in the assumptions on the size distribution of the droplets for various intensities of rain, which is a function of various climatic conditions. Without attempting to give a universal rain attenuation spectrum, Fig. 7 is to some extent a combination of the results by HAROULES and BROWN (1969), from a report by the World Meteorological Organization (TREUSSART et al., 1970), and by BETTENCOURT (1973). These investigations are normally presented for horizontal propagation, therefore the attenuation values are given per unit of distance. Because of the fitting of the results from three complementary but not completely corresponding sources, the curves have to be used with caution (Compare GAUDISSART, 1971).

It is important to note that the size of the droplets or hailstones are comparable to the wavelengths and therefore the scattering cross-section can neither be approximated by the geometric optics as in the infra red range (equal

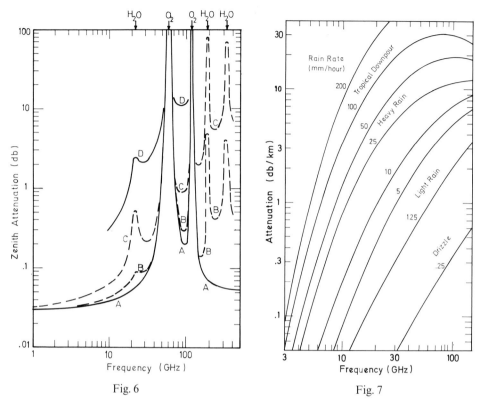

Fig. 6

Fig. 7

Fig. 6. Zenith attenuation at sea level of clear atmosphere. *A* without water vapor. *B* water vapor concentration 0.5 g/m³ at the earth's surface (1 mm equivalent depth), *C* water vapor concentration 7.5 g/m³ at the earth's surface (1.5 cm equivalent depth), both *B* and *C* are average values (Blum, 1974); curve *D* gives the effect of a cloud 1 km above earth, 2 km thick, water content 1.69 g/m³ (Edison, 1966)

Fig. 7. Horizontal path attenuation at various rain rates. (Compiled from data by Haroules and Brown, 1969; Treussart et al., 1970, and Bettencourt, 1973)

to the geometric cross-section) nor by the Rayleigh approximation as in the lowest part of the microwaves (scattering cross-section for wave intensity proportional to a^2/λ^4, where a is the diameter of the spherical droplet). Because of the broad distribution of droplet sizes within a shower, no exact relation between attenuation and rain intensity has been established so far (Treussart et al., 1970; Ott, 1972). But from the experimentally verified attenuation spectrum Fig. 7 a broad maximum around a wavelength about twice the diameter of the most abundant droplets can be recognized. For heavier rain the large drops are more abundant, and therefore the attenuation is peaked at a longer wavelength then for a light rain. An expression for the size distribution of rain droplets as used by Deirmendjian (1963) is $n_p(r) = pr \exp(-q\sqrt{r})$ where r is the radius and the constants p and q have to be adjusted for the droplet number density and for the radius at the maximum value of the distribution (e.g. $p = 5.333 \cdot 10^5$ and $q = 8.944$ for 1000 droplets per m^3 and a modal value of 50 microns).

Concerning the attenuation by hail the following relation between the attenuation α in decibles per kilometer and the precipitation rate R in mm (fluid) water per hour has been given by RYDE (1946)

$$\alpha = HR \tag{12}$$

where H can be found in the following table as a function of the wavelength and the diameter of the hailstones.

Table 2. Values of H for (12). (After RYDE, 1946)

	Diameter of the hailstones (cm)				
	0.25	0.5	1	1.5	2.0
λ (cm) 1	2.7 10^{-2}	1.1 10^{-1}	7.3 10^{-2}	2.8 10^{-2}	1.0 10^{-2}
3	3.7 10^{-4}	1.5 10^{-3}	8.6 10^{-3}	1.7 10^{-2}	1.7 10^{-2}
10	2.2 10^{-5}	2.7 10^{-5}	7.5 10^{-5}	1.8 10^{-4}	3.6 10^{-4}

There is a maximum of attenuation at a wavelength of about twice the diameter of the hailstones compared to about π times as given by the theory of Mie-scattering on a conducting sphere.

The droplets in clouds have sufficiently small diameters for the validity of Rayleigh's approximation even into the millimeter wave range (HAROULES and BROWN, 1969; TREUSSART et al., 1970; BENOIT, 1968). From these considerations the following approximate relation for the attenuation by a water cloud at a temperature of 20° C can be derived (in decibels per kilometer)

$$\alpha \approx 0.6 \cdot 10^{-3} v^{1.9} M \tag{13}$$

where v is the frequency in GHz and M the water density in the cloud in g/m^3. A size distribution of the water droplets in the cloud can be formulated (DEIRMENDJIAN, 1963) as $n_c(r) = cr^6 \exp(-d \cdot r)$ where r is the radius of the droplets and c and d have to be adjusted for number density and radius of the maximum value of the distribution (e.g. for 100 droplets per cm^3 and a modal value of 4 microns: $c = 2.373$, $d = 1.5$; the same droplet number but a modal value of 12 microns yields $c = 1.085 \cdot 10^{-3}$, $d = 0.5$, KREISS, 1968).

Measurements at sea level by WRIXON (1971) at 90 GHz under elevation angles of about 20° (path length about 2.5 times zenith path length) total attenuations of typical 2.1 db for individual cumulus clouds to typical 5.2 db for heavily overcast sky between periods of rain were noted which gives a 20–25 times higher (linear) attenuation then at 16 GHz. At 230 GHz (WRIXON, 1974a), a strong temperature dependence of the cloud attenuation (higher temperatures giving stronger attenuations) has been noted. Additionally the clear sky zenith attenuation at 230 GHz has been determined experimentally as $A (\text{db}) = 0.35 + 0.95 \, W$

where W is the precipitable water in cm. The zenith brightness temperature at 92 GHz has been observed to increase from the clear sky value (about 50–60 K) to about 100 K for high cirrus clouds, to a range of 120 K to 200 K for various medium and low height cumulus, and strato cumulus and finally to a range of 200–240 K for dark thunderstorm clouds (Schaerer and Schanda, 1974).

6.2 Instrumental Aspects of Microwave Radiometry

6.2.1 The Sensitivity of a Microwave Radiometer

A radiometer antenna pointing on an object senses the brightness temperature of this object and presents the power due to this brightness temperature as a signal to the input of the receiver. The main problems in detecting this input power sufficiently reliably and accurately are the weakness and the noise-character of this "signal". It has to be distinguished from the noise power produced by the receiver itself, which is in most practical cases much more intense. The receiver noise has various sources partly of pure thermal, partly of quantum-like character of the amplifying and detection devices. Regarding the analytical expression which describes this noise, for most receivers the same type of approximation can be made which led to (1). The differentiation between the (microwave) thermal character of the noise power ($P_T = kTB$) and the (infrared) quantum character ($P_Q = h\nu B$, with $h = 6.626 \cdot 10^{-34}$ Ws^2 the Planck constant) is given by the relation $h\nu/kT \approx 1$. Therefore a receiver at a physical temperature of 300 K produces predominantly thermal-type noise up to about $6 \cdot 10^{12}$ Hz (0.05 mm wavelength), but for a Helium cooled receiver input stage this limit is shifted to a few millimeters wavelength. If a sensitivity of a radiometer is demanded to discriminate a change of only one degree of the brightness temperature, this means a change of the input power

$$\Delta P = kB\Delta T \tag{14}$$

of about $1.4 \cdot 10^{-17}$ Watt for a 1 MHz receiver bandwidth.

The noise power produced in low-noise receivers is about hundred to several thousand times higher. Therefore several measures are necessary to warrant the needed intensity resolution:

— The use of a low pass (integrating) filter after the detection device smoothes the output voltage. Only changes of the input level slower than the response time of this low pass filter cause changes of the d.c. output voltage and fluctuations faster than the response time are eliminated. In fact the low pass filter integrates over many measurements, thus yielding an improved accuracy by the root of the number of measurements (response time).

— As becomes obvious from (14), a larger radio frequency (input) bandwidth B results in a larger input power change ΔP which makes the signal easier to detect. But by a larger bandwidth the receiver noise power also increases, and therefore the improvement is partly cancelled.

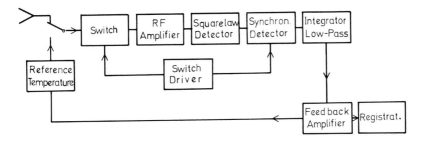

Fig.8. Block diagram of a null-balancing switched radiometer

— The lower the noise produced by the receiver, the easier to discriminate is the input power. But a practical limitation is set because receivers with very low noise contributions (as e.g. MASERS) are complicated structures which need to be cooled cryogenically and are therefore scarcely practicable in an operational automatic satellite.

— The amplifiers of a highly sensitive receiver exhibit always some drift and gain variations which prevent any quantitative determination of the input radiation. The periodic comparison of the unknown input power with the radiation of a controlled source and the adjustment of this reference source to the power level of the input eliminates essentially the effect of the drift.

These measures have already been introduced to a certain extent by DICKE (1946) in his original radiometer design, which has been refined and modified many times (see e.g. TIURI, 1964 and 1966). Figure 8 shows a block diagram of a modified switched (Dicke-type) radiometer as proposed by MACHIN et al. (1952). The output signal is proportional to the difference between antenna and reference temperature. By balancing these two inputs via a feedback loop, the effect of the gain variations is eliminated to a first order. The feedback voltage controlling the reference temperature is recorded or used for other ways of presentation. The sensitivity of this type of radiometer is approximately given (TIURI, 1964) by the minimum change of antenna temperature which can be discriminated against all other noise sources in the sensed scenery and within the receiver (with a "signal" to noise ratio of unity)

$$\Delta T_{\min} \approx 2 \frac{T_N}{\sqrt{B\Sigma}}, \tag{15}$$

where T_N is the total systems noise temperature (receiver noise and background radiation of the scenery), the integration time $\Sigma = 1/2\,b$ with b the bandwidth of a square pass-band integrating filter and B the high frequency (input) bandwidth. A typical assumption as $T_N = 1000$ K, $B = 4$ MHz, $\Sigma = 1\,s$ yields a temperature resolution $\Delta T_{\min} = 1$ K.

A more careful investigation of the remaining effect of gainfluctuations on the sensitivity by MAGUN and KÜNZI (1971) has shown that even for the null-balanced version, a deteriorating factor exists which depends mainly on the (low frequency) spectrum of the gain fluctuations and the specific switch frequency (between antenna and reference) within this spectrum. If the reference temperature T_0 is not adjusted to the received antenna temperature T_R, the minimum detectable temperature including the effect of gain fluctuations (GF) becomes

$$\varDelta T_{\text{min, }GF} = \varDelta T_{\text{min}} \left[1 + 2 \left(\frac{T_R - T_0}{\varDelta T_{\text{min}}} \right)^2 \left(\frac{\varDelta G}{G} \right)^2 \right]^{\frac{1}{2}}$$

where $\varDelta T_{\text{min}}$ is given by (15), G is the nominal power gain and $\varDelta G$ is the frequency integral of the gain fluctuations. For most amplifier and detection devices the gain fluctuation spectrum is strongly decreasing with frequency in the range of 0.1 Hz to 100 Hz (YAROSHENKO, 1964; KÜNZI and MAGUN, 1971).

6.2.2 Other Types of Radiometers

The most widely used microwave radiometers are those based on the switching (Dicke) type. But for specific applications other principles can be useful. Only two will be discussed here very briefly:

The first type (Fig. 9a) is specifically advantageous when a strongly frequency dependent phenomenon—spectral line radiation—has to be studied by many jointly aligned narrow frequency channels and mainly the relative intensities of radiation entering the various spectral channels are of importance. In the second type (Fig. 9b) the correlation receiver is specifically useful for interferometric purposes but it is also advantageous in single antenna applications (splitting the

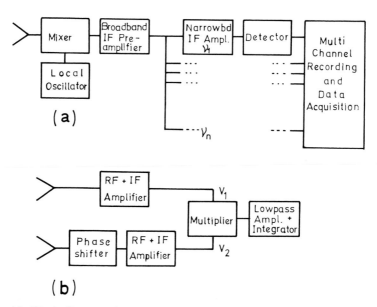

Fig. 9a and b. Block diagram of (a) a multichannel superheterdyne radiometer for narrowband spectral investigations; (b) a correlation radiometer

signal into two arms) for the elimination of the noise of the preamplifier stages. The multiplication of statistical signals (expectation value of the product) yields zero if the signals are statistically independent (noise of the two input arms), but it gives the mean square value of identical signals. Signals received in both antennas from the same source and in correct phase (integer multiples of the wavelength) are interfering to a positive product. By means of a phase shifter the direction of positive interference can be adjusted. Another application of the correlation receiver is the measurement of direction and velocity of lateral motions (e.g. motion of water surface or of precipitation or in navigational applications) by delaying the signal in one arm compared to the other one for maximizing the correlation.

The theoretical sensitivities of these types of radiometers are roughly the same and their application depends on the specific problem to be investigated. For more details on the performance of these and several other types of radiometers can be referred to the literature (e.g. TIURI, 1966).

6.2.3 Angular Resolution and Range of Passive Microwave Sensing

The antennas used in the microwave part of the radio spectrum are in most cases radiating apertures. Only in special applications (Radio Astronomy) large arrays of arials are utilized for achieving extremely high angular resolution, or arrangements of many small radiating elements are applied for electronic beam scan by using a differential phase shift between the interfering elements or large arrays are "synthetically" simulated in the synthetic aperture radar for achieving high resolution due to a holographic treatment of the coherent radar waves.

The angular resolution of a single circular aperture as used in most passive microwave sensing systems is given by the solid angle of the main lobe (half power beam width), which is approximately

$$\Omega_A \approx \lambda^2/A_E,\qquad\qquad\qquad\qquad (16)$$

where A_E is the effective antenna area. A_E is reduced compared to the geometric area (in general by a factor 0.5–0.7) due to a non-uniform illumination of the aperture. At a distance R (assumed to be in the far-field zone where the simple Fraunhofer diffraction theory is valid), the main beam covers a cross-sectional area of approximately

$$A = \lambda^2 \cdot \frac{R^2}{A_E}.$$

For an antenna area of 1 m^2 and a wavelength of 1 cm the covered area A at 1 km distance is 100 m^2.

Antennas with a well-defined directional beam pattern and negligible contributions from directions outside the beam allow us to assume that the radiometer receives only radiation from directions within the beam. Therefore the condition of an enclosed antenna for the validity of the antenna temperature concept [leading to Eq.(6)] may now be reduced to the distant cross-sectional area A covered by the beam.

Let us regard now the ability of a radiometer to detect a remote object (distance R) exhibiting a brightness temperature T_0 differing from the surroundings (T_s) by ΔT_0. The cross-sectional area of the object A_0 subtends a solid angle

$$\Omega_0 = A_0/R^2 \tag{17}$$

At a great distance only very large objects will present a homogeneous brightness temperature over an area larger than that subtended by the antenna beam ($\Omega_0 > \Omega_A$) and thus yield a radiometer input temperature $T_R = T_0 e^{-\alpha R} + (1 - e^{-\alpha R}) T_c$ with T_c the temperature of the atmosphere and α as in Section 1.2. In general, distant objects will subtend smaller solid angles than the antenna beam ($\Omega_0 < \Omega_A$) and the surroundings (T_S) will contribute to the antenna temperature as

$$T_R \approx \left[\frac{\Omega_0}{\Omega_A} T_0 + \left(1 - \frac{\Omega_0}{\Omega_A} \right) T_s \right] e^{-\alpha R} + T_c (1 - e^{-\alpha R}).$$

Sweeping the antenna over the object embedded in a homogeneously assumed surrounding, the change of the antenna temperature is

$$\Delta T_R = \frac{\Omega_0}{\Omega_A} \Delta T_0 e^{-\alpha R}, \qquad \Delta T_0 = T_0 - T_s. \tag{18}$$

Regarding the radiometer as being able to detect the object if the signal equals the noise of the receiver, we may equate (18) with (15). Substituting (16) and (17) the maximum range R up to which an object of area A_0, exhibiting a brightness temperature difference ΔT_0 compared to the background, can be detected is given by

$$R^2 \approx (A_0 \Delta T_0) \left(\frac{A_E}{\lambda^2} \right) \left(\frac{\sqrt{B \Sigma}}{2 T_N} \right) e^{-\alpha R}. \tag{19}$$

The right hand side of (19) has been grouped due to the characteristics of the object, the antenna, the receiver and the atmosphere respectively. The explicit appearance of the radiometer properties, in particular the response time Σ in the range formula, allows detailed design considerations for demanded performance specifications. For negligible atmospheric attenuation the relation (19) is shown graphically in Fig. 10. The effect of the attenuating atmosphere on T_R and thus on R can be estimated with the aid of Fig. 3.

For large scale terrain mapping as needed in surveying earth resources or meteorological features, an airborne or space borne scanning radiometer has to be used which measures a strip beneath the vehicle with a certain scan angle Θ_s transverse to the flight direction. The necessity of a finite measuring time equal or longer than Σ on every spot (footprint of the antenna beam) on the ground along a scan line in order to reliably resolve temperature changes ΔT_R puts a

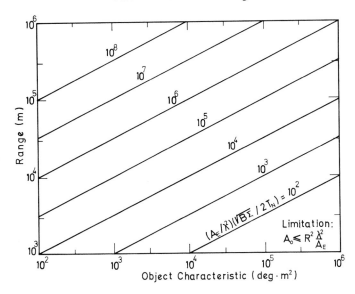

Fig. 10. Maximum detection range versus object characteristic $\Delta T_0 \cdot A_0$ for various receiver and antenna characteristics (SCHANDA, 1971)

limitation on the scan speed. The highest scan speed (for a signal to noise ratio of unity) is given by $d\Theta/dt \approx \sqrt{\Omega_A}/\Sigma$ if $\sqrt{\Omega_A}$ is taken as the angular width of a spot and Σ is calculated from (15) using the required temperature resolution ΔT_R for ΔT_{min}. During a scan of duration t the vehicle travels a certain distance, and if a certain coverage of the ground by succeeding scans (e.g. by half of the antenna beam width $\sqrt{\Omega_A}$) is required, the flight velocity enters as a new parameter. The maximum velocity for given system parameters is determined by $t \cdot v_{max} = h\sqrt{\Omega_A}$ where h is the height of the vehicle above ground. Because of $\Theta_s/t \approx d\Theta/dt \approx \sqrt{\Omega_A}/\Sigma$ we find

$$v_{max} \approx \frac{h}{\Sigma}\frac{\Omega_A}{\Theta_s}, \qquad\qquad (20)$$

where the units of Θ_s and Ω_A are radian and radians squared respectively.

Figure 11 shows the object characteristic as a function of the flight speed detectable with a signal to noise ratio of unity, assuming the instrumental response time Σ adjusted for the flight velocity and for h equal to the maximum range according to (19). The scan angle is taken 1 radian, four different heights and five different angular resolutions have been used. The receiver properties are characterized by $2 T_N/\sqrt{B}=0.1$, the same as for Fig. 10 (e.g. $T_N=1000$ K and $B=400$ MHz). Attention should be drawn to the fact that the above relations are derived along a very simplified approach and that all instrumental parameters are optimized for nadir ($h=$ maximum range with $\Sigma =$ adjusted to resolve that object characteristic within the response time with signal to noise ratio of unity), as it can only be attained on a well stabilised circularly orbiting satellite. For

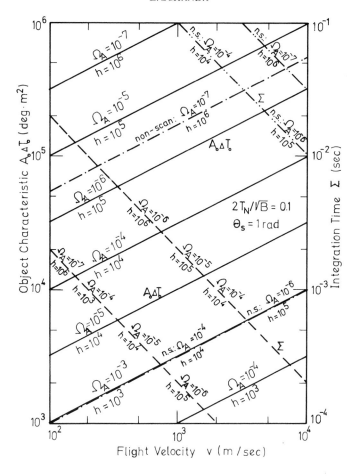

Fig. 11. Minimum object characteristic $\Delta T \cdot A_0$ for detection at flight velocities v (solid lines) with adjusted integration time (dashed lines) of a transversely scanning air- or space borne radiometer. The height above ground h is taken equal to the maximum range ($S/N = 1$), the scan angle is 1 radian, various antenna solid beam angles Ω_A. Non-scanning (Nadir-viewing) situations are given: $- \cdot - A_0 \cdot \Delta T_0$, $- \cdot \cdot - \Sigma$. The ratio $2T_N/\sqrt{B}$ of the receiver noise temperature and bandwidth is fixed at 0.1 deg. sec$^{\frac{1}{2}}$

any other practical sensing system the degradations of the performance have to be considered individually.

In a more exact analysis the effect of smoothing out small features due to a more realistic antenna beam (SCHANDA, 1966) by the convolution of the real antenna pattern with the brightness distribution on the ground (KRAUS, 1966) has to be regarded. The notion of antenna transfer function is used sometimes to describe these facts (ULABY, 1975). But the much more simplified description which has been used in this section for the sake of clarity gives fairly accurate results in most practical applications.

6.2.4 Realisations of Microwave Radiometers

Microwave radiometry is not yet as established as a few other techniques of remote sensing and therefore only very few attempts have been made by industrial firms to develop standard types of radiometers or parts of them for series production. But many experimental microwaves sensors have been brought into extensive use either for basic laboratory investigations, for ground based use or in air borne or satellite borne mission—not to speak of the highly sensitive radiometers for radio-astronomical purposes.

The performance characteristics of the radiometers are strongly dependent on their specific sensing application. The three main factors determining the sensitivity: the system noise temperature, the high-frequency (input) bandwidth and the integration time (post-detector bandwidth) are used in a trade-off to achieve the optimum performance for a particular application; the temperature resolution is in general in the range between 0.1 K and 10 K. The angular resolution, more affected by the technical constraints than by the applicational demands, are realized so far between about 0.1 degree for high resolution imagery (e.g. SCHANDA et al., 1972) and more than 10 degrees for large scale sensing of oceanographic features (e.g. LOVE and HIDY, 1971).

Figure 12 may illustrate the geometric recognizability of a thermal image produced by a scanning radiometer with 10 arcminutes angular resolution.

As an example of a satellite borne radiometer, the main specifications of the Electronically Scanning Microwave Radiometer (ESMR) on board of Nimbus 5 (altitude 1100 km, 81° inclination in sun-synchronous orbits, since December 1972) can be given (MIX, 1974):

center frequency	19.35 GHz
high frequency bandwith	300 MHz
absolute temperature accuracy	2 K
(integration time 47 msec)	
dynamic range	50 K to 330 K
antenna beam halfwidth	1.4° to 1.6°
resolution at nadir	30 km
stepwise scan over	$\pm 50°$
(perpendicular to flight path)	
physical size	$94 \times 92 \times 11 \ cm^3$
weight (without deployment mechanism)	30.6 kg
power requirement	42 W

This instrument is in use for synoptic mapping of the emissions from the surface of the earth and from meteorological features.

A comparable radiometer at a center frequency of 37.0 GHz has been developped for Nimbus F.

The second experimental microwave radiometer on board of Nimbus 5 is the Nimbus E microwave spectrometer (NEMS) primarily intended for the determination of the atmospheric temperature profile up to about 20 km and of the

Fig. 12. Radiometric image of a single building. Wavelength 3 mm, antenna beam width
10 arcmin. (SCHAERER and SCHANDA, 1972)

total water vapor content of the air. Its main specifications are (STAELIN
et al., 1973 a):

five channels at center frequencies of	22.235, 31.4, 53.65, 54.9, and 58.8 GHz
high frequency bandwidths of each channel	200 MHz
absolute temperature accuracy (integration time 2 sec)	2 K
sensitivity (r.m.s.) for 16 sec averaging	0.1 K to 0.2 K
resolution (nadir)	200 km

Another satellite—Skylab 1, launched 1973—also carries two microwave
sensors (NASA, 1973), one is a 1.42 GHz radiometer with a nadir oriented beam
of 15° width and the other is a combined radiometer/scatterometer and altimeter
system at 13.9 GHz with a two-axis mechanically scanned beam of 1.5° width
over 52° off nadir; for both radiometers a sensitivity of approximately 1 K is
claimed. The Radar/Radiometer combination has been proposed and its advantages for the applications and the required specifications have been described
by MOORE and ULABY (1969).

Various technical achievements needed for the high performance of recent
radiometers are concerned with the reference radiation (temperature) source,
e.g. HARDY (1973), who claims a long-term stability of 0.1 K, or with the filtering
for spectrometric purposes (MCLEISH, 1973) and correlation techniques for
attaining highest spectral resolution (BALL, 1973). The problem of a satisfactory

sensitivity, particularly at the shortest wavelengths, has been attacked along various approaches. Among the best receivers in use in the 1 to 2 mm wavelength range are those by WRIXON (1974 b) based on mixers with ultra low capacitance, low noise Schottky-barrier diodes in a stripline design incorporating high frequency matching elements, low pass filter and intermediate frequency output transmission line. Noise temperature contributions of less than 1000 K at 2.1 mm and about 5000 K at 1.3 mm wavelength have been achieved. Cryogenic cooling of mixers (WEINREB and KERR, 1973) seems to offer even better performance values for future designs.

6.3 Emissive Properties of Materials on the Surface of the Earth

The parameters of the ground which determine the amount of emitted radiation are essentially the complex permittivity of the material or mixture of materials and the shape of its surface. Therefore we have to deal with the effects of the

— dielectric properties (for the sake of simplicity: on flat surfaces)
— heterogeneous mixtures of various materials (only two components)
— roughness of the surface.

6.3.1 Plane Surfaces of Homogeneous Materials

Because of the conservation of energy for optically thick (non-transparent) perfectly flat media, the relation (4) holds

$$e_i(\Theta, \varphi, v) + r_i(\Theta, \varphi, v) = 1 \tag{4a}$$

for any given viewing angle Θ, φ, frequency v and any one of two orthogonal polarisations i. The expressions "emissivity" and "reflectivity" are in use for the special situation of vertical incidence ($\Theta = 0$, therefore independent on the polarisation) to define the—frequency dependent—emissive properties. Because of (4a) the emission coefficient can simply be expressed by the (power) reflexion coefficient, and the Fresnel relations of the reflected and penetrating part of a plane wave inpinging on a flat interface of two media can be applied (see textbook as BORN and WOLF, 1964 or STRATTON, 1941). The propagation behavior of waves is determined by the complex relative permittivity (dielectric constant) $\varepsilon = \varepsilon' - i\varepsilon''$ where the imaginary part $\varepsilon'' = \sigma/\omega\varepsilon_0$ is expressed by the conductivity σ [unit: $(\Omega m)^{-1}$], the vacuum permittivity $\varepsilon_0 = 8.854 \cdot 10^{-12}$ As/Vm, and the radial frequency $\omega = 2\pi v$. The index of refraction n is related to the relative permittivity by $n^2 = \varepsilon$ assuming non-magnetic media. The intensity reflection coefficients for horizontal (electric field of the incident wave parallel to the surface i.e. perpendicular to the plane of incidence) and vertical polarization (electric field parallel to the plane of incidence) on an air-dielectric interface ($\varepsilon_{air} = 1$) are respectively

$$r_h(\Theta) = \frac{(p - \cos\Theta)^2 + q^2}{(p + \cos\Theta)^2 + q^2}$$

$$r_r(\Theta) = \frac{(\varepsilon' \cos\Theta - p)^2 + (\varepsilon'' \cos\Theta - q)^2}{(\varepsilon' \cos\Theta + p)^2 + (\varepsilon'' \cos\Theta + q)^2}$$

with the abbreviations   (21)

$$p = \frac{1}{\sqrt{2}} \{[(\varepsilon' - \sin^2\Theta)^2 + \varepsilon''^2]^{\frac{1}{2}} + [\varepsilon' - \sin^2\Theta]\}^{\frac{1}{2}}$$

$$q = \frac{1}{\sqrt{2}} \{[(\varepsilon' - \sin^2\Theta)^2 + \varepsilon''^2]^{\frac{1}{2}} - [\varepsilon' - \sin^2\Theta]\}^{\frac{1}{2}}$$

where Θ is the incidence angle. For the reflectivity ($\Theta = 0$, $r_r = r_h$) Eq.(21) is simplified to

$$r = \frac{(p-1)^2 + q^2}{(p+1)^2 + q^2}$$

with   (22)

$$p = \frac{1}{\sqrt{2}} [(\varepsilon'^2 + \varepsilon''^2)^{\frac{1}{2}} + \varepsilon']^{\frac{1}{2}}, \qquad q = \frac{1}{\sqrt{2}} [(\varepsilon'^2 + \varepsilon''^2)^{\frac{1}{2}} - \varepsilon']^{\frac{1}{2}}.$$

For lossless medium (22) becomes the well-known expression $r = (\sqrt{\varepsilon'} - 1/\sqrt{\varepsilon'} + 1)^2$.

Because of the finite conductivity of all real media a wave inpinging on the surface is penetrating only into a limited depth, or in other words the radiation emitted through a surface is originating from a surface layer of finite depth. The propagation of a radiowave in a medium is characterized by the complex wave number $k = i\omega\sqrt{\tilde{\varepsilon}\tilde{\mu}} = \zeta + i\xi$, where ζ is the attenuation constant and ξ the phase constant. The amplitude attenuation constant ζ is related to the previously introduced volume absorption coefficient α of the intensity (flux density) by $\alpha = 2\zeta$. For most natural media $\tilde{\mu} = \mu_0 = 1.257 \cdot 10^{-6}$ Vs/Am the vacuum permeability and the complex permittivity $\tilde{\varepsilon} = \varepsilon_0(\varepsilon' - i\sigma/\omega\varepsilon_0)$ is expressed by ε' and σ. The amplitude of a wave propagating in a medium will be attenuated to $1/e \approx 0.37$ of its original value after a distance $1/\zeta$, which is called the penetration depth. Therefore the essential part of the radiation emitted by a medium originates from a surface layer of thickness $1/\zeta = 2/\alpha$.

Figure 13a gives the emissivity (viewing perpendicularly to the surface, $\Theta = 0$) and the normalized penetration depth $\delta = 1/\zeta\lambda$ as a function of the complex relative permittivity, and a number of media measured at wavelengths of 3 cm and 3 mm are indicated on the grid of lines of constant emissivities and constant

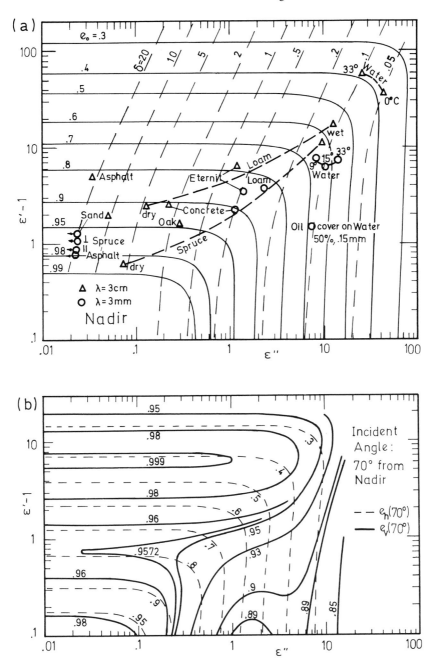

Fig. 13. (a) Emissivity and penetration depth as a function of the complex permittivity. Various natural and manmade media at wavelenghts of 3 cm and 3 mm. Loam and spruce at varying humidity, water at various temperatures (SCHANDA and HOFER, 1974b); (b) emission coefficients for two orthogonal polarizations at 70° incident angle as related to the complex permittivity (HOFER, 1974)

penetration depths. For an almost grazing viewing angle of $\Theta = 70°$ (measured from nadir), the lines of constant emission coefficients for vertical and horizontal polarizations respectively as drawn into the same plane of complex permittivities are given in Fig. 13b. The emission coefficients are not only split up due to the different polarizations, but the lines of vertical polarization are strongly determined by the Brewster effect, which will be discussed below. Various media exhibit a wavelength dependent permittivity (e.g. due to a molecular relaxation as for water) and are therefore shifting on this diagram as a function of the wavelength. Note that most solid media are characterized by an emissivity which is exclusively determined by the real part of the permittivity, but as soon as water is involved, the narrow range of $\varepsilon'' \approx \varepsilon' - 1$ becomes very important.

If the viewing angle Θ of the sensor is different from zero, the complete formula (21) is needed to determine the reflection or emission coefficient respectively. For lowloss media ($\varepsilon'' \ll \varepsilon'$) the reflection coefficients for horizontal and vertical polarizations can be approximated by

$$r_h = \left| \frac{(\varepsilon' - \sin^2 \Theta)^{\frac{1}{2}} - \cos \Theta}{(\varepsilon' - \sin^2 \Theta)^{\frac{1}{2}} + \cos \Theta} \right|^2$$

(23)

$$r_v = \left| \frac{\varepsilon' \cos \Theta + (\varepsilon' - \sin^2 \Theta)^{\frac{1}{2}}}{\varepsilon' \cos \Theta + (\varepsilon' - \sin^2 \Theta)^{\frac{1}{2}}} \right|^2 .$$

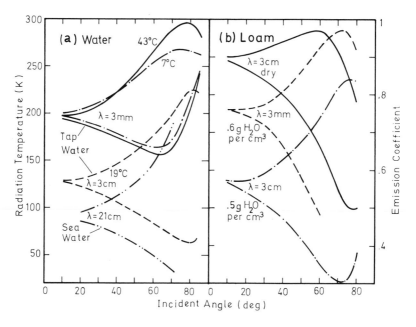

Fig. 14. (a) Radiation temperature of water at various wavelengths as a function of the incident angle ($\lambda = 21$ cm, after Hollinger, 1971), $\lambda = 3$ mm: zenith sky temperature 45 K during measurement; (b) emission coefficient of loam at 2 wavelengths and various degrees of humidity

Assuming $\varepsilon' > 1$ in all cases, the numerator of r_v vanishes at the Brewster angle which is given by $\tan^2 \Theta_B = \varepsilon'$. Also for lossy media a minimum of r_v can be found which is determined by ε' and ε''. According to (4a) a minimum of reflection is equivalent to a maximum of emission. Therefore even a material with a very low emissivity may exhibit an almost black-body behavior close to the Brewster angle for vertical polarization. Also the difference of the emission coefficients of both polarizations is a maximum close to the Brewster angle. The dependence of the emission coefficient of water on the viewing angle is given in Fig. 14a.

The depth below the surface from which the main part of the radiation is emitted is reduced for increasing nadir angle Θ as $\cos\Theta_t/\alpha$ where Θ_t is the direction of wave propagation in the medium measured from nadir, according to Snell's law of refraction.

From the emissive behavior of water (Figs. 13a and 14a) as a function of wavelength and temperature the relaxation phenomenon is evident. Water exhibits three important properties:

— a comparatively high dielectric constant at microwaves and therefore an emissivity very different from other natural media,
— a dipole relaxation occuring in the frequency range of about 10^{10} Hz, causing a strong frequency dependence of ε and thus of $e(\Theta)$,
— a low-frequency permittivity which is highly temperature dependent, causing in particular a high sensitivity of the penetration depth on the temperature.

The behavior of water as a polar liquid is described by the Debye relaxation formula (VON HIPPEL, 1954)

$$\varepsilon = \varepsilon_r + \frac{\varepsilon_s - \varepsilon_r}{1 + iv\Gamma}, \tag{24}$$

where $\varepsilon_r \approx 5.5$ is the high-frequency limit of ε. The relaxation time Γ and the low-frequency (static) limit ε_s of ε are given for water temperatures between $0°$ C and $80°$ C (HASTED, 1961) in Fig. 15. In the case of water (24) is a sufficiently accurate

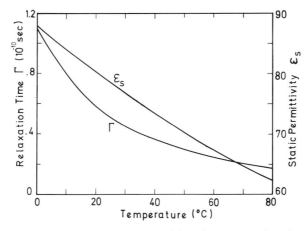

Fig. 15. Relaxation time and static (d.c.) permittivity of water as a function of temperature

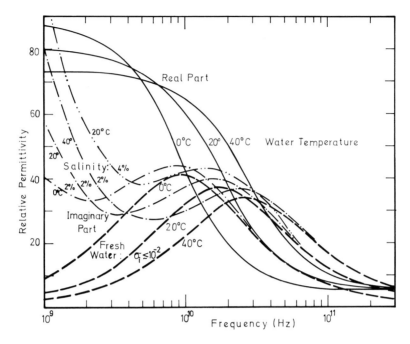

Fig. 16. Spectral dependence of the complex permittivity of water at various temperatures
and various degrees of salinity

approximation of the more general permittivity-frequency relation (COLE and
COLE, 1941), which describes the dispersive behavior of liquids and solids of more
complicated molecular structure.

The resulting spectral behavior of the permittivity of fresh and saline water
is presented in Fig. 16. The effect of the temperature is very pronounced, and
real and imaginary parts of ε are of comparable magnitude in the upper
frequency range. The frequency of the maximum dispersive losses ($v_m = 1/\Gamma$) is
shifted by approximately a factor of three within a temperature change from
$0°$ C to $40°$ C.

The effect of ionic conductivity of the salt solution of fresh water and—many
times more—of sea water can be taken into account (LANE and SAXTON, 1952)
by adding to the imaginary part of ε in (24) a term $\sigma_i/\omega\varepsilon_0$, where σ_i is the
conductivity due to ions [about 10^{-2} $(\Omega m)^{-1}$ for fresh water and several $(\Omega m)^{-1}$
for typical sea water salinities].

6.3.2 Heterogeneous Media

Most media in the natural environment are heterogeneous i.e. the scale of
the internal inhomogeneities is between a small fraction of a wavelength up to
several wavelengths.

Porous soil and gravel are only two examples of this type of texture. The depolarizing effect of macroscopic (size comparable to the wavelength) inclusions causes an effective relative permittivity ε_e which is in general a complicated function of the respective permittivities of the component media, their mixing ratio, distribution and shape (preferred orientations).

A very much simplified situation with ellipsoidal inclusions (permittivity ε_2, axes a, b, c) immersed in a medium with a different permittivity (ε_1), thus constituting a two-component mixture, has been treated by many authors (e.g. VAN BEEK, 1967, with many references) and yields practicable results.

An approximative expression of the effective (overall) permittivity ε_e of the mixture of a medium (ε_1) with ellipsoidal inclusions (ε_2) occupying a small volume fraction v_2 and randomly distributed and oriented, is due to FRICKE, (1953).

$$\varepsilon_e \approx \varepsilon_1 \left[1 + \frac{v_2}{3} \sum_{i=1}^{3} \frac{\varepsilon_2 - \varepsilon_1}{\varepsilon_1 + (\varepsilon_2 - \varepsilon_1) A_i} \right] \tag{25}$$

where A_i are the depolarizing coefficients of the ellipsoids along the three main axes. For volume fractions $v_2 > 0.1$, a modified $\bar{\varepsilon}_1$ due to the changed mean permittivity around an inclusion has to be applied (DE LOOR, 1956). In the case of a preferred orientation of all ellipsoidal inclusions, (25) becomes

$$\varepsilon_{e,a} \approx \varepsilon_1 \left[1 + v_2 \frac{\varepsilon_2 - \varepsilon_1}{\varepsilon_1 + (\varepsilon_2 - \varepsilon_1) A_i} \right] \tag{26}$$

where A_i is the depolarizing coefficient of the inclusion in the direction of the applied electric field. Table 3 gives the values of A_i for various ratios of the main axes a, b, c of the ellipsoids, using Δ as a dimensionless quantity small compared to unity. Also in Table 3 the effective permittivities for mixtures with inclusions in preferred orientations and with the assumption $\Delta \approx 0$ (extremely flat discs

Table 3

Spheres	$a = b = c$	$A_a = A_b = A_c = \frac{1}{3}$		$\varepsilon_2 \approx \varepsilon_1 \left[1 + 3v_2 \dfrac{\varepsilon_2 - \varepsilon_1}{\varepsilon_2 + 2\varepsilon_1} \right]$
Oblate ellipsoids	$a = b \gg c$	$A_a = A_b = \Delta,$ $A_c = 1 - 2\Delta$	Electric field perpend. to disks:	$\varepsilon_e \approx \varepsilon_1 \left[1 + v_2 \left(1 - \dfrac{\varepsilon_1}{\varepsilon_2} \right) \right]$
			Electric field parallel to disks:	$\varepsilon_e \approx \varepsilon_1 (1 - v_2) + v_2 \varepsilon_1$
Prolate ellipsoids	$a \gg b = c$	$A_a = 2\Delta,$ $A_b = A_c = \frac{1}{2} - \Delta$	Electric field perpendicular to needles:	$\varepsilon_e \approx \varepsilon_1 \left[1 + 2v_2 \dfrac{\varepsilon_2 - \varepsilon_1}{\varepsilon_2 + \varepsilon_1} \right]$
			Electric field parallel to needles:	$\varepsilon_e \approx \varepsilon_1 (1 - v_2) + v_2 \varepsilon_2$

and extremely thin needles) are given, basing on (26). For randomly oriented ellipsoidal inclusions of conducting or high permittivity material ($|\varepsilon_2| \gg |\varepsilon_e|$) the effective permittivity (assuming a small volume fraction v_2 of the inclusions) becomes

$$\varepsilon_{ec} \approx \varepsilon_1 \left[1 + \frac{v_2}{3} \sum_{i=1}^{3} \frac{1}{\varepsilon_1/\varepsilon_2 + A_i} \right] \tag{27}$$

in agreement with DE LOOR (1956).

This yields for conducting spheres $\varepsilon_{e,c} \approx \varepsilon_1(1 + 3v_2)$, and for conducting rods parallel to the electric field $\varepsilon_{e,c} \approx \varepsilon_1 \left(1 + \dfrac{v_2}{\varepsilon_1/\varepsilon_2 + 2\Delta} \right)$ and perpendicular to the electric field $\varepsilon_{e,c} \approx \varepsilon_1(1 + 2v_2)$.

A more pragmatic and widely practised approach is due to WIENER (1910), who calculates a "mean" electric field $E = (1 - v_2)E_1 + v_2 E_2$ due to the fields in the components 1 and 2 respectively, a "mean" electric displacement $D = \varepsilon_1(1 - v_2)E_1 + \varepsilon_2 v_2 E_2$ and uses a ratio $E_1/E_2 = (\varepsilon_2 + F)/(\varepsilon_1 + F)$ with a Form-zahl F to be determined empirically in order to find the effective permittivity by

$$\frac{\varepsilon_e - 1}{\varepsilon_e + F} = (1 - v_2) \frac{\varepsilon_1 - 1}{\varepsilon_1 + F} + v_2 \frac{\varepsilon_2 - 1}{\varepsilon_2 + F}. \tag{28}$$

The Formzahl F can be determined by the boundary conditions in simple situations and turns out to be zero for thin layers perpendicular to the electric field, infinity for thin structures parallel to the field and attains the value of two for spheres and values between 10 and 25 for irregular flakes (e.g. of snow).

Extensive analytical and numerical studies of the effects of vertical structuring and of subsurface spherical scatterers on the surface emissivity have been published by STOGRYN (1970) and more recently by WONG (1974).

When losses are present in one or both of the components of the mixture, their complex permittivities have to be used in (25). If the losses are of polar origin (relaxation losses) as for pure water, the mixture causes a shift of the relaxation frequency ($v_R = 1/\Gamma$) to higher values. The permittivity of the polar medium (e.g. water) according to (24) has to be inserted for ε_2 in (25), and a analoguous Debye type relaxation formula for the mixture can be deduced. The low and high frequency permittivities and the relaxation times of the mixture attain according to (24) and Table 3 the following values respectively for:

1. thin threadlike structures (ε_2, $a \gg b \approx c$) along the electric field:
$\varepsilon_{sm} \approx v_2 \varepsilon_{s2} + (1 - v_2)\varepsilon_1$, $\varepsilon_{\infty m} \approx v_2 \varepsilon_{\infty 2} + (1 - v_2)\varepsilon_1$, $\Gamma_m \approx \Gamma_2$;

2. thin disklike structures (ε_2, $a \approx b \gg c$) with the electric field in these planes: the same values as for 1;

3. thin disks perpendicular to the electric field

$$\varepsilon_{s,m} \approx \varepsilon_1 \left[1 + v_2 \left(1 + \frac{1 + v^2 \Gamma_2^2}{\varepsilon_{s2} + \varepsilon_{\infty 2} v^2 \Gamma_2^2} \right) \right], \quad \varepsilon_{\infty m} \approx (1 + v_2)\varepsilon_1,$$

$$\Gamma_m \approx \Gamma_2 \cdot \frac{\varepsilon_{s2} - \varepsilon_{\infty 2}}{\varepsilon_{s2} + \varepsilon_{\infty 2} v^2 \Gamma_2^2}.$$

Other shapes as e.g. spheres yield more complicated expressions but can easily be deduced. In the third example the frequency dependent relaxation time for a thin layer of water at $20°$ C on some substrate appears at $v = 10$ GHz to be $\Gamma_m = 0.91 \cdot \Gamma_2$ (just below the original relaxation time of pure water $(1/\Gamma_2 \approx 17$ GHz) and $\Gamma_m = 0.77 \cdot \Gamma_2$ at 30 GHz.

The situation of a thin layer of a lossy dielectric on a substrate has to be treated more carefully, taking into account the interference of multiple reflections on the interfaces. The simplest case of a loss-free dielectric layer (ε_2) between two infinitely extended dielectrics (ε_1 and ε_3 resp.) yields an oscillating reflection coefficient as a function of the layer thickness and of the viewing angle (BORN and WOLF, 1964). Values of ε_2 between the limits $\varepsilon_1 \leqq \varepsilon_2 \leqq \varepsilon_3$ (assuming arbitrary $\varepsilon_3 > \varepsilon_1$) cause a reduction of the reflection coefficient at layer thicknesses roughly equal to odd multiples of $\lambda_2/4$ (λ_2 the wavelength in the layer medium) compared to that by a single interface between media 1 and 3. For $\varepsilon_2 = \sqrt{\varepsilon_1 \varepsilon_3}$ the reflection vanishes at the odd multiples of $\lambda_2/4$. Values outside of the above mentioned limits, however, cause an enhancement of the reflection for the same layer thicknesses.

A specific situation of an oil film on sea water has been supposed in a computation of the frequency dependence of the microwave emission coefficient (HOLLINGER, 1973) following a solution of three layered lossy media by BREKHOVSKIKH (1960). The feasibility of unambiguous determination of oil layer thicknesses in the range of a few tenths to several millimeters on seawater by using the distinctly phased emissivities at three wavelengths (between about 0.4–1.5 cm) has been shown convincingly.

The effect of very thin layers of oil—only patches of a few centimeter diameter covering a part of the surface of tap water—on the brightness temperature as a function of the viewing angle at a single, very short wavelength is presented in Fig. 17a (HOFER and SCHANDA, 1974).

A heterogeneous material with a preferred orientation is wood. Figure 17b shows the brightness temperature for two polarizations of the electric field, one perpendicular to the fibers, one with a parallel component to them. The scale of the fiber structuring is about equal to the shorter wavelength used, but the anisotropic behavior is characteristic for both wavelengths. Humidity is enhancing the anisotropy at both wavelengths (SCHANDA and HOFER, 1974a and b; KÜNZI et al., 1971).

Finally, sand of a grain size up to two millimeter is an example of a more irregularly shaped heterogeneous medium (Fig. 17c). At the various degrees of humidity, more or less of the air between the grains is replaced by water; at about 0.5 g/cm^3 content, water is dominating the emissive behavior. A considerable amount of uncertainty is entering due to drying of the surface layer during the measuring cycle of a few minutes and an inaccurate reproduction of the humidity over the measured area (about 2 m^2). The penetration depth of 3 mm waves in dry sand has been measured to be about 10 wavelengths, therefore one may state that not the surface roughness but the heterogeneity below is dominating the emission properties (SCHANDA and HOFER, 1974). Road asphalt with a constructional texture as used in Switzerland has also a penetration depth of many wavelengths and may be regarded as a medium with an emission behavior

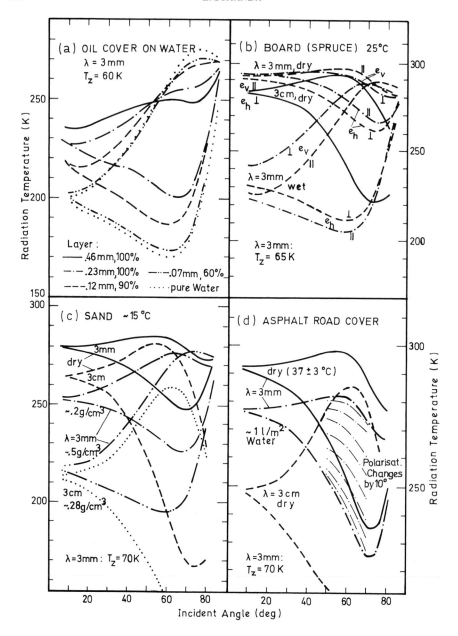

Fig. 17 a–d. Radiation temperatures as a function of the incident angle at two wavelengths (zenith temperatures T_z during the $\lambda = 3$ mm measurement). (a) Thin layers of oil (0.07 to 0.46 mm thickness) on tap water, increasing layer thickness causes increasing percentages of coverage, (b) board cut out of spruce, demonstrating the anisotropic emission behaviour, $\perp$ fibres perpendicular to incidence plane, $\parallel$ parallel, (c) sand (mean grain size about 1 mm) at various degrees of humidity, (d) asphalt road cover, roughness scale about 3 mm; effect of changing polarisation from vertical to horizontal between 50° and 70° incident angle of the humidified surface dominated by the internal heterogeneity above surface roughness (Fig. 17 d). The significant difference between horizontal and vertical polarization may originate from this fact; the inhomogeneity causes just a change of the effective permittivity but the Fresnel-Type emission behavior is still effective.

The influence upon the emissivity by volume scattering on inclusions within an otherwise rather transparent medium has been analyzed recently by ENGLAND (1975). Already very small volume fractions of spherically assumed scatterers (e.g. about 1 percent of ice cristals in dry snow) cause essential changes of the surface brightness temperature if the diameter of the scatterers equals 0.1 of the free-space wavelength.

6.3.3 Rough Surfaces

Surface roughness can be classified owing to its character and its degree measured in wavelengths. The different resulting roughness models allow different approximations in the analytical treatment, and they are based on the assumptions of randomly distributed deviations from a mean plane and of randomly distributed slopes of the surface of an intrinsically homogeneous medium (constant values of ε and a sufficient conductivity below the surface to avoid confusions with the effect of heterogeneity). The following classification is after FUNG (1971):
— slightly rough surface: the height variations are small compared with the radio wavelength and the surface slopes are small compared with unity,
— smoothly undulating surface: the height variations are comparable to or larger than the wavelength and the surface is locally flat relative to the radio wavelength (Kirchhoff tangent plane approximation),
— two-scale composite rough surface: a large scale roughness satisfying the tangent plane approximation superimposed by a small scale roughness obeying the assumptions of a perturbation method (sea surface with large gravity waves and small capillary waves).

Additionally to these continuous stationary random surfaces are those representable by a distribution of objects of assumed shape:
— protuberances uniformly and randomly distributed on a ground plane (TWERSKY, 1957),
— distribution of long, thin, lossy cylinders covering a ground plane (PEAKE, 1957 and 1959),
both models are particularly useful to represent vegetation.

The deviations from a plane surface are assumed to be somehow comparable to the radio wavelength in all the above models of "rough" surfaces which are therefore qualitatively different from both the limiting situations of:
— the perfectly smooth surface, which can be treated according to the Fresnel formulae (Section 6.3.1) and for which the emission coefficient e is related to the reflection coefficient of the intensity r in any plane of the polarization i by

$$e_i(\Theta, \varphi, v) = 1 - r_i(\Theta, \varphi, v), \qquad (4a)$$

— the Lambertian (perfectly rough) surface, for which the heigh variations are much larger than the wavelength (PEAKE et al. (1966) for microwave experimental

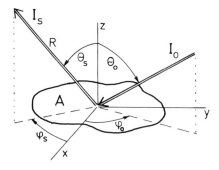

Fig. 18. Scattering geometry

verification) and produces a backscatter coefficient independent on the incident angle. Using the symbols of Fig. 18, the differential scattering coefficient γ for a given frequency and two orthogonal states of polarization i and j is given by (PEAKE, 1959)

$$\gamma_{ij}(\Theta_0, \varphi_0, \Theta_s, \varphi_s) = \frac{4\pi R^2 I_s}{I_0 A \cos \Theta_0} \tag{29}$$

and because of Kirchhoff's radiation law the emissivity of A viewed from (Θ_0, φ_0) is

$$e_i(\Theta_0, \varphi_0) = 1 - \frac{1}{4\pi} \int_{\Theta_s = -\pi}^{\pi} \int_{\varphi_s = 0}^{2\pi} [\gamma_{ii} + \gamma_{ij}] d\Theta_s d\varphi_s \tag{30}$$

and after using Peake's substitution $\gamma_{ii} + \gamma_{ij} = \gamma_0 \cos \Theta_0$ due to the projection of A into the viewing direction, the integration yields

$$e_i(\Theta_0, \varphi_0) = 1 - \frac{\gamma_0}{4} \tag{31}$$

independent of the angles Θ_0 and φ_0.

The integral term on the right hand side of (30), replacing the reflection coefficient of a plane surface [r_i in (4a)], is called the albedo of the rough surface for the particular polarization i of the sensor.

There are essentially two approaches to predict the emission coefficient of rough (non-Lambertian) surfaces treated in the literature: a geometrical optics approach by STOGRYN (1967a, b) and a physical optics approach by ULABY et al. (1970, based on earlier investigations by FUNG, 1967), which is claimed to contain the geometrical optics model as a special case. Both approaches result in rather complicated formulae, which are practicable only in computer aided numerical analyses and the reader is referred to the original literature and in particular to the most recent exact formulation by FUNG et al. (1974).

Returning to the more general formulation (30), the radiation temperature received at Θ_0, φ_0 from a rough ground T_r (omitting the attenuation of the air between ground and sensor) is composed of two contributions due to emitted radiation and due to scattering of diffuse external radiation respectively in either of two orthogonal polarisations (COSGRIFF et al., 1960)

$$T_{r,i} = T_{g,i} + T_{s,i} \tag{32}$$

where $T_{g,i}(\Theta_0, \varphi_0) = e_i(\Theta_0, \varphi_0)(T_{g0,i} + T_{g0,j})$ with e_i from (30) and T_{g0} the physical temperature of the ground, while the contribution by the scattering of the external, distributed radiation $T_{s0}(\Theta_s, \varphi_s)$ is given by

$$T_{s,i}(\Theta_0, \varphi_0) = \frac{1}{4\pi} \iint [\gamma_{ij}(\Theta_0, \varphi_0, \Theta_s, \varphi_s) T_{s0,i}(\Theta_s, \varphi_s)$$

$$+ \gamma_{ij}(\Theta_0, \varphi_0, \Theta_s, \varphi_s) T_{s0,j}(\Theta_s \varphi_s)] d\Theta_s d\varphi_s .$$

The relation (32) is a generalization of (5). In Fig. 19 the emission coefficients as a function of the viewing angle measured at 3 mm wavelength are presented in a few examples which may be regarded as surface roughness dominated emissions. Figure 19a shows the behavior of the very irregular surface of humus for various thicknesses (height variations equal to wavelength). The penetration depth in dry humus is roughly 2λ, therefore the roughness effect is dominating the emission, this is also confirmed by the vanishing difference of the emission in the two polarisations (SCHANDA and HOFER, 1974b). Lawn (Fig. 19b) is a typical example of a "long, thin, lossy cylinders" situation and even a considerable amount of water does not split up the curves for vertical and horizontal polarisations. The complex permittivity has been determined by a forward scatter measurement (SCHANDA and HOFER, 1974a). The emission coefficient of the dry lawn calculated with this permittivity compares fairly well with the measured one.

A surface of stone fragments (Serpentine, Fig. 19c) with a mean size of about 3 cm (10 wavelengths) should be expected to act as a Lambertian surface. The difference between the horizontally and vertically polarized emission leads to the conclusion that the specific shapes of these fragments caused some preferentially directed ordering. An artificial rough surface agreeing to some extent with the model of randomly distributed protuberances has been realized by metal turnings of a mean size comparable to λ. Figure 18d compares its emission with that of a flat metal plate. The higher radiation temperature will be due to two reasons: the scattering on the turnings introduces contributions of the external distributed radiation (high sky temperature near to the horizon and objects in the surroundings), and the turnings exhibit a much higher effective electrical resistance than the flat plate.

6.3.4 Various Investigations on Emissivities

One of the first comprehensive presentations of the scattering behavior of natural and man-made materials at microwaves, which can be used also for estimations of the emissivity is by COSGRIFF et al. (1960). Measurements on rough surfaces have been reported by PEAKE et al. (1966), who tried to verify the Lamber-

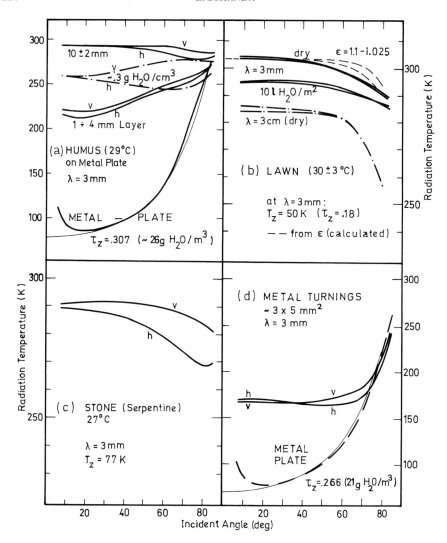

Fig. 19a–d. Radiation temperature as a function of the incident angle of rough surfaces at 3 mm wavelength, (a) humus on a perfectly reflecting plate, the penetration depth is about 8 mm; (b) lawn; (c) serpentine; (d) metal turnings as an artificially rough surface; increasing radiation temperature of the metal plate at small incidence angles is due to reflections of the measurement equipment; the thin line represents the sky temperature directly measured (without reflection on the metal plate)

tian assumption on large blocks (0.5–1 m) of pumice with low density at wavelengths between about 1 cm and 15 cm. The rough-surface emission behavior of aggregates of blocks of dolomitic limestone and of granite in the same wavelength range has been investigated in great detail by EDGERTON and TREXLER (1970).

The surface of the sea is able to present almost every type of roughness and has therefore been studied extensively and used to verify the particular roughness

models (STOGRYN, 1967a; POE et al., 1972). The strong increase of the emissivity of
the sea surface with higher wind speeds (NORDBERG et al., 1968 and 1971), which
is due the formation of foam, has been further investigated by HOLLINGER (1970)
and STOGRYN (1972a). GLOERSEN et al. (1974a) determined the spectral variation
of the seawave-, foam-, streak bubbles- and breaking wave-dependent emissivity with the wind speed in the 0.8 cm–21 cm wavelength range and they propose a
microwave anemometry for all wind speeds of interest. An investigation of the
permittivity of sea-ice by HOEKSTRA and CAPPILLINO (1971) revealed a very low
loss behavior at temperatures below about $-30°$ C ($\varepsilon'' < 0.2$) and a sudden increase around $-20°$ C to about $\varepsilon'' \approx 1$ at $-10°$ C for a measuring frequency of
9.8 GHz and a salinity of 8 parts per thousand, resulting in an attenuation of more
than 100 db/m. The same authors were able to show that in a controlled experiment with an NaCl-water mixture the jump of ε'' from less than 0.1 to about
1 occurred between $-22°$ C and $-20°$ C due to the eutectic point at $-21.2°$ C
below which NaCl·$2H_2O$ crystallizes out. The more gradual change in sea water
is claimed to occur because other salts as KCl and $MgCl_2$ remain in solution
below $-22°$ C, and a certain amount of liquid brine remains in the ice. Extensive
investigations of the emission characteristics in the wavelength range of 3 mm–
21 cm and the comparison among the various mixture formulae has been reported by POE et al. (1972).

Some of the rather scarce investigations on the emission properties of snow
are due to EDGERTON et al. (1971) on controlled snowpacks with groundbased
instruments covering the wavelength range of 0.8–21 cm, or by GLOERSEN et al.
(1974b) from analyses of aircraft-radiometer data, who conclude that in contrast
to soil not the dielectric losses, but the volume scattering (Mie) of the ice particles
is the dominant mechanism determining the emissivity. Concerning ice and snow
many authors refer back to a study by CUMMING (1952) on the dielectric properties under varying weather conditions. Investigations on different types of soil
have been reported by MELENTYEV and RABINOVICH (1972), and GEIGER and
WILLIAMS (1972), among many others. The signatures of vegetations have almost
exclusively been studied so far by radar back- and forward-scatter methods (e.g.
MOORE, 1973; ULABY, 1974; ATTEMA et al., 1974), and therefore the emission
properties of crops have to be deduced from these results.

A more general approach to formulate radiometric signatures of complex
bodies is made by HAMID (1974), and a survey of wave scattering from rough
surfaces has been presented by HÖJER (1971).

6.4 Remote Determination of Atmospheric Constituents by Their Microwave Spectra

6.4.1 The Absorption Coefficients of the Line Spectra

In Section 6.1.3 the effects of the atmospheric phenomena have been discussed
as imposing limitations on the sensing of remote objects with millimeter wave
instruments. On the other hand, the atmospheric absorption can be utilized to

determine the presence, the quantity and eventually the local distribution of an atmospheric constituent. Only the effect of the line radiation due to the molecular transitions shall be discussed briefly within this section.

Radiation measurements for the remote quantitative determination of the atmospheric structure are in use since the ultra-violet scattering spectrometry of ozone by Götz et al. (1934). Dicke et al. (1946) were the first to propose the remote measurement of the water-vapor content of the atmosphere by a micro-wave radiometer. Meeks and Lilley (1963) gave a detailed discussion on the use of the microwave emission of oxygen at 5 mm wavelength for remotely sensing the atmosphere. Reviews on the potentialities of microwave passive probing of atmospheric gases are published by Staelin (1969) and recently in a very comprehensive way by Waters (1975). The most intensively studied spectral lines for remote probing are those of water vapor at 22 GHz and 183 GHz, oxygen (O_2) near 60 GHz and at 118 GHz and of Ozone above 100 GHz.

The problem of evaluating the distributions of the various atmospheric constituents from their radiometrically sensed contributions to the brightness temperature is the correct inversion of the equation of the radiative transfer (9). In the microwave approximation ($hv/kT \ll 1$) and with the assumption of pure thermal radiation (in the situation of local thermodynamic equilibrium of the microwave absorption by the terrestrial atmosphere, $j/4\pi K = 2kT/\lambda^2$) Eq.(9) can be rewritten (see Fig. 2)

$$T_v(z_c) = T_v(0) \exp[-\tau_v(0, z_c)] + \int_0^{Z_c} T(z) \exp[-\int_0^z K_v \varrho(z')dz'] K_v \varrho(z) dz \qquad (9\,c)$$

where the subscript v indicates the respective values at the particular frequency v of these strongly frequency dependent quantities. Emission or absorption of an electromagnetic radiation (photon) with a system of quantized energy states ($E_l, E_m, \ldots$), such as an atom or a molecule, is causing a transition between two quantum states (l, m) of this system. In terms of quantum mechanics the rate of change of the radiation flux density (Section 6.1.2) can be expressed by

$$\frac{dF_v(z)}{dz} = -K_v \varrho(z) F_v(z) = -\frac{hv}{c} [N_l B_{lm} - N_m B_{ml}] f(v, v_{lm}) F_v(z) \qquad (33)$$

where B_{lm} and $B_{m,l}$ are the probabilities for induced transitions between the states l to m and m to l respectively (the rates of the spontaneous transitions are negligible for $hv/kT \ll 1$), N_l and N_m are the number densities in the initial states l and m respectively and $f(v, v_{lm})$ is a line shape factor. For a gas in the state of thermal equilibration, the quantities within the brackets can be specified for allowing accurate computations of the volume absorption coefficient $\alpha_v(z) = K_v \varrho(z)$, which is defined by (33). The reader is referred to the more specialized literature (e.g. Waters, 1975) for the derivations and more details of the application of (33).

The line shape factor is determined by the collisional disturbance of the inter-actions between the molecule and the radiating field as the dominating line broadening effect. Doppler (thermal) broadening becomes important only for the low densities at mesospheric altitudes and the natural linewidth resulting

Table 4. Center frequencies (in GHz) of absorption lines of O_2 and H_2O. (After WATERS, 1975)

H_2O^{16}	22.2351	183.3101
O_2^{16}	The bulk of (30) lines between 50 and 66 GHz	118.7503

Table 5. Concentration of trace gases at the surface of the earth, line frequencies and calculated maximum absorption coefficients of the strongest lines below 300 GHz.[b] (After WATERS, 1975)

Molecule	Relative volume concentration (earth surface)	Center frequencies of lines	Maximum absorption coefficient in cm^{-1}
Nitrous oxide (N_2O)	(2.5 to 6.0) 10^{-7} [a]	25.12325 and multiples upt zo 301	4.2 10^{-7} up to 4 · 10^{-4}
Carbon monoxide (CO)	(1 to 2) 10^{-7}	115.2712 230.5380	2.19 10^{-4} 1.68 · 10^{-3}
Ozone (O_3)	(0 to 5) 10^{-8} (maximum concentration about 10 times greater at 20 to 30 km altitude	101.7369 110.8360 124.0875 142.1751 165.7844 184.3783 195.4305 208.6424 231.2812 235.7096 237.1460 239.0930 242.3186 243.4537	1.85 10^{-3} 2.84 10^{-3} 4.02 10^{-3} 5.37 10^{-3} 6.86 10^{-3} 6.36 10^{-3} 8.66 10^{-3} 5.45 10^{-3} 9.90 10^{-3} 14.30 10^{-3} 14.30 10^{-3} 13.70 10^{-3} 13.90 10^{-3} 12.90 10^{-3}
		and several more with comparable absorption coefficient	
Ammonia (NH_3)	(0 to 2) 10^{-8}	10 strongest lines between 21 and 28 GHz	2 · 10^{-4} up to 8 · 10^{-4}
Sulfur dioxide (SO_2)	(0 to 2) 10^{-8}	31 measured lines below 300 GHz	
Hydrogen sulfide (H_2S)	(0.2 to 2) 10^{-8}	168.7625 216.7104	
Formaldehyde (CH_2O)	(0 to 1) 10^{-8}	39 measured lines below 226 GHz	
Nitrogen dioxide (NO_2)	(0 to 3) 10^{-9}	measured lines close to 16, 27, 40 GHz	

[a] (SCHÜTZ, K., et al., 1970).
[b] Recent calculations of absorption coefficients and atmospheric brightness temperatures due to O_2, H_2O, O_3, N_2O, NO, NO_2, CO, SO_2 and H_2S are reported by J. FULDE (Ph. D. thesis, Univ. of Berne, 1976).

from the finite lifetime of the particular molecular state is completely negligible. The collisional dominated line shape function has been derived by Ben-Reuven (1965, 1969), and its application to the microwave spectrum of atmospheric molecules for the determination of the resulting absorption coefficients has been discussed in detail by Waters (1975).

Of the main constituents of the terrestrial atmosphere Nitrogen, Oxygen and Water Vapor, the latter two exhibit strong absorption lines in the microwave spectrum. The water vapor molecule has an electric dipole moment because of its non-linear, asymmetric structure and the transitions between the rotational states are causing line absorptions in the microwave and the far infra-red spectra. The oxygen molecule has no electric, but it has a magnetic dipole moment due to the combined spins of two unpaired electrons. The changes of the orientation of the electronic spin relative to the orientation of the molecular rotation produces a band of spin-rotation spectral lines near 60 GHz and a single line at 118 GHz, while transitions between different rotational states cause absorption lines at frequencies higher than 300 GHz. Table 4 gives the absorption lines of the H_2O and O_2 molecules below 300 GHz.

A considerable number of the molecules of the minor atmospheric constituents exhibits spectral lines due to transitions of their rotational states. Table 5 gives the concentrations of trace constituents near the surface of the earth, the center frequencies of the most intense absorption lines and if available the related maximum absorption ciefficient α in cm^{-1} calculated for environmental temperatures at 220 K (O_3) and at 273 K (N_2O, CO).

Ozone is a special candidate for the radiometric probing of its locally and temporarily varying density profile. The calculated brightness temperature toward the zenith of an oxygen-ozone mixture due to the 1962 US standard atmospheric

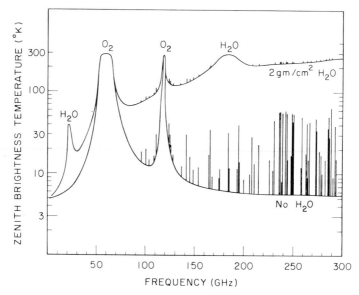

Fig. 20. Zenith brightness temperature due to O_2, H_2O and ozone. (After Waters, 1975)

model is given in Fig. 20 (WATERS, 1975) for an ideally dry atmosphere and one with $2 \, g/cm^2$ total water vapor pressure (resulting from a 2 km scale height exponential decrease from a density of $10 \, g/m^3$ at the surface up to 15 km and a constant mixing ratio of 2 parts per million above that height), corresponding to about 55% relative humidity at a temperature of $20° C$ near the surface. The feasibility of remotely detecting Ozone lines or even determining the Ozone density profiles from the ground is drastically reduced by an increasing humidity of the air.

6.4.2 Determinations of Height Profiles

For the most important constituents of the atmosphere, in particular for those with a certain environmental (e.g. meteorological) relevance, it is important to measure not only the total quantity within a column above the surface of the earth, but also to determine the eventual temporarily varying height distribution or the height profile of the temperature. A narrow band spectral filtering along the line shape of sufficiently intensive lines can yield a certain height resolution.

For the inversion of the remotely sensed line shape in the nonuniform atmosphere, the concept of weighting functions (STAELIN, 1966) has been developed. The pressure dependent contributions to the total optical thickness τ_v at a specific frequency along the line shape of a particular constituent can be determined by the frequency and pressure (height z) dependent weighting function $W(v, z)$ which replaces the previously used mass absorption coefficient $K(v)$.

$$\tau_{cv} = \int_0^\infty \varrho_c(z) W_c(v, z) dz . \tag{34}$$

The constituents density profile is $\varrho_c(z)$ and the weighting function can be expressed by $W_c(v, z) = \alpha_c(v, z)/\varrho_c(z)$ with the volume absorption coefficient α_c. Thus an extraneous constituent (e.g. water vapor) adds the amount τ_{cv} to the optical depth of a semitransparent atmosphere against a well-defined background. If the undisturbed level is subtracted, the remaining optical thickness of the desired constituent is yielded. The changes of the atmospheric pressure as a function of time and of temperature are only very small and have little effect on the measured τ_v compared to the effect of the varying content of a constituent as water vapor. Therefore the atmospheric pressure profile (approximately given by $p \, (mbar) = 1013 \exp(-z/7)$ with z in kilometers up to the stratopause at about 50 km) may be assumed constant with time, and the weighting function may be calculated by the knowledge of the line shape as a function of the pressure.

Figure 21a shows the (downward looking) weighting functions of water vapor around the 22 GHz line as calculated by STAEHLIN (1966). As long as the optical depth of the desired constituent is smaller than say 0.5 and simultaneously the background temperature is sufficiently different from the kinetic temperature of the atmosphere, the weighting functions are very similar for observation from the ground or from space, because the line center is already absorbed very effectively at high altitudes (where the pressure is low) and the edges of the line are absorbed only at low altitudes. Therefore observations from the ground against the high radiation temperature of the sun (in absorption) or against the "cold" sky (in emission) as well as observation from space against the sea surface are feasible.

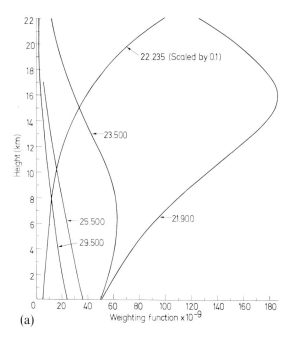

(a)

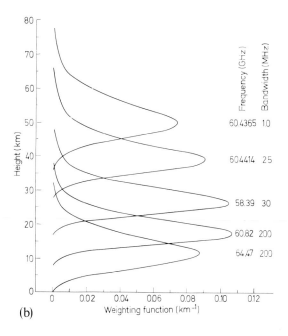

(b)

Fig. 21a and b. Weighting functions of (a) water vapor, (STAELIN, private communication),
(b) oxygen (LENOIR, 1965)

The lines of molecules with well-known and time-invariant mixing ratios throughout the atmosphere can be used to determine the height profile of such an important physical parameter as the temperature. The band of the oxygen spectral lines at about 60 GHz absorbs sufficiently intensive and the mixing ratio of O_2 is quite uniform and constant in time to make it ideally useful for remotely probing the temperature profile with a surprisingly high discrimination. With the concept of the weighting function integral, introduced by MEEKS and LILLEY (1963) as a special way to solve the equation of radiative transfer, the measured brightness temperature at a measuring frequency v along the shape of the resonance line can be expressed (STAELIN, 1969) by

$$T_v(\infty) = \int_0^\infty T(z) W_T(v, z)dz + T_v(0) \exp[-\tau_v(\infty)]. \tag{35}$$

The optical thickness $\tau_v(\infty)$ and the weighting function W_T can be found by comparison with (9c). $T(z)$ is the profile of the kinetic temperature of the atmosphere, W_T depends on the particular frequency, on the line spectrum and on the pressure (height z). The background temperature $T_v(0)$ is attenuated due to the total optical thickness of the intervening atmosphere at the frequency v. The weighting functions disclose the sensitivity to resolve the atmospheric (kinetic) temperature distribution.

LENOIR (1965 and 1968) has calculated weighting functions for the space borne radiometric measurement of the temperature profile by a multifrequency sensing along the oxygen line spectrum near 60 GHz (Fig. 21 b).

A relatively large width of the weighting functions as low altitudes is a disadvantage compared to radiosonde measurements, but the continuous measuring capability, the global coverage and the non-interference of the remote measuring eqipment with the atmosphere are considerable advantages.

Because of the statistical character of the measurement of the atmospheric parameters, special procedures are needed to minimize the errors of the conversion of the measured data into the estimates of the desired atmospheric parameters when inverting the equation of radiative transfer. The most advanced procedure has originally been proposed by RODGERS (1966) for the inversion of temperature profiles from infrared measurements. A calculus for minimizing the mean square errors between estimated and true parameter values has been developed; this method has been applied to the inversion of microwave line measurements by WESTWATER and STRAND (1968) and further discussed by STAELIN (1969), and WESTWATER (1972).

6.5 Passive Microwave Remote Sensing of Water, Ice and Snow

As far as the present experimental state of passive microwave sensing allows investigations of significance for the environmental sciences or managements, these are dealing with the following topics:
— water surfaces (sea state, pollution, salinity),
— ice (on sea and land), snow,
— soil (bare, moisture, temperature),

— very few topics on vegetation and geology,
— atmosphere (temperature profile, constituents),
— meterological features.

In the following sections only some representative examples but certainly not an exhaustive treatment of these investigations will be given.

6.5.1 Investigations of Water Surfaces

A flat water surface exhibits a low emissivity compared to all other natural surfaces, therefore it is extremely well discernible from soil or vegetation. SCHAERER (1974) investigated the effect of the polarization on the discernibility in ground-based mapping experiments at 3.3 mm wavelength. Figure 22a shows one of his images (horizontal polarization) of a lake viewed under a depression angle of about 8–15 degrees. Inspecting the emission coefficient of water (Fig. 14a), one finds that this range of angles is close to the Brewster angle. It turned out that with a temperature resolution of about 3–5 K, the map of the same scenery in vertical polarization did not yield any discernability of the lake against the vegetated surroundings.

Airborne mapping of lakes at 8 mm wavelength has been reported on by VOGEL (1972), and PREISSNER (1974).

An important application of measuring the microwave radiation temperature of water surfaces is the determination of the sea state through clouds and even through precipitation.

The possibility of a world-wide sea state sensing has been anticipated in a study of a composite radiometer—scatterometer (RADSCAT) by CLAASSEN et al. (1973). Various roughness models for the scattering properties at various sea states and their dependence on the wind speed have been developed and experimentally tested. A power law has been found for the scattering coefficient σ at given incidence angles as a function of the wind speed w (in knots)

$$\sigma \doteq w^n \tag{36}$$

with n ranging between 1.0 and 1.5 (Incidence angles 30° and 60° resp.) and lower values for horizontal, higher ones for vertical polarization. The range of validity is for windspeeds between 8 and 40 knots.

The relationship between the true and the mean slopes of the sea waves—an important parameter for a correct interpretation of the remotely sensed radiation temperatures—has been studied by MATUSHEVSKIY (1969).

The importance of foam forming on the microwave emissivity and thus on the windfield determination has been recognized and studied by NORDBERG et al. (1968), WILLIAMS (1969), EDGERTON et al. (1970), HOLLINGER (1970), and others. Investigations by STOGRYN (1972a, b) yielded a semi-empirical expression for the emission coefficient of the totally foam-covered surface as

$$e_i(v,\theta) = e(v,0)F_i(\theta) . \tag{37}$$

Fig. 22. (a) Thermal image at $\lambda = 3$ mm of a natural scenery with a lake, distance to the mountain peak 9.5 kms (SCHAERER, 1974), (b) optical identification of (a) the radiometer is situated several hundred meters above the water surface, 1 km distant from the shore, (c) the same mountain scanned from a place close to the shore with meteorological features appearing in the radiometric image. (SCHAERER and SCHANDA, 1974)

As a function of the frequency v and the look angle from nadir θ for both polarizations (i either horizontal or vertical). The factors of (37) have the meaning

$e(v, o) = 0.722 \pm 0.0045\,v$
(v in GHz) and $F_i(\theta)$ for both polarisations:
$F_h(\theta) = 1 - 1.748 \cdot 10^{-3} \cdot \theta - 7.336 \cdot 10^{-5} \cdot \theta^2 - 1.044 \cdot 10^{-7} \cdot \theta^3$
$F_v(\theta) = 1 - 9.946 \cdot 10^{-4}\,\theta + 3.218 \cdot 10^{-5} \cdot \theta^2 + 1.187 \cdot 10^{-6} \cdot \theta^3 + 7 \cdot 10^{-20} \cdot \theta^{10}$

(θ in degrees), where the fit to the measuring data is performed up to $\theta = 70°$.

Wu and Fung (1972) developed a two-scale scattering model for the emission and backscatter properties of the undulated sea-surface to take into account the small irregularities superimposed on the large scale wave structure. A bistatic two-scale noncoherent scattering concept, extended from Semyonov's (1966) theory, yields the differential scattering coefficients and from these the emission is derived. A Gaussian surface height distribution and Gaussian correlation functions are assumed for both scales of roughness. The correction to the ordinary Fresnel behavior of the emission depends on the r.m.s. slope of the large undulations, on the standard deviation and on the correlation length of the small irregularities. A remarkable coincidence of the observed (8.36 GHz) and theoretical brightness temperatures over incident angles between 20 and 70 degrees and both polarization has been shown for wind speeds of 15 and 25 knots.

Multifrequency experiments on the wind speed determination have been reported by Hollinger (1971) and by Au et al. (1974). The sensitivity to wind speed increases with the observational frequency (between 1.4 GHz and 19 GHz), and is most pronounced for horizontal polarization at larger incidence angles as measured from nadir. The effects by small scale wave structure at wind speeds below about 15–20 m/sec and by the increasing foam coverage at higher wind speeds on the brightness temperature have separately been determined. The emissivity of foam varies from 0.57 at the lowest to 0.85 at the higher (14.5 GHz) frequency and the polarization has little effect.

Melentyev et al. (1972a and b) found the nadir radiation temperature of a sea surface with 0.5 m waveheight to be 165 K at 1.35 cm wavelength, decreasing to 105 K at 3.2 cm. A clear increase of the nadir radiation temperature by about 20 K due to a sea wave increase from 0.5 m to 3 m was noted, and a better discrimination between wave heights was achieved at large nadir angles.

Mappings of surface wind fields are reported by Strong (1971), using an airborne 1.55 cm wavelength scanning radiometer (swath ± 30 degrees from nadir). A rate of the brightness temperature change due to increasing wind velocities of 1.7 K sec/m earlier found by Hollinger (1970) has been verified, and wind speed maps have been produced with 5 km/sec steps between the iso-velocity lines.

The dependence of the radiation temperature of seawater on the salinity (roughly 0.5 K per promille change of the salinity at 21 cm wavelength) has been used in experiments by Thomann (1973) to trace the salinity variations in estuary environments of coastal waters. Salinity accuracies of 3–5 $^0/_{00}$ in a 5–35 $^0/_{00}$ range have been obtained in that particular experiment, but accuracies better than 1 $^0/_{00}$ are certainly attainable for quiet sea surfaces by an improved radiometer.

The effect of oil slicks on ocean water is two-fold: the oil cover damps the small scale surface roughness and it changes the surface emissivity due to the much lower permittivity of the oil layer. In fact the layer of a thickness between few tenths of a millimeter and several millimeters causes multiple reflections by the boundaries air-oil and oil-water respectively, thus exhibiting a radiation temperature oscillating as a function of thickness and being therefore strongly frequency dependent. HOLLINGER (1973), who studied this effect, used it for the design of a multifrequency sensor which is able to map the thickness distributions of oil slicks (HOLLINGER and MENNELLA, 1973).

EDGERTON et al. (1971a) were able to show that a sensing wavelength of 8 mm is close to optimum for the detection of oil slicks because of the increases of the brightness temperature signatures and of the angular resolution of an imaging system and the decrease of the atmospheric transmissivity with frequency. The emission characteristics vary with the oil type and film thickness, but the modification of the sea state by reducing the small scale roughness seems to be largely independent of the film thickness.

A prototype airborne oil pollution surveillance system has been developed by EDGERTON and WOOLEVER (1974), where a 37 GHz passive microwave imaging instrument is utilized for adverse weather spill mapping and approximate spill thickness determination together with a side looking radar and visible and infrared mapping instrumentation.

The absolute determination of the thermal emission from the sea at 2.69 GHz with an accuracy of ± 0.3 K is quoted by GRAY et al. (1971), where the remaining uncertainties in determining the kinetic temperature of the sea are due to surface contamination (oil slicks), spray and foaming, salinity variations and surface waves. During the process of oil spreading a very large temperature contribution was observed which is thought to be associated with some lens forming of oil and water with intermediate dielectric constants of this "quasi-emulsificated" surface region, while homogeneous oil films contribute only about $1 \,^{\circ}$K at this low frequency.

The interference between the boundaries of a layer on a substrate has also been used by BASHARINOV et al. (1971) to detect thickness and dielectric properties of floating ice. The spectral dependence of the effective radiation on layer thickness and degree of polarization allows the determination of the physical conditions as temperature and density of the surface cover and to accomplish remote sensing of depths to ten meters. Discrimination of floating and (shelf) continental ice at and around the antarctica was possible (BASHARINOV et al., 1972).

The large combined US—SU remote sensing expedition to the Bering sea has revealed the potentialities of the microwave methods for a complex earth science project. In a preliminary report by the WOEIJKOV group (1973)
— the process of sea-ice formation,
— the distribution and properties of sea-ice,
— salinity, density and temperature of the sea-ice,
— the sea state,
as measured with microwave methods together with extensive "ground" truth observations are discussed in some detail.

6.5.2 Investigations of Ice and Snow

Several satellite and airborne sensing experiments have been devoted partly to the study of the ice cover of the polar regions.

A laboratory type of measurement on ice-samples during the above mentioned Bering sea expedition has been reported by MOROZOV and CHOCHLOV (1973). This investigation of salinity (varying roughly between $2^o/_{oo}$ and $8^o/_{oo}$, density (between 0.69 and 0.915), index of refraction (roughly from 1.7 to 2.2) and

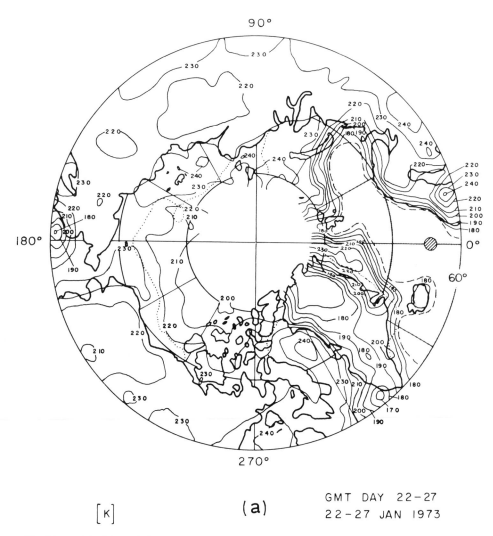

[K]   (a)   GMT DAY 22−27
                                             22−27 JAN 1973

Fig.24a and b. Map of the North pole region as obtained by the Nimbus 5 Microwave Spectrometer at the same season as Fig.23a. (a) Mean brightness temperature, (b) spectral gradient of the brightness temperature, both by using the channels of 22 GHz and 31 GHz. (KÜNZI et al., 1976, copyright by American Geophysical Union)

absorption coefficient (roughly from 0.4 to 4 db/cm) delivered the "ground truth" for the remotely sensed (airborne or satellite borne) emission data of that area.

Investigations on the formation and development of ice covers on water and on land by WILHEIT et al. (1972), GLOERSEN et al. (1973a), and SCHMUGGE et al. (1973) accumulated a considerable amount of knowledge on the properties, distribution and history of ice as measured by microwave radiometry. This allowed a fair understanding of the radiometric maps of the polar regions as obtained with the 19 GHz scanning radiometer on board of the Nimbus 5—satellite (CAMPBELL

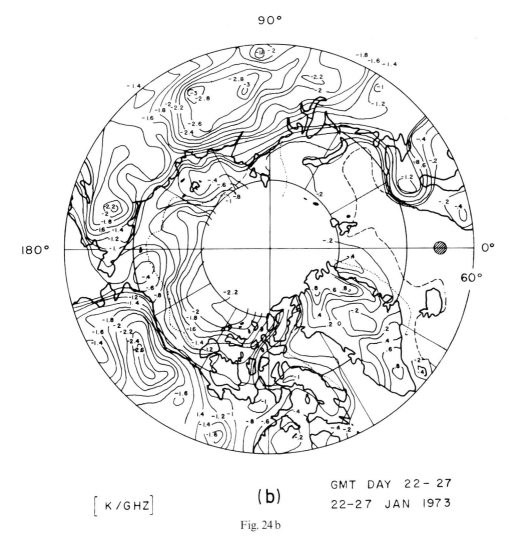

90°

180°   0°

60°

[K / G H Z]   **(b)**   GMT DAY 22- 27
                                       22-27 JAN 1973

Fig. 24 b

et al., 1973; GLOERSEN et al., 1973b). Figure 23 (see p. 364) shows two false color images of the Arctic region (resolution 30 km × 30 km) at two different stages of development of the ice shields around Greenland and other near polar areas. One

can observe a very low brightness temperature of the Greenland ice cap comparable to that of the open sea. The strong variation (amounting to 50 degrees) across
the ice cap with the highest emittance corresponding to the highest elevations, are
thought to be due to emissivity variations probably induced by variations of the
small scale ice stands, analogous to the differences of first year and multiyear ice
(CAMPBELL et al., 1973). The lower emissivity of the latter may partly be caused by
recristallization due to seasonal thawing and refreezing. The central Arctic Ocean
is covered by multiyear ice stretching broad tongues into the eastern Beaufort
Sea. A more detailed study on the discrimination of first- and multi-year sea-ice in
these regions is given by RAMSEIER et al. (1974). As the winter progresses, the
formation of the broad ice area east of Greenland can be observed. The zones of
first-year ice north of Asia are believed to result from the large fresh water influx
in a shallow structured sea yielding low surface salinities (CAMPBELL et al., 1973).

The whole scale of snow and ice conditions has also been observed by the
second Nimbus 5 microwave experiment with a coarse nadir viewing ground
resolution of 200 km × 200 km but a multispectral receiving facility (KUENZI et al.,
1976). A discrimination between the signatures of sea ice, land ice and snow
covered land has been derived by using the mean brightness temperature and the
spectral gradient of the brightness temperatures of the two receiver channels at
22.2 GHz and 31.4 GHz, respectively. Sea ice and land snow generally exhibit low
spectral gradients (0 to +2.5 K/GHz) strongly correlated with the mean brightness temperature of 190–250 K, while land ice exhibits spectral gradients of −0.5–
+1.2 K/GHz and mean brightness temperatures between 160 K and 230 K,
which are in general not correlated.

Figure 24 a and b present the mean brightness temperature and the spectral
gradient respectively of the same polar region (down to 60° latitude) as in the
maps of Fig. 23, taken at the same season as Fig. 23 a. The central region is missing
because of the inclination of 81° of the Nimbus 5 orbit and the limitation to Nadir
of the microwave Spectrometer.

In ground-based multi-wavelength experiments, EDGERTON et al. (1971 b) used
the "snowpack water equivelant" (total mass per area)

$$M_a = \int_0^d \varrho(z) dz$$

with $\varrho(z)$ the mass density and d the depth of snow above the underlying media,
as the characteristic parameter for run-off forecasting. Their experiments in the
wavelength range between 0.8 cm and 21 cm on increasing amounts of snow over
a frozen soil revealed a large penetration depth into dry snow (e.g. 1.6 m at 8 mm
wavelength and many meters at 21 cm wavelength), a pronounced decrease of the
brightness temperature (by 50 K and 8 K at 8 mm and 6 cm respectively for a
snowpack water equivalent increasing from 0 to 40 cm), thus confirming a direct
relationship between brightness temperature and water equivalent of dry snow.
On moist snow the effect of liquid water causes oscillations of the brightness
temperature with increasing water equivalents (most pronounced at the longest
wavelengths). These oscillations are presumably due to interferences within the
snowpack.

It is important that the inset of the melting process results in a very distinct increase of the brightness temperature (e.g. by about $70°$ at $\lambda = 2.2$ cm with only about one percent liquid water), followed by a slight decline with increasing water content (MEIER and EDGERTON, 1971). Thus the cyclic melting and freezing due to the diurnal temperature variations are detected very clearly (see also KENNEDY and SAKAMOTO, 1966).

Radiometric images taken at 9 mm wavelength of a rural snowcovered site show that the spring melting snow image appears to be a negative of the spring frozen-snow and the winter images, due to the presence of liquid water on the terrain and the particular response in the image presentation (MOORE and HOOPER, 1974).

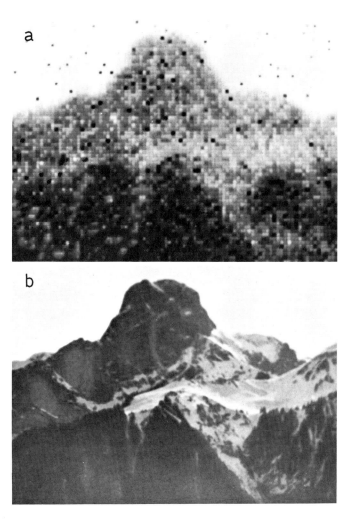

Fig. 25. (a) Thermal image (wavelength 3 mm) and (b) optical identification of a partly snowcovered mountain at about 7.5 km distance. (SCHAERER and SCHANDA, 1974)

Figure 25 demonstrates the feasibility of mapping areas of fresh snow of modest humidity as fallen in the early winter.

The usefulness of a combination of active and passive microwave methods in snow determinations is implicitly demonstrated by WAITE and MACDONALD (1970). In interpreting their radar returns they show that the radiative behavior of snow packs depend not only on the humidity but to a comparable amount on the granular texture, which results from the combination of snow accumulation, melting, compaction and refreezing. These processes are particularly important in alpine environments where granular snow (or firn) comprises the major portion of the snow fields and contributes significant amounts of water for stream flow. Therefore the determination of their volume and areal distribution are important to the hydrologist.

6.6 Investigations of the Soil, Vegetation and Geological Features

The treatment of the emissive properties of natural and man-made materials in Section 6.3 can only give some basic idea of the relations between the natural parameters of a given land area and its emissive properties. Due to the complexity of natural surfaces (various media, mixtures and degrees of roughness respectively) which have to be discriminated, even a big library of emission parameters (as: spectral and polarisation behavior and dependance on the look angle of the emission coefficients) would hardly be sufficient to characterize unambiguously all natural and man-made media and their combinations encountered in the environment. Therefore the determination of the natural parameters (in terms of the natural-science man) can only be realized by preselecting an expected range of possible parameter values or/and by an adaptive measuring procedure.

In the following a few particular fields of investigations and some—more or less isolated—individual studies will briefly be reviewed.

The first satellite (Cosmos 243) measurements of land surfaces, in particular the humidity of the soil at innundated areas of big rivers, were reported by BASHARINOV et al. (1971) and in a later version (Cosmos 384) the microwave experiments have been combined with a thermal infra-red channel (BASHARINOV et al., 1972).

A laboratory measurement of the complex permittivity of sandy and loamy soils (LESHCHANSKII et al., 1973) yielded the following information: the permittivity increases from about 3 for dry soil to about 11 at 18% moisture content at 8 mm wavelength, or to about 28 at 25% moisture for 226 cm wavelength. At the same time the attenuation factor is quoted to become about 5000 db/m at 8 mm wavelength, about the same for sandy as for loamy soil, and about 60 db/m at 226 cm wavelength for sandy and only about 8 db/m for loamy soil. It is thought that in sandy soils the water molecules are determining the attenuation, while in loamy soils the ions are taking over and the attenuation due to water becomes less important.

The moisture dependence of the brightness temperature at wavelengths of one to a few centimeters is given by BASHARINOV et al. (1974a, b) to be 3–4 K per %

moisture content for sandy and loamy soils, near 2 K per % for arable areas and 1 K per % or less for corn covered or mown areas.

The determination of the vertical moisture profiles by the brightness temperatures at various wavelengths has been achieved by POE and EDGERTON (1972), using the different penetration depths of the different wavelengths at various look angles, and calculating the effective permittivity from the Wiener mixing formula (Section 6.3). Good agreement with the direct moisture measurements has been achieved.

Various Form numbers F are to be used (POE and EDGERTON, 1972) in the mixing formula:

For the frequencies 1.42 and about 5 GHz applies the same range of Form numbers $F = 32$–64, while for 37 GHz the range $F = 16$–36 has to be used; only the lower frequencies 1.4 and 5 GHz exhibit a statistical meaningful agreement between measured data and model predictions, while less agreement of the shorter wavelength can be explained by near surface moisture changes, vegetational effects and roughness. For substantial amounts of moisture (in excess of 14–21%), penetration through alfalfa and wheat (25 cm and 15 cm tall, respectively) was noted at the longer wavelengths, but not observed with discernible consistancy at higher frequencies.

A soil moisture experiment at four wavelengths between 0.8 and 21 cm on a spectrum of samples starting off with sand, sandy loam, clay loam, until silty clay loam tracing through the soil-texture triangle (SCHMUGGE et al., 1974) revealed that the longer wavelengths (6–21 cm) have greater sensitivity to soil moisture. There is little change in the emission from soils with moisture contents less than 10–20%, and above this point there appears to be a linear decrease at a rate of about 2 K/% soil moisture.

The value of the knee in the curve depends on the soil type. Surface roughness and vegetative cover decrease the ability to sense soil moisture at the shorter wavelengths. The laboratory microwave method for determining the complex permittivity as a function of moisture content in various soils which can be used for the calculation of the emission has been described by GEIGER and WILLIAMS (1972).

Among the considerable attention that remote detection of soil moisture on a synoptic basis has received because of its importance to hydrologists and agriculturists, ROUSE et al. (1974) acquired measured data for developing analytical models of microwave emission from bare and vegetated soil. The factors: soil type, moisture profile, surface roughness and vegetative cover affect the emissivity e.g. moisture content decreases the brightness temperature at a rate of 1.5–2 °K/%. But vegetation cover masks the soil moisture dependence by a proper transmission coefficient, assuming the canopy as a homogeneous layer. Soil plots of 15×50 m^2 of controlled conditions were used for the experimental verifications at 1.4 and 10.7 GHz. Moisture and temperature profiles have been measured to a depth of 18 cm.

The moisture dependence of the brightness temperature is more pronounced at 1.4 GHz than at 10.6 GHz, and more for horizontal than for vertical polarization, e.g. decrease from 300 K to about 200 K for a moisture contents' increase from zero to about 30%.

For the investigation of vegetation a combined Radiometer-Scatterometer (MOORE and ULABY, 1969; ULABY et al., 1972) has been developed for achieving more experimental parameters (backscatter in addition to emission, a large number of frequency channels, dual polarisation and all measurements as a function of the look angle) for a wider range of interpretational possibilities. Agricultural targets as corn, sorghum, soybeams and alfalfa were recorded during their development and under all natural conditions (ULABY, 1974b). The RADSCAT system at a single frequency of 13.9 GHz has been used as an experimental space borne sensing system on board of Sky-Lab (MOORE et al., 1974), and soil moisture variations of a large test area have been investigated (ULABY et al., 1974).

For the discrimination of various rough surfaces or various degrees of roughness the careful investigation of the polarization dependence is of great importance; not only a pair of orthogonal linear polarizations, but the degree of smoothness of this dependence gives indications on the shape of the deviation (plants) from an emitting flat plane (MARTIN, 1972).

A balloon borne application of an 8 mm wavelength radiometer for the investigations of peats and forest burning zones have been reported by ARTEMOV et al. (1973). The temperature increase of burning peat piles and burning areas of peat swamps reached 300 and 400 degrees, respectively. The feasibility of detecting areas of burning forests, swamps etc. covered by a heavy smoke screen has been demonstrated convincingly. An investigation of the radiation temperatures of various types of soil and canopies in the wavelength-range of 0.4–1.2 mm has been reported by ISKHAKOV et al. (1974).

The microwave radiometry as an integral part of a system sensing the biosphere (composition, distribution and condition of plants and soil cover; the development and use of plants, soil and water-resources for agricultural and forestry management and protection) from space has been studied by VINOGRADOV (1973). It is the medium to large scale all weather mapping of vegetation and its dynamics, of the primary productivity of pasture vegetation and of agricultural crops, of the soil-moisture, forest fires and wind erosions of the soil which is regarded as the field of potential microwave contributions. The feasibility considerations of microwave investigations of the biosphere are preferably published with the exclusive regard of Radar techniques (e.g. MOORE and SIMONETT, 1967).

The usefulness of passive microwave imaging for urban and rural terrain analysis has been studied by a combined analysis of careful ground-truth determination and comparison of the microwave imagery with conventional photography and satellite imagery (BRUNNELLE et al., 1974). The most promising application results from the ability to generate information on water related resources and soil moisture. Its use for detecting urban features and structures also shows considerable promise, while for the differentiation of crop and vegetation types as well as monitoring water quality more detailed ground investigations are needed.

In experimental mapping of urban sceneries with ground based radiometers at wavelengths of 3 cm and 3 mm with angular resolutions well below one degree (KUNZI and SCHANDA, 1968; SCHANDA et al., 1972; MENDIS, 1972), the feasibility of detecting and recognizing the large variety of features and details of single buildings and of the street traffic has been demonstrated.

The spectral discriminations offered by the Nimbus 5—microwave spectrometer has been used partly (the 22.2 and the 31.4 GHz channels) for the determination of the surface brightness temperatures over land areas where the total atmospheric absorption can be neglected, as over the large deserts of central Australia and the Sahara (KUENZI et al., 1974). The mean brightness temperatures are between 270–300 K at daytime and 10 degrees less for night time. The spectral gradient of the brightness temperatures between 31.4 and 22.2 GHz is close to zero, typically 0.1 at daytime and decreases for nighttime by approximately 0.3. For calculated penetration depths in sand of 9–11 cm and 5–7 cm for the lower higher frequency channel resp. and assuming that surface reflectivities (about 0.05) do not change from day to night, one can explain the day to night variation by a change of the soil temperature profile. The night-time temperature gradient between about 6 and 10 cm depth can thus be estimated at 0.6 K/cm, while the daytime temperatures are about the same at these depths. Figure 26 (see p. 365) is a world map of the radio brightness as obtained by the Nimbus 5 scanning radiometer. In central Africa a surprisingly marked line is separating the warmer and vegetated south (midsummer in the southern hemisphere) from the desert and savannah regions. However, this particular presentation does not permit an unambiguous interpretation of the surface features (equal presentation of the Australian desert and the South American tropical forest), mainly due to the arbitrary gradiation. Rain bands along the intertropical convergence zone (light blue above the equator) can be recognized convincingly.

Because of the limited penetration depths of microwaves into soil—in particular into moist soil—the detection of geological phenomena by passive radiometry is restricted to few specific applications and to the long wavelength range (decimeter and even meterwaves) where the angular resolution becomes extremely poor.

HRUBY and EDGERTON (1971) reported on the feasibility of detecting subsurface voids such as caverns and tunnels of the near-surface soil moisture. Large microwave temperature anomalies were observed due to the moisture patterns, but no unambiguous correlation between the microwave temperature contours and the subsurface voids was determined.

Field observations—supplemented by laboratory investigations—of the emission by igneous and sedimentary rocks, and both metallic and non-metallic ore deposits at wavelengths between 0.8 and 21 cm (EDGERTON and TREXLER, 1970) yielded correlations between the computed emissivities and the physical characteristics of the minerals as surface roughness (microrelief, texture etc.), moisture content and specific gravity. No convincing direct association was noted between chemical compositions and emissivities. Multiwavelength data permit an evaluation of the effects of surface roughness. For outcrop materials a qualitative correspondence between decreasing emissivity at low antenna view angles and increasing specific gravity—in the 21 cm and despite the rough nature of outcrops even in the 0.8 cm, data—has been observed. The emissivities at 30° from nadir for 8 mm wavelength are between 0.8 and 0.97 for vertical polarization and between 0.7 and 0.96 for horizontal polarization of the various rocks and minerals of the field tests. Figure 27 shows the small differences of the emission coefficients among a considerable number of minerals.

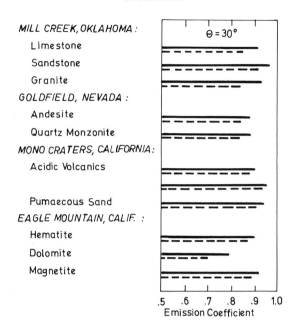

Fig. 27. Emissivities of various rocks and minerals at 0.81 cm wavelength, calculated from radiometric measurements (EDGERTON and TREXLER, 1970). Solid and dashed bars for vertical and horizontal polarization, respectively

Radiometric profiles across a portion of the San Andreas Fault Zone show close correlation between low emissivities and high near-surface moisture content values associated with seepage along the fault zone. This result illustrates the potential use of microwave radiometry for interfering surface structure from the resultant moisture maps. BLINN et al. (1972) have investigated the ability of detecting minerals of various kinds and other geologic materials from their emissions at 21, 2.8, and 0.9 cm wavelength using the differences of the respective penetration depths in controlled circumstances.

Recommendations for feasibility studies of passive microwave sensing of geological features have been formulated by PARASNIS (1971). From the Geophysicist's point of view, supplementary field investigations have to include:

— pre-cambrian shield areas, including lakes in these areas,
— outcropping contacts between rocks and outcropping faults in sedimentary and volcanic areas,
— phosphatic, bauxitic and lateritic deposits,
— dried-up deltaic tracts,
— areas of thermal springs,
— arid regions and permafrost regions and glaciated regions.

Additionally, an overview of European test areas where the above features can be studied are given.

6.7 Investigations of the Atmosphere and of Meteorological Features

6.7.1 Sounding of Atmospheric Constituents from Molecular Line Radiation

As explained in Section 6.1.3 the line radiation in the millimeter and submillimeter spectrum due to molecular transitions—mainly between rotational states—offers the possibility for remotely detecting the presence of trace gases or for determining the height profiles of the stronger constituents up to the mesosphere. The advantages of remotely probing the atmosphere—notably the upper atmosphere—are the avoidance of the difficult access for *in situ* measurements, the avoidance of any disturbing effect on the atmospheric conditions by the presence of equipment and the possibility of continuous measurements.

The line complex of O_2 at about 6 mm wavelength has been utilized by various investigators for ground-based experiments to determine the temperature profile of the lower atmosphere (up to about 10 km), as e.g. WESTWATER (1971) and (1972), MINER et al. (1972) and SNIDER (1972). The techniques of either multifrequency sensing or angular scanning as a tool for the determination of the low altitude temperature profile due to the inversion method have been proven to deliver r.m.s. accuracies of about 1 K as compared to radiosonde measurements. In particular, a combination of both methods yields a profile recovery of high accuracy. The analysis of several sets of fixed-frequency variable angle radiometer data has indicated the potential use of ground-based microwave measurements in airpollution forecasting (WESTWATER, 1971).

THOMPSON (1971) used the effect of water vapor on the velocity of electromagnetic signals through the atmosphere which causes differential phase delays of two signals at different, properly chosen frequencies, to develop and demonstrate a radio-optical dispersometer. With operational wavelengths of 3.1 cm and 6328 Å he claims to measure the variations of the averaged water vapor content along a path of several hundred km.

The measurements of stratospheric parameters or mixing ratios is hardly possible from the ground.

Balloon borne experiments for the observation of the microwave emission by molecular oxygen of the terrestrial atmosphere have been reported by BARRETT et al. (1966) and by LENOIR et al. (1968). Another balloon-borne radiometer with nine frequency channels on the O_2-line complex pointing downward from an altitude of 40 km to the ground has been used by MISME and ROBERT (1968) to sound the temperature profile in the height range of 17.8–37 km. By a proper choice of the frequency channels, the weighting functions allowed only the measurement of the stratospheric sections and avoided any significant effect due to the radiation temperature of the ground. The accuracy of this downward sounding experiment is about 1–5 degrees K for the various bands (taking a signal to noise ratio of 4). A special property of a balloon flight—which introduces some uncertainty to the experiment—is the un-controlled horizontal motion.

A more promising step in the sense of a global surveillance of the upper atmosphere with a well-defined repetition rate of the measurement is the Nimbus-5 experimental microwave spectrometer (NEMS), (STAELIN et al., 1973a, b).

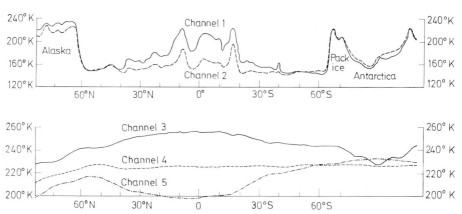

Fig. 28. Traces of the 5 channels of the Nimbus 5 Microwave Spectrometer during one half orbit on December 23, 1972, from Alaska through the Pacific Ocean to Antarctica (Staelin, private communication)

The five channels are centered at: the water vapor line (22.235 GHz), in the "window" range at 31.4 GHz as a reference channel and at three positions on the edge of the oxygen absorption band 53.65, 54.9, and 58.8 GHz, related to layers centered at 4, 11, and 18 km respectively, according to the weighting functions. The measurements are made continuously towards nadir with a 200 km resolution. Some general features of the global pattern of atmospheric temperatures are shown in the orbit given in Fig. 28. The temperature of the troposphere (channel 3) is cooler at the poles, whereas the stratospheric channel 5 indicates lower temperatures in the equatorial regions except for a cold region near Alaska. A surprising result is the observed lack of significant lateral discontinuities in the brightness temperatures of the channels 3, 4, 5 attributable to the terrestrial temperature or to clouds. Only in the intertropical convergence zone, high dense clouds decrease the brightness temperature of channel 3 by 0–4 K. These variations on an otherwise smooth temperature curve induced by this zone are extending over 3–5° of latitude and are sometimes spatially unresolved. Channels 4 and 5 respond exclusively to the atmospheric temperature profile according to their weighting functions and are unaffected by clouds or the terrestrial surface.

The atmospheric water vapor content is sensed by channel 1. Because of the small optical depth of this line, the surface structuring measured with this channel is comparably significant as it is in the "window" channel 2. Clearly marked are the sea-land boundaries and the regions of high humidity over the oceans. The radiation temperature of the sea surface, 150 K according to an emissivity of about 0.5, allows detection of the much warmer atmospheric water vapor in emission. The emissivities of the landmasses of 0.7 to 1 prevent the unambiguous detection of atmospheric water vapor over these regions because of the warm background. Liquid water in clouds and rain exhibits nonresonant absorption approximately proportional to the square of the frequency and appears also in emission over the sea. These two phases of water can be distinguished because water vapor produces a two to three times larger response in the water vapor

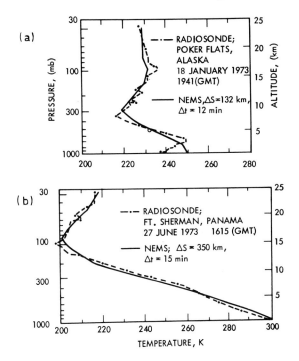

Fig. 29 a and b. Atmospheric temperature profiles as obtained by the inversion of the Nimbus 5 Microwave Spectrometer (Oxygen channels) measurements over (a) a polar and (b) a tropical latitude. Comparison with Radiosonde measurements (WATERS et al., 1975 a)

channel 1 than in the window channel 2, whereas liquid water produces a response about two times greater in the higher frequency channel 2 than in 1.

The increased water vapor in the tropical zone over the Pacific Ocean is clearly distinguished (see also Fig. 31).

The liquid water content of the atmosphere, which is roughly proportional to the deviation of the window channel 2 from its nominal ocean values, reaches high values in two rain bands at the edges of a humid region (see also radio brightness map Fig. 26). The inferred abundance of water vapor typically ranges from 0 to 5 g/cm^2 and compares well with nearby radio sonde data (errors less than 0.4 g/cm^2), and the liquid water values are ranging up 0.2 g/cm^2. The reversed temperatures over central Antarctica as compared to central Alaska (over central Greenland the same reversal has been observed) should provide new information about the distribution of various types (ages) of ice.

Figure 29 presents the temperature profiles over an arctic (winter) place and a tropical place respectively, as obtained by the inversion of the three oxygen-channels data.

In a payload consideration for spacelab missions, BOLLE (1974) is reviewing the state of the art for the monitoring of trace constituents by space techniques, and more specifically he reviews the experiments to be used in Nimbus G and gives quantitative estimates of the detection limits.

The significance of the stratospheric ozon shield for the biological equilibrium on earth has been stressed (RATNER and WALKER, 1972) because of the potential threat by supersonic air planes (CRUTZEN, 1972). Investigations on the production and annihilation processes of ozone under stratospheric conditions need not only a careful observation of the vertical and lateral distributions and the seasonal changes, but also involve the correlated study of various stratospheric gases which are participating in the photochemical reactions or are potential candidates for those processed (DÜTSCH, 1973).

A successful detection and line-profile resolution of the $4_{0.4} \rightarrow 4_{1.3}$ transition of atmospheric ozone at 101.737 GHz has been reported by CATON et al. (1968) in an earth-based remote probing experiment. The detection of the $6_{0.6} \rightarrow 6_{1.5}$ transition at 110.836 GHz and some estimation of the height distribution by the measured line profile has been given by SHIMABUKURO and WILSON (1973). An earth-based experiment using the 101 GHz line to detect mesospheric ozone, described by RUDZKI (1972) has been without success. This result does not support the prediction by HUNT (1965) of a night-time increase of the mesospheric ozone density to around $5 \cdot 10^{10}$ cm^{-3} from a day time concentration of about 10^9 cm^{-3}. It is concluded that the reaction rates are so poorly known that a positive statement on the night-time increase of mesospheric ozone based on the equations of the photochemical theory is unjustified. This experiment also indicates that the water vapor mixing ratio might be about $2 \cdot 10^{-4}$ at 68 km, dropping to $5 \cdot 10^{-6}$ at 80 km (HESSTVEDT, 1964) and correspondingly yielding an ozone number density of only $2 \cdot 10^6$ at 70 km without significant diurnal variation. But before more measurements are made of the water vapor and the ozone concentrations above 50 km, it is only speculation to say that the low water vapor mixing ratios (around 10^{-5}) at 70 km account for the low mesospheric ozone concentration.

An approach to resolve this question is the proposal by WATERS (1972) to observe the water vapor and the ozone due to lines at 183.3 GHz ($2_{2.0} \rightarrow 3_{1.3}$ transition) and 184.4 GHz ($10_0 \rightarrow 9_1$ transition), respectively. These lines have sufficient intensity for the determination of the height profiles and their frequencies are close enough to allow a single high frequency part of the radiometer front end. But because of the opacity of the lower atmosphere around these frequencies, the measurement is proposed from an airborne platform at 12 km altitude. For the various assumptions of the mixing ratio profiles at altitudes up to 100 km, sufficiently high brightness temperatures due to these lines are predicted for various elevation angles to allow accurate determinations of the profiles.

A feasibilty study for the measurement of the stratospheric and mesospheric carbon monoxide concentration due to the 115 GHz line of the J_{0-1}-transition by FULDE and SCHANDA (1974), basing on the mixing ratios by WOFSY et al. (1972) yielded a rather unfavorable prediction. However the successful measurement of this line recently by WATERS et al. (1975b) seems to push the mesospheric mixing ratio up by about one order of magnitude. Another calculation of the brightness temperatures due to the strongest O_3 lines between 100 and 200 GHz, namely at 110, 142, and 195 GHz and due to N_2O at 150.7 GHz has been performed (FULDE et al., 1974). The calculations suggest that ozone, nitrous oxide, water vapor and oxygen can be studied by virtue of their microwave spectra in the above mentioned band, and they confirm that carbon monoxide would be difficult to analyze. For even shorter wavelengths—the submillimeter region—many more constituents of the atmosphere exhibit transition lines, but the atmosphere becomes very opaque and only high altitude observations have good chances. Quasi-optical techniques have to be used because of the technological limitations of microwave methods at these short wavelengths. HARRIES (1973) reports on determinations of the stratospheric concentrations of H_2O, O_2, O_3, HNO_3, N_2O, NO_2, and SO_2 by using an airborne Michelson interferometer with phase modulation together with a helium-cooled Rollin In Sb-electron-bolometer having a spectral response in the 300 μm to 3000 μm range. The interferometer has a resolution of 0.067 cm^{-1}. The following global densities (number of molecules above the flight level of the SST-"Concorde" in a column of 1 cm^2 crosssection) were obtained: $7.7 \cdot 10^{18}$ for H_2O, $1.2 \cdot 10^{19}$ for O_3, $7.2 \cdot 10^{15}$ for HNO_3 and $7.0 \cdot 10^{17}$ for N_2O measured at the tropopause.

6.7.2 Observation of Meteorological Features

The frequency dependent attenuation by clouds and precipitation in the short wavelength part of the microwave spectrum has been recognized as a tool for the remote observation of these meteorological features, and the whole field of weather Radar has emerged from it. A first comprehensive study on the feasibility of using passive microwave methods has been done by KREISS (1968), and the possibility to obtain data of the water phase, the density and the particle-distribution in clouds has been investigated. In the frequency range 15–30 GHz (around the water vapor line at 22 GHz), the increase of the brightness temperature for a radiometer looking downward (over oceans) is given to be about 20 K, 35 K and some 70–80 K for cloud thickness of 1320 m, 2380 m and 6560 m respectively assuming equal liquid water density of 1 gram per cubicmeter (KREISS, 1969).

This fairly agrees with an independent study on the increase of the zenith brightness temperature due to thickness and water content of clouds for a ground based measurement (TOONG and STAELIN, 1970), which is found to be e.g. from about 50 K to 130 K at 30 GHz for 0.6 g/m^3 density and 3 km thickness. Common to both studies is the result of the separability of the resonant absorption of water vapor and the nonresonant absorption of the liquid phase in clouds. A composite picture of a cold-front assembled from the microwave data agrees with typical frontal characteristics, demonstrating the ability of this procedure to yield reasonable estimates of water vapor and cloud parameters (TOONG and STAELIN, 1970). Except in heavy rainfall, the spectra remain relatively unaffected by scattering and absorption effects of rain.

BASHARINOV et al. (1971) are giving an accuracy of 10% for the inference of the atmospheric humidity from microwave water vapor absorption and note the association of the radio brightness temperature shapes over oceans (from space) with rain clouds and precipitation zones. An increase of the total water vapor (column) content in clouds up to 0.3 g/cm^2 has been estimated.

RABINOVITCH et al. (1972) conclude from calculations that short cm waves would be most useful to detect radiometrically hydrometeorological formations of the atmosphere and intensities of precipitations. The correspondence of passive and active microwave results with pluviographic measurements has been demonstrated in a coordinated experiment (RABINOVITCH and SHTSHUKIN, 1972).

The effect of scatter by precipitation which is estimated to be small on the brightness temperature of clouds has been studied by CRANE (1972).

In the Nimbus 5 electrically scanning microwave radiometer experiment, the frequency 19.35 GHz was chosen to make the instrument sensitive (2 K) to meteorological features. The ground resolution of 30 km at nadir which is enlarged to about 45 by 165 kms² at 50° from nadir is sufficient for the imagery of large

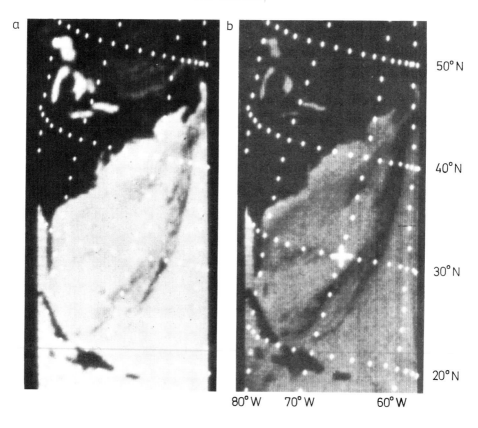

Fig. 30a and b. Maps of the east coast region of the US (Great Lakes top left, Nova Scotia top right) down to the Carribean (Puerto Rico bottom) obtained by the Nimbus 5 Scanning Microwave Radiometer. (a) Brightness temperature dynamic range 190–250 K, (b) range 130–200 K. The darker image (b) represents gradations in the moisture (liquid and vapor), the brighter image (a) (higher brightness temperature range) represents precipitation over the ocean. The rows of dots indicate lines of geographic longitudes and latitudes and are correcting for effects of varying viewing geometry (WILHEIT et al., 1973)

meteorological features. The brightness temperature over oceans can be formulated approximately (WILHEIT et al., 1973) as

$$T_B[K] = 125 + V \cdot 6.8 + L \cdot 300$$

where V is the net water vapor in a column (in g/cm^2) and L is the net non-raining cloud liquid water content in g/cm^2. Rain droplets, because of their larger size have enhanced absorption effect. This enhancement, combined with the fact that a raining cloud typically contains much more liquid water than a non-raining cloud enables the delineation of rain over the oceans. But ice clouds produce much less effect. The maps of Fig. 30 are shown in two different dynamic ranges, the darker

Fig. 31a and b. Maps of the tropical region of the earth between $\pm 33°$ latitudes (African east coeast left, African west coast right) obtained by the 22 GHz and 31 GHz channels of the Nimbus 5 Microwave Spectrometer. (a) Total water vapor above the oceans, contour lines in grams per square centimeter. (b) Total liquid water content above the oceans, contours in milligrams per square centimeter (STAELIN et al., 1975)

corresponds to roughly 130–200 K and the lighter to the range 190–250 K, and for both maps dark areas correspond to higher temperatures. In the 130–200 K map the gradiations of the moisture (liquid and vapor) of the atmosphere over the ocean can be recognized. A feature to be recognized in the lower end of the 190–250 K map requires about 10 g/cm² of water vapor or 0.2 g/cm² of liquid water which is improbable in a nonraining situation. Snow was indicated all along the coast of New England and Nova Scotia from surface weather observations on that day. The map (Fig. 30) shows rain along a front from as far as Puerto Rico and is heaviest just south of Nova Scotia.

The data of the earlier mentioned multichannel spectrometer aboard Nimbus 5 allow estimates of thickness of various atmospheric layers (STAELIN et al., 1973 b). Figure 31 are maps of the total water vapor contents and liquid water contents in the tropical zones of the world, which are derived from the data of the water vapor and the "window"-channels.

Passive radio sensing of a very different type—the observation and analysis of radio noise from thunderstorms—has been described by HORNER (1964), HEYDT et al. (1967), GUPTA et al. (1972), and HARTH (1974). They are investigating methods for the classification and location of lightnings by measuring the very low frequency pulses of the first leader of a lightning flash to the ground and by studying the propagation conditions of the atmospherics in the Earth-Ionosphere waveguide. The step lengths range between the extreme values of 3 and 200 m and the pause times lie between 30 and 125 sec. Narrow band (e.g. 400 Hz) tuned amplifiers at the frequencies in the 10 kHz range are in use to observe the spectrum of the lightning and in particular of stepped leaders.

Acknowledgement. The author wants to thank Prof. D. STAELIN (MIT) and Drs. J. W. WATERS (JPL, previously MIT), K. F. KUENZI (U. Bern, previously MIT), P. GLOERSEN, W. J. WEBSTER, T. T. WILHEIT (all NASA-GSFC) for placing material of recent investigations at his disposal. He also acknowledges herewith the numerous valuable results achieved by his collaborators G. SCHAERER, R. HOFER and J. FULDE, used in this chapter, and for the fruitful discussions, with them. The author is endebted to Prof. K. P. MEYER, head of the Institute of Applied Physics, University of Berne, for enabling the compilation of this review. Last not least many thanks to Mrs. ROTH, who typed the manuscript.

References

ARTEMOV, V. A., BASHARINOV, A. E., BORODIN, L. F., BULATNIKOV, V. L., EGOROV, S. T., MISHENEV, V. F.: Radiobrightness characteristics of various natural formations. Abstr. XXIV Internat. Astronautical Congr., 30–32, Baku, USSR (1973).
ATTEMA, E. P. W., DEN HOLLANDER, L. G., DE BOER, TH. A., UENK, D., ERADUS, W. J., DE-LOOR, G. P., VAN KASTEREN, H., VAN KUILENBURG, J.: Radar cross sections of vegetation canopies determined by monostatic and bistatic scatterometry. Proc. 9th Symp. Rem. Sens. Env., 1457–1466, Univ. of Michigan, Ann Arbor (1974).
AU, B. D., KENNEY, J. E., MARTIN, L. U., ROSS, D.: Multifrequency radiometric measurements of foam and monomolecular slick. Proc. 9th Symp. Rem. Sens. Env., 1763–1773, University of Michigan, Ann Arbor (1974).
BALL, J. A.: The Harvard Minicorrelator. Inst. El. Electron, Eng. Trans. Instr. Meas. **22**, 193–196 (1973).

BARRETT, A. H., KUIPER, J. W., LENOIR, W. B.: Observations of microwave emission by molecular oxygen in the terrestrial atmosphere. J. Geophys. Res. **71**, 4723–4734 (1966).

BASHARINOV, A. E., BORODIN, L. F., SHUTKO, A. M.: Passive microwave sensing of moist soils. Proc. 9th Symp. Rem. Sens. Env., 363–367, Univ. of Michigan, Ann Arbor (1974 a).

BASHARINOV, A. E., BORODIN, L. F., SHUTKO, A. M.: Passive microwave sensing of moist soils. Proc. URSI meeting Microw. Scatt. Emiss. from the Earth, 131–136, Univ. of Berne, Switzerland (1974 b).

BASHARINOV, A. E., GURVITCH, A. S., GORODEZKY, A. K., EGOROV, S. T., KUTUZA, B. G., KURSKAYA, A. A., MATVEEV, D. T., ORLOV, A. P., SHUTKO, A. M.: Satellite measurements of microwave and infrared radiobrightness temperature of the earth's cover and clouds. Proc. 8th Symp. Rem. Sens. Env., 291–296, Univ. of Michigan, Ann Arbor (1972).

BASHARINOV, A. E., GURVITCH, A. S., IGOROV, S. T.: Features of microwave passive remote sensing. Proc. 7th Symp. Rem. Sens. Env., 119–131, Univ. of Michigan, Ann Arbor (1971).

VAN BEEK, L. K. H.: Dielectric behaviour of heterogeneous systems. In: BIRKS, J. B. (Ed.): Progress in Dielectrics, Vol. 7, pp. 69–114. London: Iliffe Books Ltd. 1967.

BENOIT, A.: Signal attenuation due to neutral oxygen and water vapor, rain and clouds. Microwave J. **11**, No. 11, 73–80 (1968).

BEN-REUVEN, A.: Transition from resonant to nonresonant line shape in microwave absorption. Phys. Rev. Letters **14**, 349–351 (1965).

BEN-REUVEN, A.: The meaning of collision broadening of spectral lines: the classical oscillator analog. Advan. Atom. Molec. Phys. **5**, 201–235 (1969).

BETTENCOURT, J. T.: Statistics of terrestrial millimeterwave rainfall attenuation. Preprint. Symposium Inter Union Commun. Radio Meteorology, Nice (1973).

BLINN, J. C., CONEL, J. E., QUADE, J. G.: Microwave emission from geological materials: Observations of interference effects. J. Geoph. Res. **77**, 4366–4378 (1972).

BLUM, E. J.: Some aspects of observational Radio-Astronomy in the millimeter wavelength range. Proc. 4th meeting Comm. Europ. Sol. Rad. Astron. In: SCHANDA, E. (Ed.): 31–36, Univ. of Berne, 1974.

BOLLE, H. J.: Payload considerations for Space Lab missions to study air quality and atmospheric effects on earth science data. ESRO-Report (1974).

BORN, M., WOLF, E.: Principles of Optics, 2nd Ed. Oxford: Pergamon Press 1964.

BREKHOVSKIKH, L. M.: Waves in layered media. London-New York: Academic Press 1960.

BRUNELLE, D. N., ESTES, J. E., MEL, M. R., THAMAN, R. R., EVANISKO, F. E., MOORE, R. P., HAWTHORNE, C. A., HOOPER, J. O.: The usefulness of imaging passive microwave for rural and urban terrain analysis. Proc. 9th Symp. Rem. Sens. Env., 1603–1620, Univ. of Michigan, Ann Arbor (1974).

CAMPBELL, W. J., GLOERSEN, P., NORDBERG, W., WILHEIT, T. T.: Dynamics and morphology of Beaufort sea ice determined from satellite, aircraft and drifting stations. Preprint X-650-73-194, NASA-GSFC, Greenbelt Maryland (1973).

CATON, W. M., MANNELLA, G. G., KALAGHAN, P. M., BARRINGTON, A. E., EWEN, H. J.: Radio measurement of the atmospheric ozone transition at 101.7 GHz. Astrophys. J. **151**, L 153–L 156 (1968).

CLAASSEN, J. P., FUNG, A. K., WU, S. T., CHAN, H. L.: Toward Radscat measurements over the sea and their interpretation. NASA Contr. Rep. 2328 (1973).

COLE, K. S., COLE, R. H.: Dispersion and absorption in dielectrics. J. Chem. Phys. **9**, 341–351 (1941).

COSGRIFF, R. L., PEAKE, W. H., TAYLOR, R. C.: Terrain Scattering properties for sensor system design. Engin. Exper. Station Bull., Ohio State Univ. **29** (1960).

CRANE, R. K.: Virginia precipitation scatter experiment: data analysis. NASA-GSFC Rep. X-750-73-55, Greenbelt, Maryland (1972).

CRUTZEN, P. J.: SST's—A threat to the earth's ozone shield. Ambio **1**, 41–51 (1972).

CUMMING, W. A.: The dielectric properties of ice and snow at 3.2 centimeters. J. Appl. Phys. **23**, 768–773 (1952).

DEIRMENDJIAN, D.: Complete microwave scattering and extinction properties of polydispersed cloud and rain elements. Report R-422-PR. The Rand Corporation (1963).

DICKE, R. H.: The measurement of thermal radiation at microwave frequencies. Rev. Sci. Instr. **17**, 268–275 (1946).

Dicke,R.H., Beringer,R., Kyhl,R.L., Vane,A.B.: Atmospheric absorption measurements with a microwave radiometer. Phys. Rev. **70**, 340 (1946).

Duetsch,H.U.: Recent developments in photochemistry of atmospheric ozone. Pure & Appl. Geophys. **108**, 1361–1384 (1973).

Edgerton,A.T., Meeks,D., Williams,D.: Microwave emission characteristics of oil slicks. Proc. Joint Conf. Sensing of Environm. Pollutants, AIAA paper No.71–1071, Palo Alto (1971a).

Edgerton,A.T., Stogryn,A., Poe,G.: Microwave radiometric investigations of snowpacks. Rep. 1285 R-4. Aerojet General Corp., El Monte, Calif. (1971b).

Edgerton,A.T., Trexler,D.T.: A study of passive microwave techniques applied to geologic problems. Report No 1361R-1. Aerojet-General Corp., El Monte, Calif. (1970).

Edgerton,A.T., Trexler,D.T., Poe,G.A., Stogryn,A., Sakamoto,S., Jenkins,J.E., Meeks,D., Soltis,F.: Passive microwave measurements of snow, soils and oceanographic phenomena. Techn. Rep. 6 SD 9016-6. Aerojet-General Corp., El Monte, Calif. (1970).

Edgerton,A.T., Woolever,G.: Airborne oil pollution surveillance system. Proc. 9th Internat. Symp. Rem. Env., 1791, Univ. of Michigan, Ann Arbor (1974).

Edison,A.R.: Calculated cloud contribution to sky temperature at millimeter wave frequencies. Nat. Bur. Standards, US Dept. Commerce, NBS Rep. 9138 (1966).

England,A.W.: Relative influence upon microwave emissivity of fine-scale stratigraphy, internal scattering and dielectric properties. Submitted to Pure and Appl. Geophys. (1975).

Fricke,H.: J. Phys. Chem. **57**, 934 (1953).

Fulde,J., Kuenzi,K.F., Staelin,D.H.: The microwave spectrum of the atmosphere between 100 and 200 GHz. Quart. Progr. Rep. 114, Res. Lab. Electronics, 68–73, Mass. Inst. Techn., Cambridge, Mass. (1974).

Fulde,J., Schanda,E.: On the detectability of atmospheric carbon monoxyde by microwave remote sensing. Proc. 9th Symp. Rem. Sens. Env., 465–469, Univ. of Michigan, Ann Arbor (1974).

Fung,A.K.: Theory of cross polarized power returned from a random surface. Appl. Sci. Res. **18**, 50–60 (1967).

Fung,A.K.: A review of rough surface scattering theories. The University of Kansas, CRES Techn. Rep. 105–10, Lawrence, Kansas (1971).

Fung,A.K., Axline,R.M., Chan,H.L.: Exact scattering from a known randomly rough surface, Proc. URSI meeting Microwave Scattering and Emission from the Earth, 227–238. University of Berne, Switzerland (1974).

Gaudissart,E.: Propagation tropospherique aux hyperfrequences. Hyp. Frequ. **8**, 173–182 (1971).

Geiger,F.E., Williams,D.: Dielectric constants of soils at microwave frequencies. NASA-GSFC Preprint X-652-72-283, Greenbelt, Maryland (1972).

Gloersen,P., Nordberg,W., Schmugge,T.J., Wilheit,T.T.: Microwave signatures of first-year and multi-year sea ice. J. Geophys. Res. **78**, 3564–3572 (1973a).

Gloersen,P., Schmugge,T.J., Chang,T.C.: Microwave signatures of snow, ice and soil at several wavelengths. Proc. URSI meeting Microw. Scatt. & Emiss. from the Earth, 101–112, Univ. of Berne, Switzerland (1974b).

Gloersen,P., Webster,W.J., Wilheit,T.T., Ross,D.B., Chang,T.C.: Spectral variation in the microwave emissivity of the roughened seas. Proc. URSI meeting Microw. Scatt. & Emiss. from the Earth, 11–16, Univ. of Berne, Switzerland (1974a).

Gloersen,P., Wilheit,T.T., Chang,T.C., Nordberg,W., Campbell,W.J.: Microwave maps of the polar ice of the earth. Preprint X-652-73-269. NASA-GSFC, Greenbelt, Maryland (1973b).

Goetz,F.W.P., Meetham,A.R., Dobson,G.M.B.: The vertical distribution of ozone in the atmosphere. Proc. Roy. Soc. London **A 145**, 416 (1934).

Gray,K.W., Hall,W.F., Hardy,W.N., Hidy,G.M., Ho,W.W., Love,A.W., van-Melle,M.J., Wang,H.: Microwave measurement of thermal emission from the sea. Proc. 7th Symp. Rem. Sens. Env., 1827–1845, Univ. of Michigan, Ann Arbor (1971).

Gupta,S.P., Rao,M., Tantry,B.A.P.: VFL spectra radiated by stepped leaders. J. Geophys. Res. **77**, 3924–3927 (1972).

Hamid,M.A.K.: Radiometric signatures of complex bodies. Proc. AGARD Conf., The Hague 21/1-21/27 (1974).

HARDY, W. N.: Precision temperature reference for microwave radiometry. Inst. El. Electron. Eng., Trans. **MTT-21**, 149–150 (1973).

HAROULES, G. G., BROWN, W. E.: The simultaneous investigation of attenuation and emission by the earth's atmosphere at wavelengths from 4 centimeters to 8 millimeters. J. Geophys. Res. **74**, 4453–4471 (1969).

HARRIES, J. E.: Measurements of some hydrogen-oxygen-nitrogen compounds in the stratosphere from Concorde 002. Nature (Lond.) **241**, 5392, 515–519 (1973).

HARTH, W.: The propagation of atmospherics. 5th Internat. Conf. Atmosph. Electricity, Garmisch-Partenkirchen, FRG (1974).

HASTED, J.: The dielectric properties of water. In: BIRKS, J., HART, J. (Eds.): Progress in Dielectrics, Vol. 3. New York: Wiley & Sons 1961.

HESSTVEDT, E.: On the water vapor content in the high atmosphere. Geofisiske publikasjoner, Geophys. Norveg. **25**, 3, p. 1 (1964).

HEYDT, G., FRISIUS, J., VOLLAND, H., HARTH, W.: Beobachtung entfernter Gewitterzentren mit dem Atmospherics-Analysator des Heinrich-Hertz-Instituts. Kleinheubacher Berichte **12**, 103–110 (1967).

VON HIPPEL, A. R.: Dielectrics and Waves. Cambridge, Mass.; MIT Press (1954).

HOEIJER, S.: A survey of wave scattering from rough surfaces. Rep. C 2486-El, FOA-2, Stockholm (1971).

HOEKSTRA, P., CAPPILLINO. P.: Dielectric properties of sea and sodium chloride ice at UHF and microwave frequencies. J. Geophys. Res. **76**, 4922–4931 (1971).

HOFER, R.: Streu- und Emissionseigenschaften natürlicher Materialien bei 3 mm Wellenlänge. Diploma Thesis, Univ. of Berne, Switzerland (1974).

HOFER, R., SCHANDA, E.: Emission properties of water surfaces at 3 mm wavelength, Proc. URSI meeting Microwave Scattering and Emission from the Earth, 17–23, Univ. of Berne, Switzerland (1974).

HOLLINGER, J. P.: Passive Microwave measurements of the sea surface. J. Geophys. Res. **75**, 5209–5213 (1970).

HOLLINGER, J. P.: Remote passive microwave sensing of the ocean surface. Proc. 7th Symp. Rem. Sens. Env., 1807–1817, Univ. of Michigan, Ann Arbor (1971).

HOLLINGER, J. P.: The determination of oil slick thickness by means of multifrequency passive microwave techniques. Naval Res. Lab., Interim Rep. 7110-1 (1973).

HOLLINGER, J. P., MENELLA, R. A.: Oil Spills: Measurements of their distributions and volumes by multifrequency microwave radiometry. Science **181**, 54–56 (1973).

HORNER, F.: Radio noise from thunderstorms. In: SAXTON, J. A. (Ed.): Adv. in Radio Research, Vol. 2, pp. 121–204. London-New York: Academic Press 1964.

HRUBY, R. P., EDGERTON, A. T.: Subsurface discontinuity detection by microwave radiometry. Proc. 7th Symp. Rem. Sens. Env., 319–325, Univ. of Michigan, Ann Arbor (1971).

HUNT, B. G.: A non-equilibrium investigation into diurnal photochemical atomic oxygen and ozone variations in the mesosphere. J. Atmos. Terr. Phys. **27**, 133 (1965).

ISKHAKOV, I. A., SOKOLOV, A. V., SUKHONIN, E. V.: Estimation of the apparent temperature of local objects and some earth's covers in the range of $6.66–25\,cm^{-1}$, Proc. URSI meeting Microwave Scattering and Emission from the Earth, 137–140. Univ. of Berne, Switzerland (1974).

KENNEDY, J. M., SAKAMOTO, R. T.: Passive microwave determination of snow wetness factors. Proc. 4th Symp. Rem. Sens. Env., 161–171, Univ. of Michigan, Ann Arbor (1966).

KISLYAKOV, A. G.: On the atmospheric transparency spectrum in the millimeter waveband. Infrared Phys. **12**, 61–63 (1972).

KRAUS, J. D.: Radio Astronomy. New York: McGraw Hill 1966.

KREISS, W. T.: Meteorological observations with passive microwave systems. Ph.D. Dissert., Univ. of Washington, Boeing Scient. Res. Lab. Doc. D1-82-0692 (1968).

KREISS, W. T.: The influence of clouds on microwave brightness temperatures viewing downward over open seas. Proc. Inst. El. Electron. Eng. **57**, 440–446 (1969).

KUENZI, K., MAGUN, A.: Statistical gain fluctuations of microwave amplifiers measured with a Dicke-radiometer. Z. Angew. Math. Phys. **22**, 404–412 (1971).

KUENZI, K., MAGUN, A., MAETZLER, C., SCHAERER, G., SCHANDA, E.: Passive microwave remote sensing at the University of Berne. Proc. 7th Symp. Rem. Sens. Env., 1819–1826, Univ. of Michigan, Ann Arbor (1971).

KUENZI, K., SCHANDA, E.: A microwave scanning radiometer, Inst. El. Electron. Eng. Trans. **MTT-16**, 789–791 (1968).

KUENZI, K. F., STAELIN, D. H., WATERS, J. W.: Earth surface emission measured with the Nimbus-5 microwave spectrometer. Proc. URSI meeting Microwave Scattering and Emission from the Earth, 113, Univ. of Berne, Switzerland (1974).

KUENZI, K., STAELIN, D. H., WATERS, J. W.: Snow and Ice surfaces measured by the Nimbus 5 Microwave Spectrometer, to be published in J. Geophys. Res. (1976).

LANE, J. A., SAXTON, J. A.: Dielectric dispersion in pure polar liquids at very high radio frequencies, the effects of electrolytes in solution. Proc. Roy. Soc. London **214**, 531–545 (1952).

LENOIR, W. B.: Remote sounding of the upper atmosphere by microwave measurements. Ph.D. dissertation M.I.T., Cambridge, Mass. (1965).

LENOIR, W. B.: Microwave spectrum of molecular oxygen in the mesosphere. J. Geophys. Res. **73**, 361–376 (1968).

LENOIR, W. B., BARRETT, J. W., COSMO PAPA, D.: Observations of microwave emission by molecular oxygen in the stratosphere. J. Geophys. Res. **73**, 1119–1126 (1968).

LESHCHANSKII, YU. I., LEBEDEVA, G. N., SHUMILIN, V. D.: The electrical parameters of sandy and loamy soils in the range of centimeter, decimeter and meter wavelengths. Rad. Phys. Qu. El. **14**, 445–451 (1973).

DE LOOR, G. P.: Dielectric properties of heterogeneous mixtures. Thesis, Leiden (1956).

LOVE, A. W., HIDY, G. M.: Sea temperature by microwave radiometry. NASA-AFFE Conf. (1971).

MACHIN, K. E., RYLE, M., VONBERG, D. D.: The design of an equipment for measuring small radio frequency noise powers. Proc. IEE **99**, 127–134 (1952).

MAETZLER, C.: Messung der Wärmestrahlung der Erdoberfläche im Mikrowellengebiet. Diploma-Thesis, Univ. of Berne, Switzerland (1970).

MAGUN, A., KUENZI, K.: Influence of statistical gain fluctuations of the high-frequency amplifier on the sensitivity of a Dicke-Radiometer. Z. Angew. Math. Phys. **22**, 392–403 (1971).

MARTIN, D.: Polarization effects with a combined Radar-Radiometer. 4th Ann. Earth Res. Progr., Vol. II, 54/1–54/12, NASA-MSC Houston, Texas (1972).

MATUSHEVSKIY, G. V.: Relationship between true and mean slopes of the wave covered sea surface. Izvestiya Ak. Nauk SSSR, Fizika Atmosferiy i Okeana, 404–415 April 1969.

MC LEISH, C. W.: A total power filter spectrometer for radio astronomy. Inst. El. Electron. Eng., Trans. Instr. Meas. **22**, 279–281 (1973).

MEEKS, M. L., LILLEY, A. E.: The microwave spectrum of oxygen in the earth's atmosphere. J. Geophys. Res. **68**, 1683–1703 (1963).

MEIER, M. F., EDGERTON, A. T.: Microwave emission from snow—a progress report. Proc. 7th Symp. Rem. Sens. Env., 1155–1163, Univ. of Michigan, Ann Arbor (1971).

MELENTYEV, V. V., RABINOVICH, Y. I.: Remote sounding of water surface conditions from aboard artificial satellites. Proc. COSPAR Space Res. Meeting, 105–106, Madrid (1972a).

MELENTYEV, V. V., RABINOVICH, Y. I.: Emission properties of natural surfaces at microwave frequencies. Proc. 8th Symp. Rem. Sens. Env., 217–227, Univ. of Michigan, Ann Arbor (1972b).

MELENTYEV, V. V., RABINOVICH, Y. I., SHUKIN, G. G.: Airborne measurements of the radio radiation of a wavy sea surface. Voeikov Main Geophys. Obs., Report No. 291, 34–39 (1972).

MENDIS, F. V. C.: Some observations of urban terrain and buildings by X-band radiometry. Dept. Electronic & Electrical Eng., Memorandum 436, Univ. of Birmingham, England (1972).

MINER, G. F., THORNTON, D. D., WELCH, W. J.: The inference of atmospheric temperature profiles from ground-based measurements of microwave emission from atmospheric oxygen. J. Geophys. Res. **77**, 975–991 (1972).

MISME, P., ROBERT, A.: Mesure de temperature de l'atmosphère par radiométrie en ondes millimétriques. Compt. Rend. Acad. Sci. Paris, Ser. B., 1434–1436 (1968).

MIX, R. F.: Electronically scanning microwave radiometers. Proc. 9th Symp. Rem. Sens. Env., 1649–1655, Univ. of Michigan, Ann Arbor (1974).

MOORE, R. K.: Physics of remote sensing. Proc. ESRO Summer School, Earth Res. (1973).

MOORE,R.K., et al.: Simultaneous active and passive microwave response of the earth.—The Skylab Rad Scat experiment. Proc. 9th Symp. Rem. Sens. Env., 189–217, Univ. of Michigan, Ann Arbor (1974).

MOORE,R.K., SIMONETT,D.S.: Radar remote sensing in biology. Bioscience **1967**, 384–390.

MOORE,R.K., ULABY,F.T.: The radar radiometer. Proc. Inst. El. Electron. Eng. **57**, 587–590 (1969).

MOORE,R.P., HOOPER,J.O.: Microwave radiometric characteristics of snow-covered terrain. Proc. 9th Internat. Symp. Rem. Sens. Env., 1621–1632, Univ. of Michigan, Ann Arbor (1974).

MOROZOV,P.T., CHOCHLOV,G.P.: The physical-chemical and electrical characteristics of ice within the polar circle of the Bering sea. In: WOEJKOV,Obs. Rep.: Preliminary Results of the Expedition "Bering". Leningrad (1973).

NASA (Office of manned space flight). Skylab Earth Resources Investigations, Superintendent of Documents. US Gov. Print. Off., Washington (1973).

NORDBERG,W., CONAWAY,J., ROSS,D., WILHEIT,T.: Measurements of microwave emission form foam covered, wind driven sea. J. Atmos. Sci. **28**, 429 (1971).

NORDBERG,W., CONAWAY,J., THADDEUS,P.: Microwave observations of sea state from aircraft. NASA preprint X-620-68-414 (1968).

OTT,R.H.: Scattering, propagation and refraction of electromagnetic and acoustic waves. In: DERR,V.E. (Ed.): Remote Sensing of the Troposphere, 8/1–8/22. Wave Propag. Lab., Boulder, Colo. (1972).

PARASNIS,D.S.: Identification, description and justification of passive microwave radiometric applications to Geology. In: HOPPE,G. (Ed.): Microwave radiometry and its potential applications to earth resources surveys, ESRO Report RAC-0-3-R 17 (1971).

PEAKE,W.H.: Simple models of radar return from terrain. Ohio State Univ. rep. 694-3 and 694-7, Columbus, Ohio (1957).

PEAKE,W.H.: Interaction of electromagnetic waves with some natural surfaces. Inst. Rad. Eng. **AP-7**, (special suppl.), 5324–5329 (1959).

PEAKE,W.H., RIEGLER,R.L., SCHULTZ,C.H.: The mutual interpretation of active and passive microwave sensor outputs. Proc. 4th Symp. Rem. Sens. Env., 771–777, Univ. of Michigan, Ann Arbor (1966).

PENZIAS,A.A., BURRUS,C.A.: Millimeter-Wavelength Radio-Astronomy techniques. In: GOLDBERG,L. (Ed.): Ann. Rev. Astron. and Astrophys., Vol.11, 51–72. Annual Review Inc., Palo Alto (1973).

POE,G., EDGERTON,A.T.: Soil moisture mapping by ground and airborne microwave radiometry. Proc. 4th Ann Earth Res. Progr. Rev., Vol.4, 93/1–93/23, NASA-MSC, Houston Texas (1972).

POE,G., STOGRYN,A., EDGERTON,A.T.: Microwave emission characteristics of sea ice. Report 1749R-2. Aerojet Electro Systems, Azusa, Calif. (1972).

PREISSNER,J.: Airborne microwave radiometric measurements at DFVLR Oberpfaffenhofen. Proc. URSI meeting Microwave Scattering and Emission from the Earth, 159–162, Univ. of Berne, Switzerland (1974).

RABINOVICH,J.I., SALMAN,E.M., SHIFRIN,K.S., SHTSHUKIN,G.G.: The possibility of determining the intensity of precipitations by the thermal radio radiation. Proc. Main Geophys. Obs. Voeikov, No.**291**, 57–62 (1972).

RABINOVICH,J.I., SHTSHUKIN,G.G.: Results of measurements of the radio radiation of precipitation. Proc. Main Geophys. Obs. Voeikov, No.**291**, 63–71 (1972).

RAMSEIER,R.O., GLOERSEN,P., CAMPBELL,W.J.: Variation in the microwave emissivity of sea ice in the Beaufort and Bering seas. Proc. URSI meeting Microw. Scatt. & Emiss. from the Earth, 87–94, Univ. of Berne, Switzerland (1974).

RATNER,M.I., WALKER,J.C.G.: Atmospheric Ozone and the history of life. J. Atmos. Sci. **29**, 803–808 (1972).

RODGERS,C.D.: Satellite infrared radiometer, a discussion of inversion methods. Clarendon Lab., Univ. of Oxford, England, Memo 66.13 (1966).

ROUSE,J.W., NEWTON,R.W., LEE,S.H.: On the feasibility of remote monitoring of soil moisture with microwave sensors. Proc. 9th Symp. Rem. Sens. Env., 725–738, Univ. of Michigan, Ann Arbor (1974).

RUDZKI, J. E.: Remote sensing of mesospheric ozone. Proc. 8th Symp. Rem. Sens. Env., 487–504, Univ. of Michigan, Ann Arbor (1972).

RYDE, J. W.: The attenuation and radar echoes produced at centimeter wavelengths by various meteorological phenomena. Proc. of Conf. Meteorological factors in radio wave propagation, Phys. Society, 169–189, London (1946).

SCHAERER, G.: Passive sensing experiments and mapping at 3.3 mm wavelength. Rem. Sens. Env. **3**, 117–131 (1974).

SCHAERER, G., SCHANDA, E.: Thermal imaging by a 3 mm-wavelength radiometer. Z. Angew. Math. Phys. **23**, 507–508 (1972).

SCHAERER, G., SCHANDA, E.: Deteriorating effects on 3 mm wave passive imagery. Proc. 9th Symp. Rem. Sens. Env., 1593–1602, Univ. of Michigan, Ann Arbor (1974).

SCHANDA, E.: Auflösung von Kontraststrukturen ausgedehnter Strahler durch Radiometer-Antennen mit Gauss'scher Charakteristik. Arch. El. Ueb. **20**, 569–572 (1966).

SCHANDA, E.: Messung des Emissionsvermögens mit einem Mikrowellen-Radiometer. Arch. El. Ueb. **22**, 133–140 (1968).

SCHANDA, E.: Graphs for Radiometer Applications. Elektronica en Telecommunicatie **4**, 51–55 (1971).

SCHANDA, E., HOFER, R.: Emissivities and forward scattering of natural and man-made material at three millimeter wavelength. Proc. 9th Symp. Rem. Sens. Env., 1585–1592, Univ. of Michigan, Ann Arbor (1974a).

SCHANDA, E., HOFER, R.: Scattering, emission and penetration of 3 mm waves in soil. Proc. URSI meeting Microwave Scattering and Emission from the Earth, 141–149, Univ. of Berne, Switzerland (1974b).

SCHANDA, E., SCHAERER, G., WUETHRICH, M.: Radiometric terrain mapping at 3 mm wavelength. Proc. 8th Symp. Rem. Sens. Env., 739–745, Univ. of Michigan, Ann Arbor (1972).

SCHMUGGE, T., GLOERSEN, P., WILHEIT, T., GEIGER, F.: Remote sensing of soil moisture with microwave radiometers. J. Geophys. Res. **79**, 317–323 (1974).

SCHMUGGE, T., WILHEIT, T. T., GLOERSEN, P., MEIER, M. F., DRANK, D., DIRMHIRN, I.: Microwave signatures of snow and fresh water ice. Preprint X-652-73-335. NASA-GSFC, Greenbelt, Maryland (1973).

SCHÜTZ, K., JUNGE, C., BECK, R., ALBRECHT, B.: Studies of Atmospheric N_2O. J. Geophys. Res. **75**, 2230–2246 (1970).

SEMYONOV, B.: Approximate computation of scattering electromagnetic waves by rough surface contours. Radio Eng. Electron. Phys. **11**, 1179–1187 (1966).

SHIMABUKURO, F. I., WILSON, W. J.: Observations of atmospheric ozone at 110.836 GHz. J. Geophys. Res. **78**, 6136–6139 (1973).

SNIDER, J. B.: Ground-based sensing of temperature profiles from angular and multi-spectral microwave emission measurements. J. Appl. Meteorology **11**, 958–967 (1972).

STAELIN, D. H.: Measurement and interpretation of the microwave spectrum of the terrestrial atmosphere near 1 centimeter wavelength. J. Geophys. Res. **71**, 2875–2881 (1966).

STAELIN, D. H.: Passive remote sensing at microwave wavelength. Proc. Inst. El. Electron. Eng. **57**, 427–439 (1969).

STAELIN, D. H., BARRETT, A. H., KUNZI, K. F., LENOIR, W. B., PETTYJOHN, R. L., POON, R. K. L., WATERS, J. W., BARATH, F. T., BLINN, J. C., JOHNSTON, E. J., ROSENKRANZ, P. W., GAUT, N. E., NORDBERG, W.: Meteorological measurements from space with passive microwave techniques. Internat. Symp. Meteorol. Satell., Paris (1973a).

STAELIN, D. H., BARRETT, A. H., WATERS, J. W., BARATH, F. T., JOHNSTON, E. J., ROSENKRANZ, P. W., GAUT, N. E., LENOIR, W. B.: Microwave spectrometer on the nimbus 5 satellite: meteorological and geophysical data. Science **182**, 1339–1341 (1973b).

STAELIN, D. H., KUENZI, K. F., PETTYJOHN, R. L., POON, R. K. L., WILCOX, R. W., WATERS, J. W.: Remote sensing of atmospheric liquid water and water vapor contents with the Nimbus 5 Microwave Spectrometer. To be published in J. Atmosph. Sci. (1975).

STOGRYN, A.: The apparent temperature of the sea at microwave frequencies. Inst. El. Electron. Eng., Trans. **AP-15**, 278–286 (1967a).

STOGRYN, A.: Electromagnetic scattering from rough finitely conducting surfaces. Radio Sci. **2**, 415–428 (1967b).

STOGRYN, A.: The brightness temperature of a vertically structured medium. Radio Sci. **5**, 1397–1406 (1970).

STOGRYN, A.: The emissivity of sea foam at microwave frequencies. J. Geophys. Res. **77**, 1658–1666 (1972a).

STOGRYN, A.: A study of radiometric emission from a rough sea surface. Rep. NASA-CR-2088, Aerojet-General Corp., El. Monte, Calif. (1972b).

STRATTON, J. A.: Electromagnetic Theory. New York-London: McGraw Hill 1941.

STRONG, A. E.: Mapping sea-surface roughness using microwave radiometry. J. Geophys. Res. **76**, 8641–8648 (1971).

THOMANN, G. C.: Remote Measurement of salinity in an estuarine environment. Rem. Sens. Env. **2**, 249–259 (1973).

THOMPSON, M. C.: A radio-optical dispersometer for studies of atmospheric water vapor. Rem. Sens. Env. **2**, 37–40 (1971).

TIURI, M. E.: Radio astronomy receivers. Inst. El. Electron. Eng. Trans. **AP-12**, 930–938 (1964).

TIURI, M. E.: Radio-Telescope Receivers. In: KRAUS, J. D. (Ed.): Radio Astronomy, pp. 236–293. New York: McGraw Hill 1966.

TOONG, H. D., STAELIN, D. H.: Passive microwave spectrum measurements of atmospheric water vapor and clouds. J. Atmos. Sc. **27**, 781–784 (1970).

TREUSSART, H., BECKWITH, W. B., BIGLER, S. G., OTANI, K., KOSTAREV, V. V., SCHWARZ, R.: Use of weather radar for aviation. World Meteor. Org., Techn. Note No. 110, Geneva (1970).

TWERSKY, V.: On scattering and reflection of electromagnetic waves by rough surfaces. Inst. Rad. Eng., Trans. **AP-5**, 81–90 (1957).

ULABY, F. T.: Vegetation and soil backscatter over the 4–18 GHz region. Proc. URSI meeting Microwave Scattering and Emission from the Earth, 163–176, Univ. of Berne, Switzerland (1974).

ULABY, F. T.: Microwave radiometry. In: Manual of Remote Sensing. Am. Soc. of Photogrammetry, Vol. 1 (1975).

ULABY, F. T., BARR, J., SOBTI, A., MOORE, R. K.: Soil moisture detection by Skylab's microwave sensors, Proc. URSI meeting Microwave Scattering and Emission from the Earth, 205–208, Univ. of Berne, Switzerland (1974).

ULABY, F. T., FUNG, A. K., WU, S.: The apparent temperature and emissivity of natural surfaces at microwave frequencies. The Univ. of Kansas, CRES Techn. Rep. 133–12, Lawrence Kansas (1970).

ULABY, F. T., MOORE, R. K., MOE, R., HOLTZMAN, J.: On microwave remote sensing of vegetation. Proc. 8th Symp. Rem. Sens. Env., 1279–1285, Univ. of Michigan, Ann Arbor (1972).

VINOGRADOV, B. V.: Remote sensing of biosphere from space. Abstr. XXIV Internat. Astronautical Congr., 427–428, Baku, USSR (1973).

VOGEL, M.: Microwave radiometry at the DFVLR Oberpfaffenhofen. Proc. 8th Symp. Rem. Sens. Env., 199–216, Univ. of Michigan, Ann Arbor (1972).

WAITE, W. P., McDONALD, H. C.: Snowfield mapping with K-band radar. Rem. Sens. Env. **1**, 143–150 (1970).

WATERS, J. W.: Remote sensing of atmospheric O_3 and H_2O to 70 km by aircraft measurements of radiation at 1.64 mm wavelength. Proc. 8th Symp. Rem. Sens. Env., 467–473, Univ. of Michigan, Ann Arbor (1972).

WATERS, J. W.: Absorption and emission of microwave radiation by atmospheric gases. In: MEEKS, M. L. (Ed.): Methods of Experimental Physics, Vol. 12, Part B, Radio Astronomy. Academic Press 1975.

WATERS, J. W., PETTYJOHN, R. L., POON, R. K. L., KUENZI, K. F., STAELIN, D. H.: Remote sensing of atmospheric temperature profiles with the Nimbus-5 Microwave Spectrometer. To be published in J. Atmosph. Sc. (1975a).

WATERS, J. W., WILSON, W. J., SHIMABUKURO, F. I.: Microwave measurement of mesospheric carbon monoxide. Submitted to Nature (1975b).

WEBSTER, W. J., WILHEIT, T. T., CHANG, A., GLOERSEN, P., SCHMUGGE, T. J.: A radio picture of the earth, Sky and Telescope **41**, Nr. 1, 14–16 (1975).

WEINREB, S., KERR, A. R.: Cryogenic cooling of mixers for millimeter and centimeter wavelengths. Inst. El. Electron. Eng., J. Solid State Circ. **8**, 58–63 (1973).

WESTWATER, E. R.: Microwave determination of low altitude temperature profiles. Proc. 7th Symp. Rem. Sens. Env., 585–594, Univ. of Michigan, Ann Arbor (1971).

WESTWATER, E. R.: Ground-based determination of low altitude temperature profiles by microwaves. Monthly Weather Rev. **100**, 15–28 (1972).

WESTWATER, E. R., STRAND, O. N.: The statistical estimation of the numerical solution of the Fredholm integral equation of the first kind. J. Ass. Comp. Mach. **15**, 100 (1968).

WIENER, O.: Mathematisch-Physikalische Klasse. Leipziger Berichte **62**, 256 (1910).

WILHEIT, T., NORDBERG, W., BLINN, J., CAMPBELL, W., EDGERTON, A.: Aircraft measurements of microwave emission from Arctic sea ice. Rem. Sens. Env. **2**, 129–139 (1972).

WILHEIT, T., THEON, J., SHENK, W., ALLISON, L.: Meteorological interpretations of the images from Nimbus 5 electrically scanned microwave radiometer. Preprint X-651-73-189. NASA-GSFC, Greenbelt, Maryland (1973).

WILLIAMS, G.: Microwave radiometry of the ocean and the possibility of marine wind velocity determination from satellite observations. J. Geophys. Res. **74**, 4591–4594 (1969).

WOEJKOV: Main Geoph. Obs. Report, Preliminary results of the expedition "Bering", Leningrad (1973).

WOFSY, S. C., McCONNELL, J. C., McELROY, M. B.: Atmospheric CH_4, CO and CO_2. J. Geophys. Res. **77**, 4477–4493 (1972).

WONG, P. S. K.: Effect of subsurface spherical scatterers on surface emissivity. M. Sc. Thesis, Res. Lab. Electronics, Massachusetts Inst. Techn., Cambridge Mass. (1974).

WRIXON, G. T.: Measurements of atmospheric attenuation on an earth-space path at 90 GHz using a sun tracker. Bell Syst. Tech. J. **50**, 103–114 (1971).

WRIXON, G. T.: Measurements of atmospheric attenuation on an earth space path at 230 GHz using a sun tracker. Proc. 4th Europ. Microwave Conf., 222–226, Montreux (1974a).

WRIXON, G. T.: Low noise diodes and mixers for the 1–2 mm wavelengths region. Submitted to Inst. El. Electron. Eng., Trans. **MTT-23**, (1974b).

WU, S. T., FUNG, A. K.: A noncoherent model for microwave emissions and backscattering from the sea surface. J. Geophys. Res. **77**, 5917–5929 (1972).

YAROSHENKO, V.: Influence of the fluctuating factor of amplification on the measurements of weak noiselike signals. Radiotechnica **7**, 749–751 (1964).

7. Applications of Gamma Radiation in Remote Sensing

R. L. GRASTY

7.1 The Natural Gamma-radiation Field

In 1896 Becquerel discovered that a photographic plate could be fogged by an invisible and penetrating radiation emitted by uranium salts. It was later found by Rutherford and other workers that uranium emits three types of radiation, alpha-particles, beta-particles and electromagnetic radiation. Alpha-particles are doubly positively charged helium nuclei and beta-particles are electrons carrying unit negative charge. Electromagnetic radiation is generally referred to as gamma-radiation when emitted by the nucleus, and X-radiation when originating from the electrons orbiting the nucleus. Elements that emit one or more of these radiations are termed radioactive, a derivation of the Latin word "radius", meaning ray. Alpha particles are absorbed by a few centimeters of air, whereas beta particles being smaller and travelling at higher velocities are more penetrating and may travel up to a meter or so. High energy gamma radiation is strongly penetrating and can travel several hundred meters through the air.

Both X and gamma-rays are emitted in packets or "quanta" of energy known as photons. The energy, E, of each photon depends on its characteristic wavelength (λ) or frequency (v), which is given by:

$$E = hv = hc/\lambda$$

where h is Planck's constant (6.6×10^{-34} Js)
and c is the velocity of light (3×10^{10} cm/s)

Energies of all radioactive emissions are measured in terms of electron-volts (1.6×10^{-19} J) which is the energy acquired by a particle carrying unit electronic charge when it falls through a potential difference of one volt. Considerable overlap exists between the energy range of X and gamma-rays. X-rays have energies from a few tens of electron-volts (eV) to around 100 KeV (10^5 eV) for elements of high atomic number, corresponding to wavelengths between 10^{-5} cm and 10^{-9} cm. Gamma-rays can have energies as low as a few KeV up to several MeV (10^6 eV) i.e. wavelengths between 10^{-8} cm and 10^{-11} cm.

X and gamma-rays are absorbed by matter through three main processes:

1. the photo-electric effect, in which the photon gives up all its energy in ejecting a bound electron. This process predominates at low energy, especially in matter with a high atomic number.

2. Compton scattering, in which the photon gives a part of its energy to an electron and is scattered at an angle to the original direction of the incident photon. This process predominates for moderate energies in a wide range of material.

3. Pair production, in which the energy of the incident photon is completely absorbed through the creation of an electron-positron pair in the electrostatic field of a nucleus. The photon must have an energy greater than 1.02 MeV, since each member of the pair has a mass equivalent of 0.51 MeV. This interaction predominates at high energies particularly in materials of high atomic number.

Because most materials (rocks, air and water) encountered in airborne radioactivity surveys have a low atomic number and because the natural gamma-ray energies are moderate to low (<2.62 MeV), Compton scattering is the predominant absorption process occurring between the ground and the detector.

When a beam of collimated monoenergetic photons passes through an absorber, the intensity, I, of the beam, measured as photons per square centimeter per second, will depend on the distance, x, travelled by the photon into the material and will be given by:

$$I = I_0 e^{-\mu x}$$

where I_0 is the original intensity at $x=0$,
and μ is the total absorption coefficient and is the sum of the absorption coefficients for all three processes.

It is useful to consider the half-thickness, $x_{\frac{1}{2}}$ of an absorbing material, defined as the thickness of the material which reduces the intensity of the beam to half its initial value. For this half-thickness

$$I/I_0 = \tfrac{1}{2}, \text{ and } x_{\frac{1}{2}} = \log_e 2/\mu .$$

Table 1 shows the half-thickness at various energies calculated from total absorption coefficients given by GRODSTEIN (1957), for water, air at 20° C and 76 cm Hg and rock of density 2.35 g/cm^3. At aircraft altitudes of 100 metres or more, the intensity of gamma-rays below 0.10 Mev, emitted by rocks and soils in the ground will be considerably reduced and dominated by Compton-scattered high energy

Table 1. Variation of half-thickness[a] with energy for water, air and rock

Energy (MeV)	Air (m)	Water (cm)	Rock[b] (cm)
0.01	1.2	0.1	0.01
0.10	38.1	4.1	1.7
1.0	90.6	9.8	4.6
2.0	129.3	14.1	6.6
3.0	161.1	17.5	8.1

[a] The thickness of material which reduces the intensity of the beam to half its initial value.
[b] Normal composition of concrete 2.35 g/cm^3.

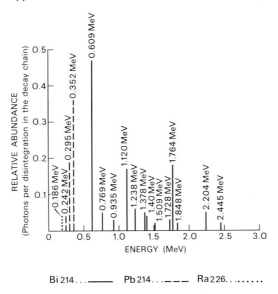

Fig. 1. Principal gamma-rays over 0.1 MeV emitted by uranium in equilibrium with its decay products

gamma radiation. It is also apparent that the measurement of high-energy gamma radiation must be carried out within a few hundred meters of the ground and only gamma-rays from a few tens of centimeters below the surface can be detected.

All rocks and soils are radioactive and emit gamma radiation of which the three major sources are:

1. Potassium-40, which is 0.012% of the total potassium, and emits gamma-ray photons of energy 1.46 MeV;

2. Decay products in the uranium-238 decay series and

3. Decay products in the thorium-232 decay series.

The gamma-ray spectrum from the uranium and thorium decay series is extremely complex. Figures 1 and 2 show the principal gamma-rays over 0.1 MeV that are emitted by uranium-238 and thorium-232 in equilibrium with their decay products. Their relative abundance, measured as photons per disintegration, is also indicated. Tables 2 and 3 show the two radioactive series together with the principal emissions and most recently determined half-lives of the decay products given by LEDERER et al. (1967). Potassium is a major constituent of most rocks, whereas uranium and thorium concentrations are in the parts per million range. Typical potassium, uranium and thorium concentrations for a variety of rock types, from ADAMS et al. (1959) and KOGAN et al. (1971), are shown in Table 4. An important characteristic of these radioactive elements is their strong correlation over a wide range of rock types (CLARK, 1966). It is only on rare occasions that some uncommon process results in the enrichment of one of these elements unaccompanied by enrichment in the others. The thorium/potassium ratio and the uranium/potassium ratios tend to show larger variations than the uranium/thorium ratio.

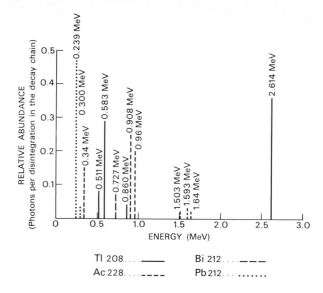

Fig. 2. Principal gamma-rays over 0.1 MeV emitted by thorium in equilibirum with its decay products

Table 2. Uranium-238 decay series[a]

Isotope	Half-life	Principal emission
Uranium-238 ↓	4.51×10^9 years	α
Thorium-234 ↓	24.10 days	β
Protactinium-234 ↓	6.75 hrs	β
Uranium-234 ↓	2.47×10^5 years	α
Thorium-230 ↓	7.5×10^4 years	α
Radium-226 ↓	1602 years	α
Radon-222 ↓	3.82 days	α
Polonium-218 ↓	3.05 min	α
Lead-214 ↓	26.8 min	β
Bismuth-214 ↓	19.9 min	β
Polonium-214 ↓	1.64×10^{-4} sec	α
Lead-210 ↓	22 years	β
Bismuth-210 ↓	5.013 days	β
Polonium-210 ↓	138.40 days	α
Lead-206	Stable	

[a] Isotopes constituting less than 0.2% of the decay products are omitted.

Table 3. Thorium-232 decay series

Isotope	Half-life	Principal emission
Thorium-232	1.41×10^{10} years	α
↓		
Radium-228	6.7 years	β
↓		
Actinium-228	6.13 hrs	β
↓		
Thorium-228	1.910 years	α
↓		
Radium-224	3.64 days	α
↓		
Radon-220 (Thoron)	55.3 sec	α
↓		
Polonium-216	0.145 sec	α
↓		
Lead-212	10.64 hrs	β
↓		
Bismuth-212	60.60 min	α (Thallium) β (Polonium)
↓		
Polonium-212 36%	3.04×10^{-7} sec	α
64% Thallium-208	3.10 min	β
↓		
→Lead-208	Stable	

Table 4. Radioelement composition of some common rocks. (Adapted from ADAMS et al., 1959 and KOGAN et al., 1971)

Rock type	Potassium (%)		Uranium (ppm)		Thorium (ppm)	
	Average	Range	Average	Range	Average	Range
Basaltic	0.7	0.2–1.6	1.0	0.2–4.0	4.0	0.5–10.0
Granitic	2.5	1.6–4.8	3.0	1.0–7.0	12.0	1.0–25.0
Shales	2.2	1.3–3.5	3.7	1.5–5.5	12.0	8.0–18.0
Sandstones	1.0	0.6–3.2	0.5	0.2–0.6	1.7	0.7– 2.0
Carbonates	0.25	0.0–1.6	2.2	0.1–9.0	1.7	0.1– 7.0
Seawater	0.035		∼0.003		<0.0002	
Continental water	—		∼0.001		∼0.0002	
Plants	∼0.05		∼0.001		—	

7.2 Gamma-ray Detector Systems

In the late 1940ies and early 1950ies, increased demand for uranium as a source of fuel for nuclear reactors resulted in the rapid development of airborne prospecting techniques. The detectors first used consisted basically of a chamber containing electrodes in a gas. Gamma-rays absorbed in the gas create ion pairs

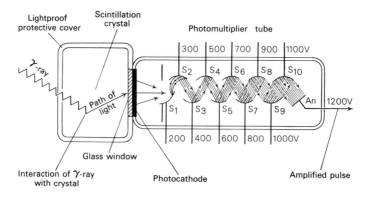

Fig. 3. Schematic diagram of a scintillation detector

which by the application of a suitable electrostatic field produce an electrical pulse at the electrodes. This pulse can be used to trigger electronic circuitry as in the Geiger-Müller counter, or integrated to produce a direct current as in the ionization chamber. Geiger-Müller counters were the only gas-filled counters used to any extent and were generally used in multiple arrays for increased sensitivity. In 1949, a large survey was carried out in the Northwest Territories of Canada using a seven tube array (GODBY et al., 1952), and in 1950, the United States Geological Survey in cooperation with the U.S. Atomic Energy Commission used a nineteen tube array for an airborne survey in northern Michigan (STEAD, 1950).

During the late 1940ies, it was discovered that crystals of certain compounds, when doped with traces of impurities, emit visible light or "scintillate" when subjected to ionising radiation. These scintillations can be converted to an electrical pulse by means of a photomultiplier tube. This assembly is the basic system for all present day radioactivity surveys. The NaI(T1), sodium iodide (thallium activated) crystal, is by far the most commonly used scintillator, although many other crystals, liquids and plastics exist which exhibit this property. The main advantages over gas filled detectors are their high density, resulting in greatly increased sensitivity per unit volume, and in particular the proportionality of the height of the output pulse to the energy expended in the crystal by the incident gamma-ray photon. Its main shortcoming is the poor resolution which results from the amplification process occurring in the photomultiplier tube. Figure 3 is a schematic diagram of a typical scintillation detector.

Scintillation detectors were first flown in conventional aircraft in 1949 and resulted in the obsolescence of gas filled detectors. Total gamma-ray activity equipment was used initially because of its simplicity and low cost, however, it was soon realised that anomalous areas could be caused by thorium which was of little interest in the atomic energy program. The need to recognise the radioactive nuclide responsible for the anomaly lead to the development of present day gamma-ray spectrometers. In a gamma-ray spectrometer, the output pulses from the photomultiplier tubes are sorted into channels according to amplitude, and a frequency distribution of the various pulse sizes is presented. Although this is not

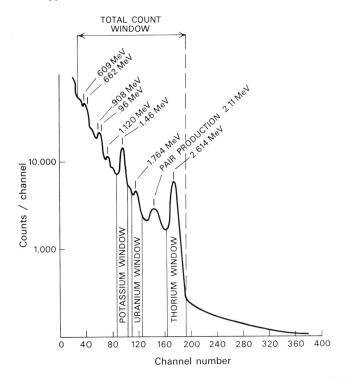

Fig. 4. Airborne gamma-ray spectrum at 120 m. (Adapted from FOOTE, 1968)

a true spectrum of the gamma radiation field, it may be interpreted and calibrated in terms of the gamma-ray energies incident on the crystal. A typical gamma-ray spectrum taken at 120 meters is shown in Fig. 4. The peaks at 2.62 MeV, 1.76 MeV, and 1.46 MeV representing thallium-208 in the thorium decay series, bismuth-214 in the uranium decay series and potassium-40 can be readily distinguished. A multichannel spectrometer records the pulses from many uniformly distributed channels, while in a window-type spectrometer broader energy bands in the vicinity of the peaks of interest are recorded. The photons of energy 1.76 and 2.62 MeV have been generally accepted as being most suitable for the measurement of uranium and thorium respectively because they are relatively abundant and high in energy and, consequently, are not appreciably absorbed in the air. They can also be readily discriminated from other gamma-rays in the spectrum. For mineral exploration and geological mapping window-type spectrometers have normally been used. Table 5 gives the window widths used by the Geological Survey of Canada (GSC) in the airborne system, described by DARNLEY and GRASTY (1971 a). These windows are also indicated in Fig. 4.

A total count window is commonly employed, however, the energy range may vary between systems, see for example KOGAN (1971) and SCHWARZER (1972). Since there is a sympathetic variation between the abundances of potassium, uranium and thorium, the total count reflects general lithological variations and is therefore useful in geological mapping. The total count technique is also used

Table 5. Spectral window widths of the GSC system

Element	Isotope	Photon energy	Window width
Potassium	K-40	1.46 MeV	1.37–1.57 MeV
Uranium	Bi-214	1.76 MeV	1.66–1.86 MeV
Thorium	Tl-208	2.62 MeV	2.41–2.81 MeV
Integral			0.41–2.81 MeV

by the Aerial Radiological Measuring System (ARMS) for monitoring the radiation from major nuclear facilities in the United States. All photons above 50 KeV are recorded and exposure rates at 1 m above the ground are inferred (Burson et al., 1972).

Early airborne gamma-ray spectrometers used small volume detectors which were incapable of sufficient sensitivity and spatial resolution to evaluate small scale anomalies. Large cylindrical detectors up to 29.2 cm in diameter are now readily available. Foote (1968) used six 29.2 cm × 10.2 cm detectors in an array; the ARMS surveys currently operate with fourteen 10.2 cm × 10.2 cm NaI(Tl) detectors, though special surveys in helicopters utilized as many as forty 12.7 cm × 5.1 cm crystals. The GSC system uses twelve 22.9 cm × 10.2 cm crystals.

In order to minimize spectral drift due to temperature fluctuations in the detector and photomultipliers, it has been customary either to maintain the entire detector assembly in a thermally insulated container at a constant temperature, or to employ an artificial radioactive source to provide an automatic gain adjustment for the high voltage supply to the photomultiplier tubes.

7.3 Operational Procedures

Spectral data are normally recorded in digital form on magnetic tape for computer data processing, together with other related information such as the output from a radar altimeter, meteorological parameters, coordinate information and any operator information. Normal survey speeds are around 200 km/hr for a fixed wing aircraft at altitudes between 50 and 150 m depending on the application. If spatial resolution is important, or the terrain is too mountainous, helicopters flown at 25–50 m and at slower speeds of 100–150 km/hr would be employed.

7.3.1 Background Radiation

In any airborne radioactive survey, three sources of "background" radiation exist. They are:
1. the radioactivity of the aircraft and its equipment,
2. cosmic ray interactions with nuclei present in the air, aircraft, or detector,
3. airborne radioactivity arising from nuclear explosions and daughter products of radon gas, from the uranium decay series.

The radioactivity of the aircraft and its equipment is found to remain constant and is due to the presence of small quantities of natural radioactive nuclides in the detector system and in the airframe. Particularly large contributions, which must be removed, can arise from luminous watches and the radium dials of the instrument panels. The cosmic ray contribution increases with altitude but shows little variation on a day to day basis (Zotimov, 1965; Dahl and Odegaard, 1970; Grasty, 1973).

By far the most difficult background radiation correction arises from the decay of radon. Radon, being a gas, can diffuse out of the ground. Furthermore it has a half-life of 3.8 days. The rate of diffusion will depend on such factors as air pressure, soil moisture, ground cover, wind and temperature. The decay products, lead-214 and bismuth-214, are charged particles and attach themselves to airborne dust, consequently their distribution is dependent to a large extent on the wind patterns. Under early morning still-air conditions, there can be measurable differences in atmospheric radioactivity at sites a few miles apart. As the day progresses it has been found that increasing air turbulence tends to mix the air to a greater extent and reduce the atmospheric background. Darnley and Grasty (1971a) report that on the average 70% of the photons detected in the uranium window arise from this source. Since accurate measurement of the uranium window is of prime importance in locating possible uranium targets, it is essential to measure this background contribution as accurately as possible. However, this background variation can be applied in studies of atmospheric mixing as has been carried out by Jonassen and Wilkening (1970) using an airborne ionisation chamber to measure vertical profiles of radon and its decay products.

Foote (1968) used a detector shielded from ground radiation by 10 cm of lead to monitor atmospheric radiation. This procedure has also been carried out in Russia and in the United States, however, the extra detectors and shielding use up valuable space and weight. The technique adopted by the Geological Survey of Canada has been to fly over a lake before the commencement of a survey flight. Since the concentrations of radioactive nuclides in water are several orders of magnitude lower than that of normal crustal material, the activity measured will be the total background contribution from all three sources. Fortunately in most of Canada lakes are abundant, and background values can be updated frequently during the course of the survey. Many experimenters have found this method satsifactory when large lakes are present and homogeneous mixing of the radioactive decay products has occurred. An alternative approach when large lakes are not available has been to sample the air by the use of filters (Burson, 1973). Reasonable estimates of the radioactivity of the air can be made from the beta or gamma activity of the dust collected on the filter papers. A disadvantage is the modification required to the air-frame necessary to sample the air. Another technique is to fly sufficiently high that ground radiation is reduced to negligible proportions through absorption in the air.

7.3.2 Height Correction

The background-corrected count rates depend on the mass of air between the aircraft and the ground and consequently on the altitude of the aircraft and the air density. A common procedure is to normalize the measurements to the nominal terrain clearance of the aircraft. BURSON (1973), DARNLEY et al. (1969), and KO-GAN et al. (1971) have found experimentally that the count rates in the spectral windows drop off exponentially with height. These exponential parameters will depend on the energy of the window considered. BURSON (1973), GRASTY (1972), and DARNLEY and GRASTY (1971a) all give a similar value for the attenuation of total radioactivitiv.

In the ARMS program for monitoring nuclear facilities it is necessary to quantify the radiation exposure rates at 1 m above the ground. This conversion from the nominal flying height of 150 m is carried out on a theoretical basis, assuming a uniform distribution of the natural emitters in the ground, and an exponential distribution of caesium-137 from nuclear fallout (BECK and DE PLAQUE, 1968). Radiation exposure rates are normally presented in microroentgens per hour, where one roentgen is that amount of radiation that will produce in one cubic centimeter of air, a total ionic charge of one electrostatic unit of electricity.

7.3.3 Calibration and Spectral Stripping

In order to determine ground level radioelement concentrations from the measurement of the photons detected in the various spectral windows, a spectral stripping procedure is carried out.

Because of the finite size of the detectors, a certain fraction of high energy thallium-208 photons that would be detected in the thorium window are incompletely absorbed in the crystal and appear as counts in the lower-energy uranium and potassium windows. Similarly, incompletely absorbed bismuth-214 photons can appear as counts in the potassium window. The fraction of counts in a high energy window that appears in a lower-energy window is known as a Compton scattering coefficient or stripping ratio. The spectra obtained during a survey are from sources which are effectively infinite in thickness and extent, however, 90% of the gamma radiation at the surface of a rock of density 2.7 g/cm^3 is emitted from the top 15–25 cm (GREGORY and HORWOOD, 1961). The Compton scattering coefficients have been determined using five radioactive calibration slabs $7.6 \times 7.6 \times 0.46$ m (GRASTY and DARNLEY, 1971), which for a detector at or very close to the surface can therefore be considered infinite in size. The Compton coefficients for three detector systems operated by the Geological Survey of Canada are shown in Table 6.

If K_B, U_B, Th_B, are the potassium, uranium and thorium background count rates and K, U, Th are the total count rates, then the corrected count rates K_C, U_C, and Th_C are related by the following equations:

$$Th_C = Th - Th_B$$
$$U_C = U - U_B - \alpha Th_C$$
$$K_C = K - K_B - \beta Th_C - \gamma U_C$$

where α = uranium counts per thorium count
       β = potassium counts per thorium count
       γ = potassium counts per uranium count

Table 6. Compton scattering coefficients

Detector size (cm)	Uranium counts per thorium count	Potassium counts per thorium count	Potassium counts per uranium count
7.62 × 7.62	0.710	0.878	1.03
12.7 × 12.7	0.426	0.622	0.908
22.9 × 10.2	0.348	0.331	0.560

To relate the corrected airborne count rates with ground level concentrations, CHARBONNEAU and DARNLEY (1970), and DARNLEY and GRASTY (1971a) used uniformly radioactive test strips and compared the airborne results with ground concentrations measured with a portable field spectrometer.

7.4 Applications

7.4.1 Mineral Exploration

By far the greatest effort in radioactivity surveying has been directed towards mineral exploration. In 1972 alone, over 390 000 line-kilometers were flown throughout the world at a cost exceeding two million dollars (Anonymous, 1974).

Early in the development of airborne uranium exploration, it was realized that it was necessary to distinguish the particular element causing an anomaly, since anomalies were frequently found to be related to thorium or potassium. This realization led to the use of gamma-ray spectrometers which produce count rates proportional to ground radioelement concentrations. However, the magnitude of an anomaly is not necessarily the best indicator of uranium mineralization. Granitic bodies tend to be more radioactive than most other rock types but generally are not of interest economically, although they may be of interest in relation to heat flow and crustal evolution. The intensity of radiation may also be low due to poor exposure or because the zone of mineralization is small in comparison to the total exposure. From a mineral exploration point of view, the ratios of the radioelements have more significance than absolute abundances. DARNLEY et al. (1970) have shown that for almost all economic uranium occurrences uranium is preferentially concentrated relative to both potassium and thorium.

The importance of ratio measurements is demonstrated in Figs. 5 and 6. These radioactivity profiles were obtained with the high sensitivity spectrometer operated by the Geological Survey of Canada. They show the integral or total radioactivity, potassium, uranium and thorium count rates in counts per second, as well as the uranium-to-potassium, uranium-to-thorium and thorium-to-potassium ratios. Figure 5 is a profile from a survey in the Uranium City area of Saskatchewan (DARNLEY and GRASTY, 1971b) and demonstrates how a very significant uranium anomaly can be lost in a total radioactivity survey because it coincides with low thorium and potassium. Figure 6 is an illustration from the Bancroft area of

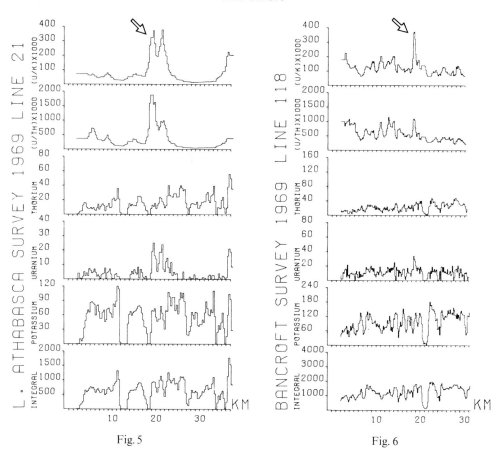

Fig. 5

Fig. 6

Fig. 5. Computer-plotted gamma-ray spectrometer profile, Lake Athabasca, Saskatchewan. (Adapted from Darnley and Grasty, 1971b)

Fig. 6. Computer-plotted gamma-ray spectrometer profile, Bancroft, Ontario. (Adapted from Darnley and Grasty, 1970)

Ontario (Darnley and Grasty, 1970) over a small uranium occurrence which is readily distinguished by a spectrometer survey but would not be found by a total radioactivity survey.

Any ore deposit represents an anomalous concentration of particular elements, and consequently gamma-ray spectrometer surveys are not confined to uranium exploration. Potassium enrichment is found to characterize porphyry copper deposits (Davis and Guilbert, 1973). High uranium to thorium ratios have also been found associated with hydrothermal copper deposits (Bennett, 1971). Radiation patterns have also been reported over oil and gas fields (Lundberg et al., 1952; Pringle et al., 1953; Alekseyev, 1960) although in some instances the patterns were believed to be related to the surface soils (Gregory, 1956).

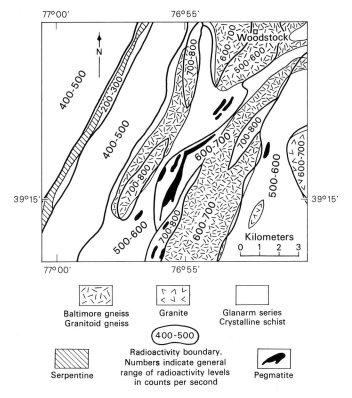

Fig. 7. Geology and total radioactivity of an area in Howard County, Maryland. (Adapted from PITKIN et al., 1964)

7.4.2 Geological Mapping

In most geological materials, there is high correlation between the radioactive elements with the result that general lithological variations are reflected in the total radioactivity. The high count rate associated with total count data also allows a greater number of significant subdivisions than would be possible from any single radioelement.

Approximately 90% of the total gamma radiation from a dry overburden of density 1.5 g/cm^3 is received from the top 30–45 cm (GREGORY and HORWOOD, 1961). Therefore, in areas of little outcrop, to be an effective mapping tool, the surface material must be reasonably representative of the underlying bedrock and be either residual or locally derived. The usefulness of radioactivity surveys in geological mapping will therefore vary from place to place, depending on the nature of the bedrock cover. In areas of the mid-western United States, thick glacial debris completely obscures any radioactivity pattern of the underlying bedrock (PITKIN, 1968). In other areas where there is a greater variation in radioactivity and only thin glacial cover of local origin, the correlation between geology and total radioactivity is excellent, as shown by Fig. 7 for an area in Maryland (PITKIN et al., 1964).

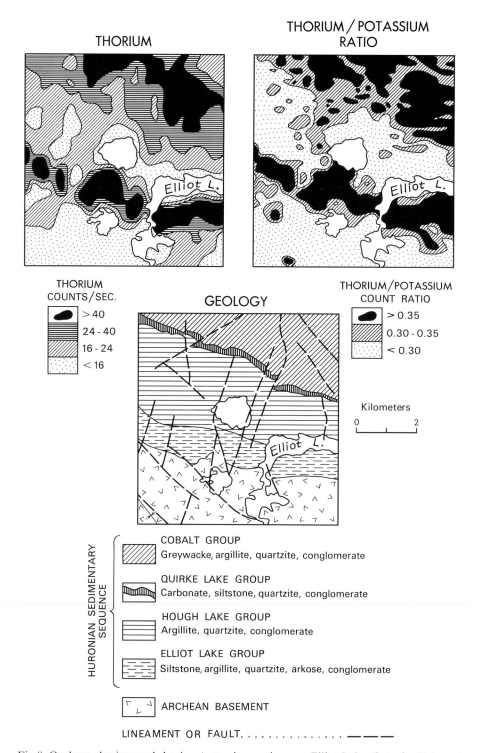

Fig. 8. Geology, thorium and thorium/potassium ratio map, Elliot Lake, Ontario. (Adapted from CHARBONNEAU et al., 1973)

At 43 test areas in the Canadian Shield, the glacial debris was found to be relatively thin and locally derived and reflected the radioactivity of the underlying geological units (CHARBONNEAU et al., 1973). However, over highly radioactive bedrock, the activity of the overburden was reduced by transported material of average radioactivity.

In general, ratios of the radioelements remain relatively constant irrespective of the rock type. Figure 8 taken from CHARBONNEAU et al. (1973) is a comparatively unusual example of where the thorium-to-potassium ratio could be useful in delineating stratigraphic units. However, a similar and more distinct pattern can be obtained from the total radioactivity or any of the single radioelements as illustrated by the thorium map.

7.4.3 Water-equivalent Snow Measurements

For forecasting run off from snowmelt, regulating water storage in reservoirs and general hydrologic studies, the water equivalent of the snowpack must be known. To obtain an accurate estimate involves considerable expense in labour and time because the distribution of the snow is highly variable and many point measurements must be made. An airborne technique, based on the attenuation by a snow layer of natural gamma radiation emitted from the ground, has proved successful in giving areal information and also has the added advantage of being independent of snowpack conditions. This method was first used in the U.S.S.R. (KOGAN et al., 1965) where it is now reported to be operational (DMITRIEV, 1972). Research has been carried out in the U.S.A. (PECK et al., 1971), Norway (DAHL and ODEGAARD, 1970) and in Canada where a large survey was flown in the winter of 1972/73 (GRASTY et al., 1973).

The radiation received by a detector is dependent on the mass of water and air between the detector and the ground. By comparison of the radiation detected for a pre-snow cover flight to one in which snow is present, the water equivalent of the snowpack can be calculated.

For a flight with no snow, the detector count rate N, is related to the altitude H of the aircraft (GODBY et al., 1952), according to:

$$N = N_0 E_2 (\mu_a H) \tag{1}$$

E_2 is an exponential function known as the exponential integral of the second kind, N_0 is the count rate at ground level, and μ_a is the linear absorption coefficient of air at the energy concerned.

For ground covered with snow of water equivalent, D, the count rate N_s is given by:

$$N_s = N_0 E_2 (\mu_a H + \mu_w D) \tag{2}$$

μ_w is the absorption coefficient of water.

If a snow course is flown first in the absence of snow and then with snow, from Eqs. (1) and (2):

$$\frac{N}{N_s} = \frac{E_2(\mu_a H)}{E_2(\mu_a H + \mu_w D)}.$$

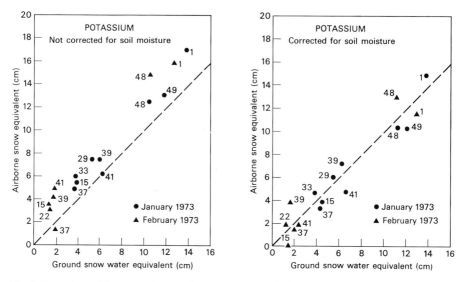

Fig. 9. Potassium airborne and ground-snow-water equivalent measurements, Southern Ontario, 1972/73. (From GRASTY, 1973)

Provided the altitude of the aircraft and the linear absorption coefficients for water and air are known, the water equivalent D can readily be calculated. The linear absorption coefficients of air for all four energy windows can be measured experimentally by flying over the snow course at different altitudes, as has been carried out by GRASTY (1973). The values for water can be calculated from the values for air, since water is known to be 1.11 times as effective as air in absorbing gamma rays (DAVISSON, 1965). PECK et al. (1971) have determined the absorption coefficients in a similar manner by flying over a snow course at different altitudes with varying snow depths of known water equivalent.

The measured count rate is dependent on soil moisture content and for a uniformly radioactive ground will decrease almost linearly with increasing soil moisture. Soil moisture changes of 20% are common, and these will change the count rate by a similar amount. These changes must be accounted for if accurate water-equivalent measurements are required. LOIJENS and GRASTY (1973) sampled the soil to a depth of 15 cm and corrected all radiation measurements to the soil moisture content of the pre-snow flight. These corrections were found to decrease the calculated snow-water equivalent an average of 1.7 cm. Figure 9 compares the airborne results using data from the potassium window with those obtained on the ground. It is clear that the airborne results are considerably higher than the ground results if no allowance is made for soil moisture changes.

Results using the total count data would be expected to be more reliable than those from potassium, in view of the greater number of photons detected. However, another environmental parameter affecting the accuracy of the technique is the variation in atmospheric background radiation due to the presence of radon and its decay products. This is found to have a more significant effect on the total count results than on those obtained using the potassium data (PECK et al., 1971;

GRASTY et al., 1973). Accuracies of 1.0 and 1.5 cm, up to water equivalents of 14 cm, are generally obtained for the potassium and total count results, respectively.

7.4.4 Soil Moisture Measurement

Another important parameter in flood forecasting is the moisture content of the soil. Because of large natural variations in the water content of the soil over an area, a large number of point measurements are required. The attenuation by soil moisture, of natural gamma radiation emitted by the ground, has the potential of permitting continuous measurement of average water content over large areas avoiding many of the problems associated with point measurements. This technique is limited to measuring variations in soil moisture content in the top 30–45 cm, since most of the radiation measured at ground level originates from this layer. ZOTIMOV (1971), reports that using this technique the average soil moisture content can be measured to an accuracy of 5–10%. BURSON (1973) reports that average soil moisture changes along a flight line as predicted by aerial survey data compared well with ground measurements.

7.4.5 Monitoring Nuclear Facilities

For many years the ARMS aircraft has been monitoring nuclear facilities in the United States, to evaluate the impact of any radiation accident or long term build-up of radioactivity. Over 100 areas covering 800 000 km^2 have been surveyed, generally at altitudes of 150 m and line spacings of 1.5 km (BURSON, 1973). Contour maps of radiation exposure rates at 1 m above the ground are produced for each area. From correlations between air and ground measurements these exposure rates are estimated to be accurate to $\pm 20\%$ (BURSON et al., 1972). This accuracy is comparable to the maximum variations generally found in natural radiation levels due to soil moisture fluctuations. Figure 10 from BURSON et al. (1972) is an example of exposure rate contours around a nuclear site and also shows a radioactive gas plume originating from the plant. Artificial isotopes are detected by examination of the spectral data. Caesium-137 concentrations can be detected when they contribute more than 10% of the total exposure rate.

The ARMS aircraft has also proved useful in locating lost radioactive sources. In 1968, a cobalt-60 source was lost from a truck and was quickly located through an airborne search along the highway (WEISSMAN and HAND, 1968). In 1971, a U.S. missile carrying two cobalt-57 sources failed in flight and crashed 650 km south of the White Sands Missile Range. Ground searches proved fruitless in locating the missile; the ARMS system using an energy window centered on the cobalt-57 peak at 122 KeV was successful within three days (DEAL et al., 1972). Rocket payloads have also been deliberately "tagged" with a radioactive source to aid in their recovery.

7.5 Future Prospects

The principal limiting factor in gamma-ray spectrometry is the lack of sensitivity. Increased sensitivity can be achieved by using larger volume sodium iodide

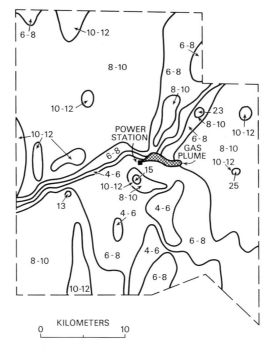

Fig. 10. Exposure rate contours around the Dresden Nuclear Power Station. (From BURSON
et al., 1972)

detectors in multiple arrays. However, this will appreciably increase the cost of
the system since the detectors constitute a large portion of the total cost. An
important factor in the cost of any survey operation is also the size of the aircraft
which is governed to a large extent by the weight and volume of the detector
assembly. It therefore seems unlikely that detector volumes will show any marked
increase in size.

One of the main disadvantages of sodium iodide detectors is their poor spectral resolution. Solid-state lithium-drifted germanium detectors have been available for several years and have the advantage of high resolution. However, this
advantage is offset by their small size and consequent low sensitivity, as well as
their high cost and need for cryogenic cooling. The use of this type of detector will
probably be limited to monitoring nuclear facilities where the identification of
artificial isotopes is difficult with conventional sodium iodide detectors.

Recording the total energy spectrum will almost certainly become standard
practice in the future. This could well solve the problem of accounting for airborne radioactivity, by monitoring the lower energy peaks from the decay products of radon. Full spectral recording also has the advantage that the spectrum
can be accurately calibrated from the position of the prominent potassium peak
at 1.46 MeV.

Identifying and correlating geological formations with their gamma-ray signature is in its infancy. In view of the energy requirements of the future, more gamma-ray spectrometer data will become available through increased exploration for uranium, and airborne gamma-ray spectrometry may well become a more powerful tool in geological mapping than it has been up to the present time.

References

ADAMS, J. A. S., OSMOND, J. K., ROGERS, J. J. W.: In: AHRENS, L. H. (Ed.): Physics and Chemistry of the Earth, Vol. 3. London: Pergamon Press 1959.

ALEKSEYEV, F. A.: Radiometric Method of Oil and Gas Exploration. Nuclear Geophysics, pp. 3–26. Moscow, 1959. Reviewed in USGS Bull. 1116-D, p. 607, 1960.

Anonymous: Geophysical Activity in 1972. Geophysics **39**, 106–107 (1974).

BECK, H., DE PLANQUE, G.: The Radiation Field in air due to distributed gamma-ray sources in the ground. Rept. HASL-195, USAEC, 1968.

BENNETT, R.: Exploration for Hydrothermal Mineralization with Airborne Gamma-ray Spectrometry. CIMM Special Vol. 11, Proc. Third-International Geochemical Exploration Symposium, Toronto, 475–478 (1971).

BURSON, Z. G.: Airborne Surveys of Terrestrial Gamma Radiation in Environmental Research. IEEE Nuclear Science Symposium 1973.

BURSON, Z. G., BOYNS, P. K., FRITZSCHE, A. E.: Technical Procedures for Characterizing the Terrestrial Gamma Radiation Environment by Aerial Surveys, Rept. EGG-1183-1559 (EG&G Inc, Las Vegas, Nevada), 1972.

BURSON, Z. G., FRITZSCHE, A. E.: Snow Gaging by Airborne Radiological Surveys—Status through September 1972, Rept. L-1078, EGG-1183-1565 (EG&G, Inc, Las Vegas, Nevada) 1972.

CHARBONNEAU, B. W., DARNLEY, A. G.: A test strip for calibration of airborne gamma-ray spectrometers. Geol. Surv. Canad. Pap. **70-1**, pt. B, 27–32 (1970).

CHARBONNEAU, B. W., RICHARDSON, K. A., GRASTY, R. L.: Airborne gamma-ray spectrometry as an aid to geological mapping Township 155 Elliot Lake Area, Ontario. Geol. Surv. Canad. Pap. **73-1**, pt. B, 39–47 (1973).

CLARK, S. P., PETERMAN, Z. E., HEIER, K. S.: Abundances of uranium, thorium and potassium. In: Handbook of Physical Constants, Revised Ed. The Geol. Soc. Amer. Mem. **97**, 521–541 (1966).

DAHL, J. B., ODEGAARD, H.: Areal measurement of water equivalent of snow deposits by means of natural radioactivity in the ground. In: Isotope Hydrology IAEA, Vienna, Austria, pp. 191–210 (1970).

DARNLEY, A. G., BRISTOW, Q., DONHOFFER, D. K.: Airborne gamma-ray spectrometer experiments over the Canadian Shield. In: Nuclear Techniques and Mineral Resources (International Atomic Agency, Vienna), pp. 163–186 (1969).

DARNLEY, A. G., GRASTY, R. L.: Airborne radiometric survey of the Bancroft area. Geol. Surv. Canad. Open File Release No. **45** (1970).

DARNLEY, A. G., GRASTY, R. L.: Mapping from the air by gamma-ray spectrometry. CIMM Special, Vol. 11. Proc. Third International Geochemical Symposium, Toronto, 485–500 (1971a).

DARNLEY, A. G., GRASTY, R. L.: Airborne radiometric survey of the uranium city area. Geol. Surv. Can., Open File Release, No. **63** (1971b).

DARNLEY, A. G., GRASTY, R. L., CHARBONNEAU, B. W.: Highlights of GSC airborne gamma spectrometry in 1969. Canad. Mining. J. **1970**, 98–101.

DAVIS, J. D., GUILBERT, J. M.: Distribution of the radioelements potassium, uranium, and thorium in selected prophyry copper deposits. Econ. Geol. **68**, 145–160 (1973).

DAVISSON, C. M.: Gamma-ray attenuation coefficients. In: α, β, γ-ray Spectroscopy, SIEGBAHN, K. (Ed.): Vol. 1, pp. 827–843. Amsterdam: North Holland 1965.

DEAL, L. J., DOYLE, J. F., BURSON, Z. G., BOYNS, P. K.: Locating the lost Athena missile in Mexico by the aerial radiological measuring system (ARMS). Health Phys. **23**, 95–98 (1972).

DMITRIEV, A. V., KOGAN, R. M., NIKIFOROV, M. V., FRIDMAN, SH. D.: The Experience and Practical Use of Aircraft Gamma-Ray Survey of Snow Cover in the USSR. WMO Symposium on Measurement and Forecasting, Banff, Alberta, 1972.

FOOTE, R. S.: Improvement in airborne gamma radiation data analyses for anomalous radiation by removal of environmental and pedologic radiation changes. In: Symposium on the Use of Nuclear Techniques in the Prospecting and Development of Mineral Resources, International Atomic Energy Meeting, Buenos Aires, 1968.

GODBY, E. A., CONNOCK, S. H. G., STELJES, J. F., COWPER, G., CARMICHAEL, H.: Aerial Prospecting for Radioactive Minerals, Atom. Energy Report 13 (1952).

GRASTY, R. L.: Snow-water equivalent measurement using natural gamma emission. Nord. Hydr. **4**, 1–16 (1973).

GRASTY, R. L., DARNLEY, A. G.: The calibration of gamma-ray spectrometers for ground and airborne use. Geol. Surv. Can. Pap. **71—17** (1971).

GRASTY, R. L., LOIJENS, H. S., FERGUSON, H. L.: An experimental gamma-ray snow survey over southern Ontario. In: Symposium on Advanced Concepts and Techniques in the Study of Snow and Ice Resources, Monterey, California, 1973.

GREGORY, A. F.: Analysis of radioactive sources in aeroradiometric surveys over oil fields. Bull. Amer. Ass. Petr. Geol. **40**, 2457–2474 (1956).

GREGORY, A. F., HORWOOD, J. L.: A Laboratory Study of the Gamma-Ray Spectra at the Surface of Rocks. Mines Branch. Res. Rept. R 85. Dept. Mines and Tech. Surv. Ottawa (1961).

GRODSTEIN, G. W.: X-ray attenuation coefficients from 10 KeV to 100 KeV. Nat. Bur. Standards Circ. 583 (1957).

JONASSEN, N., WILKENING, M. H.: Airborne measurements of radon 222 daughter ions in the atmosphere. J. Geophys. Res. **7**, 1745–1752 (1970).

KOGAN, R. M., NAZAROV, I. M., FRIDMAN, SH. D.: Gamma Spectrometry of Natural Environments and Formation. Israel Program for Scientific Translations, 5778, Jerusalem, 1971.

KOGAN, R. M., NIKIFOROV, M. V., FRIDMAN, SH. D., CHIRKOV, V. P.: Determination of water equivalent of snow cover by method of aerial gamma-survey. Soviet Hydrol. Selected Papers **2**, 183–187 (1965).

LEDERER, C. M., HOLLANDER, J. M., PERLMAN, I.: Table of Isotopes. Sixth Edition. New York-London-Sydney: John Wiley 1967.

LOIJENS, H. S., GRASTY, R. L.: Airborne Measurement of Snow-Water Equivalent using Natural Gamma Radiation over Southern Ontario, 1972-1973. Scientific Series 34, Inland Waters Directorate, Dept. of Environment, Ottawa, 1973.

LUNDBERG, H., ROULSTON, K. I., PRINGLE, R. W., BROWNELL, G. W.: Oil exploration with airborne scintillation counters. Oil Canada **1952**, 40.

PECK, E. L., BISELL, V. C., JONES, E. B., BURGE, D. L.: Evaluation of snow water equivalent by airborne measurement of passive terrestrial gamma radiation. Wat. Res. Res. **7**, 1151–1159 (1971).

PITKIN, J. A.: Airborne measurements of terrestrial radioactivity as an aid to geological mapping. Geol. Surv. Prof. Pap. **516-F** (1968).

PITKIN, J. A., NEUSCHEL, S. K., BATES, R. G.: Aeroradioactivity surveys and geological mapping. In: ADAMS, J. A. S., LOWDER, W. M. (Eds.): The Natural Radiation Environment, pp. 723–736. Chicago: Univ. Chicago Press 1964.

PRINGLE, R. W., ROULSTON, K. E., BROWNELL, G. W., LUNDBERG, H. T. F.: The scintillation counter in the search for oil. Mining Eng. **1953**, 1255–1261.

SCHWARZER, T. F., COOK, B. G., ADAMS, J. A. S.: Low altitude gamma-spectrometric surveys from helicopters in Puerto Rico as an example of the remote sensing of thorium, uranium and potassium in rocks and soils. Remote Sensing Environment **2**, 83–94 (1972).

STEAD, F. W.: Airborne radioactivity surveying speeds uranium prospecting. Eng. Min. J. **1950**, 74–77.

WEISSMAN, V. F., HAND, J. E.: ARMS Aircraft Recovery of lost Cobalt-60 source. Rept. L-901, EGG-1183-1395 (EG&G Inc., Las Vegas, Nevada), 1968.

ZOTIMOV, N. V.: A surface method of measuring the water equivalent of snow by means of soil radioactivity. Soviet Hydrol. Selected Papers **6**, 537–547 (1965).

ZOTIMOV, N. V.: Use of the gamma field of the earth to determine the water content of soils. Soviet Hydrol. Selected Papers **4**, 313–320 (1971).

8. Sonar Methods

D. J. CREASEY

8.1 Introduction

An acoustic system to detect enemy submarines was developed by the military during World War I (1914–1918), and this was given the name ASDIC (Allied Submarine Devices Investigation Committee). Since World War II (1939–1945) acoustic echo-location systems have come into ever increasing civilian use in such areas as oceanography, fishing, hydrographic surveying, siesmic surveying, geophysics, whaling, civil engineering and meteorology. Gradually the name ASDIC has been dropped in favor of sonar (SOund NAvigation and Ranging), an acronym which is similar to its younger cousin, the electro-magnetic echo-location term radar.

The majority of uses of sonar have been in the underwater field, but more recently sonar has been used for sounding the atmosphere. This chapter will concentrate upon the underwater uses of acoustics, with less emphasis on the use of acoustics in air.

In an active underwater sonar system, acoustic energy is transmitted into the water. This energy propagates through the water and any object insonified will reflect a fraction of the acoustic energy incident upon it. The reflected energy is detected by a hydrophone receiver (underwater micophone), and by measuring the time, t, that it took the energy to travel from the transmitter to the target and back to the receiver, the range, R, of the target is calculated from the simple equation $R = t/2c$, where c is the speed of sound in water. If either or both the transmitter projector and receiver hydrophone are directional, the bearing and elevation of the target may be obtained by noting the angular co-ordinates of maximum response; that is to say by noting the direction in which the received echo signal is a maximum.

Without modern electronics to amplify and process the received echo signals, sonar as developed by man would be extremely limited. Modern technology is only imitating sonar systems that occur in nature. Certain species of bat use acoustics for navigation and ranging, whilst the porpoise possesses a very efficient sonar system. Although we may not know precisely how these animals process the incoming signals, an analysis of their transmitted signals indicate that they employ sophisticated signal processing techniques. For example, the bat, *Versper-tilionidae*, emits a 1.5 millisecond pulse of acoustic energy whose carrier frequency varies linearly with time from 70 kHz at the start of the pulse to 35 kHz at the end of the pulse (Kay, 1962). This type of signal is characteristic of many modern echo location systems which use pulse compression (or chirp) techniques. It may be argued that engineers have been very slow to copy nature, especially as it was as long ago as 1794 that Spallanzani first demonstrated that a bat uses sound to avoid obstacles in its flight path.

8.2 Propagation of Acoustic Energy

8.2.1 General

Acoustic energy radiating through a medium is observed as a time varying change of pressure which travels through the medium and is measured as the difference between the instantaneous and ambient pressures. This pressure difference, called the excess or acoustic pressure, produces compressions and rarefactions in the medium and particles within the medium are displaced and given both a velocity and acceleration. The resulting disturbance is not transmitted instantaneously, but it travels at a finite speed depending upon the compressibility and density of the medium. For water, these parameters vary according to temperature, depth and salinity, and empirical formulae exist for calculating c. In the temperature range $6°$ C to $17°$ C, Wood (1949) gives

$$c = (1410 + 4.21\,T - 0.037\,T^2 + 1.14\,s + 0.018\,d) \text{ metres per second,}$$

where T = temperature in degrees centigrade, s = salinity in parts per thousand, and d = depth in metres. In examples given in the text the speed of sound will be rounded off to 1500 m/sec.

For low amplitude acoustic pressures, water is a linear medium and the time waveform of the acoustic pressure can be broken up into its frequency components. For large values of acoustic pressure, water possesses a second order non-linear characteristic. Consequently the various frequency components interact and intermodulation products are formed (Westervelt, 1963). With extremely large values of acoustic pressure the rarefactions physically break up the molecules of water and gas is formed. This phenomenon, called cavitation, limits the intensity of acoustic energy measured at some point distant from the cavitation.

If water was a lossless and infinite medium, the acoustic pressure at a distance r from the source of radiation would be

$$p(r) = \frac{p}{r} \cos\left(\omega t - \frac{\omega r}{c}\right),$$

where p is the acoustic pressure 1 m from the source, and $\omega/2\pi$ = the frequency of radiation.

In the calculation of acoustic energy, it is usual to use acoustic intensity (or power per unit area), which is proportional to the square of the acoustic pressure. Thus the acoustic intensity varies inversely as the square of range from the source. This well known inverse square law characteristic is caused by spherical spreading of radiation. For a doubling of range the intensity is reduced by one-quarter for one-way propagation, and by one sixteenth for two-way propagation as in the case of an echo signal.

Of course the sea has boundaries at the surface and the bottom and at distant ranges the propagation is bounded vertically, the acoustic energy can only spread laterally, and cylindrical spreading occurs where for a doubling of range, the intensity is halved for one-way propagation, and is reduced by one-quarter for echo signals. In practice, acoustic energy propagates in the sea starting with an inverse square law characteristic, and then as the range increases the spherical spreading gives way to cylindrical spreading.

8.2.2 Absorption Losses in Water

Losses, other than spreading losses, occur in the water itself. The displacement of particles within the body of the water as radiation passes sets up stresses which are opposed by viscous forces (RAYLEIGH, 1945). To overcome these forces energy is extracted from the acoustic field. This loss of energy is proportional to the square of the frequency of the acoustic field.

Magnesium sulfate partially dissociates in sea-water into positive and negative ions. These ions continually recombine and dissociate and the process is in a state of equilibrium (HALL, 1948). The presence of excess pressure causes a change in the number of dissociated ions, and energy is absorbed to bring about this change in an analogous manner to the charging and discharging of a capacitance through a resistance. At low frequencies there is sufficient time to bring about the change in ion dissociation, but as the frequency increases the period of the acoustic pressure waveform reduces and there is insufficient time to change the number of dissociated ions. This chemical relaxation process is temperature dependent and causes absorption anomalies at frequencies below about 100 kHz.

Apart from the frequency dependence, absorption losses vary with depth, salinity and temperature. For example, the absorption coefficient of sound in water decreases by about 2% for every 300 m increase in depth. The effect of temperature is far more complex especially in sea-water where the relaxation frequency is liable to alter. Absorption losses are used in the calculation of the performance of the sonar system. Many other terms in the resulting sonar equation are not known exactly, and as far as absorption losses are concerned the curves of Fig. 1 are usually adequate.

8.2.3 Beam Bending in Water

As mentioned earlier, the speed of sound varies in water according to temperature, depth and salinity. When working over long ranges it is particularly important to know how the speed varies with depth as this variation causes the sound to be refracted. The resulting bending of the acoustic beams occur usually in the vertical plane, and the effect can be analysed using ray theory.

The speed of sound may be determined by using a velocimeter. This meter measures the speed over a known path length, usually about 300 mm or so. A very

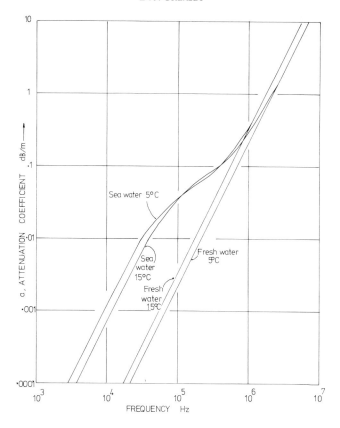

Fig. 1. Attenuation of sound in water

narrow pulse of acoustic energy of high frequency is radiated and the received signal having travelled the known path length initiates a further transmission pulse. The process is continuous and from a knowledge of the pulse repetition frequency and the path length, the speed of sound is easily computed. The speed of sound may be computed from an equation similar to that given earlier if the temperature, depth and salinity are known. However this method is less accurate, and more time consuming than that using a velocimeter.

Figure 2 illustrates how the velocity profile is related to beam bending in a practical solution. It is particularly important to know these ray paths when interpreting a sonar display, or when taking measurements of slant ranges.

Near the surface, the water is well mixed by wave action and the temperature is relatively constant. Below this surface isothermal layer, the temperature reduces with depth often with a sharp gradient which produces a thermocline. The volume of the water between the surface and the thermocline contains a complex collection of biological matter such as plankton which causes scattering of the acoustic field known as volume reverberation. Below the thermocline, the deep ocean thermal profile has a near constant slope, the water temperature getting lower

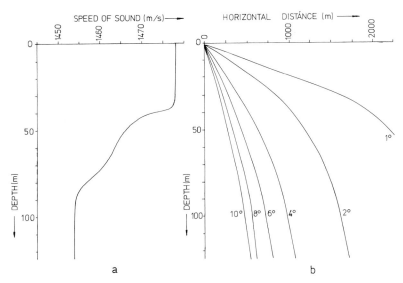

Fig. 2. (a) Typical velocity profile on water and (b) the associated ray diagrams for an array
mounted near the surface

with increasing depth until it reaches 4° C. Any further depth increase results in
no further temperature change, but the depth term in the equation for the speed of
sound, which has a positive coefficient, means that at some depth there is a
minimum speed of sound. Increasing depth beyond this point increases the speed
of sound, and a ray which enters this area may become trapped in a channel about
the speed minimum. This channel is often called the SOFAR channel and because
the spreading losses are only proportional to the reciprocal of the range, long
range propagation can be obtained.

   Another characteristic caused by beam bending is the formation of shadow
zones into which it is impossible to get acoustic energy from a fixed transducer. As
the velocity gradient increases the shadow zones approach the receiver and the
maximum range of the sonar reduces. Usually the surface temperature of the sea
warms up during the morning until in the afternoon the surface thermocline is
relatively deep and long range echo location is severely restricted. This effect is
called "the afternoon effect" in naval circles.

   Speed of sound variations can give rise to errors in the measurement of slant
ranges if care is not taken to correct for them. One simple technique used to
minimise these errors is to calculate the average speed of sound over the path and
to use this in the estimation of range. The average is calculated by measuring the
sound speed between the depths of the array and the target, and averaging this in
a vertical plane. This neglects any beam bending, but beam bending is usually
negligible except where the angle between the hydrophone and the target is
shallow. As a rule of thumb, if the depression angle between the hydrophone and
the receiver is greater than about 20° the effect of ray bending can be neglected; if
the depression angle is less than 20° the effect of ray bending should be evaluated
(MARQUET, 1973). For example, with the array at 15 m depth with a horizontal

separation of 9 km between the array and a target which is at 1.5 km depth, the neglect of ray bending gives only a 23 m error in slant range for typical oceanic conditions.

8.2.4 Propagation in Air

In air, acoustic attenuation is sufficiently small so that remote sensing is possible in the atmosphere at ranges up to 10 km. The interaction of acoustic energy with the atmosphere is very much higher than for most parts of the electromagnetic spectrum. LITTLE (1969) expressed the sensitivity of interaction as the change in velocity, and Table 1 shows that the sensitivity of interaction for sound is several orders of magnitude greater than that for electromagnetic energy.

Table 1

	Change in phase velocity (parts per million)	
	Acoustic energy	Electromagnetic
$1°$ C change in temperature	1700	1
1 m/s change in wind speed	3000	2×10^{-6}
Humidity change of 1 mbar vapour pressure	140	Radio wavelengths 4 optical wavelengths 0.04

The changes in velocity are relative to 1 atmosphere (1 mbar) pressure dry air at $0°$ C for sound, and relative to a vacuum for electromagnetic propagation.

At 1 atmosphere pressure, the speed of sound in air is given by $c = 331.5 \sqrt{(1 + T/273)}$ m/s (RANDALL, 1951), where T is the temperature in degrees centigrade. Absorption losses in air are very dependent upon frequency, and they are caused mainly by the presence of water vapor. The total absorption losses in an air sonar can be calculated from data such as that given by HARRIS (1966).

8.3 The Sonar Equation

8.3.1 Transducers and Arrays

Piezo-electric, electrostrictive and magnetostrictive materials can all be used for both the projector and hydrophone transducer elements. At frequencies above 10 kHz, the most commonly used materials are electrostrictive polycrystalline ceramic materials such as barium titanate, or lead zirconate titanate.

Arrays of these elements form the projector and hydrophone transducers, and usually the arrays are directional. This directionality is analogous to a beam of light from a handtorch, and consequently the projector and hydrophone are both regarded as having a beam pattern. This beam pattern is a measure of the angular response of the arrays: for example, an array of length d, uniformly excited along

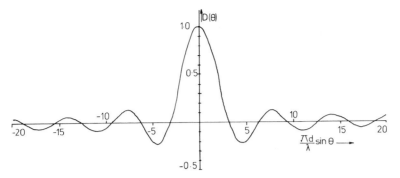

Fig. 3. Beam pattern for an untapered line array of length d metre operating at a frequency c/λ Hz

the length of the array has an angular directional response in the plane of the array given by

$$D(\Theta) = \frac{\sin\left(\dfrac{\pi d}{\lambda}\sin\Theta\right)}{\dfrac{\pi d}{\lambda}\sin\Theta} \quad \text{(Tucker and Gazey, 1966),}$$

where Θ is the angle normal to the array and λ is the wavelength.

When $D(\Theta)$ is plotted, as in Fig. 3, it is seen that there is a maximum response at $\Theta = 0$, and that it passes through a first zero value when $\pm\Theta = \sin^{-1}\lambda/d$. The negative values of $D(\Theta)$ represent a phase shift of 180° relative to the main lobe about $\Theta = 0$. The magnitude of the first major negative sidelobe has a height 0.217 relative to the mainlobe value at $\Theta = 0$. The sidelobe responses represent sensitivities in unwanted directions. Sidelobe levels can be reduced by making the excitation along the array non-uniform in amplitude and, or phase. This is known as tapering, and when applied correctly it results in a lowering of the sidelobe levels but at the expense of broadening the mainlobe width.

Although for many purposes the width of the mainlobe is a good measure of the directional response of an array, the parameter directivity factor is used for reception to indicate the extent the array discriminates against omnidirectional noise in favour of the signal. On the projector hydrophone, the directivity factor compares the acoustic intensity transmitted at the peak of the mainlobe, to that which would be transmitted by an array equally sensitive in all directions. URICK (1967) gives a table of $D(\Theta)$ and the directivity factors for several array tapers. Provided the array is untapered and the linear dimensions of the array are greater than the wavelength of the radiated signal, λ, the directivity factor is approximately $4\pi A/\lambda^2$ where A is the area of the array.

8.3.2 Noise and Reverberation

There are many sources of interference signals in the sea, these interference signals may be broadly classified as noise (those signals originating from sources

other than sonar or an active target), and as reverbation (the backscattered
energy from unwanted minor targets, the original source of energy being the
sonar transmitter). Reverberation is very complex, and is more difficult to discriminate against than noise. In fact, to the oceanographer using sonar to study
rock formations on the sea bed, fish represent unwanted targets and their
echoes are considered to be reverberation. On the other hand, the fisherman
using sonar to detect fish will find their presence often obscured by reverberation
signals received from the bottom. In other words, one man's signal is often
another man's reverberation.

Reverberation originates from three different parts of the sea. Volume reverberation results from scatterers in the main volume of the sea. One of the
main causes of volume reverberation is the so called "deep scattering layer"
a complex collection of different biological organisms whose scattering strength
is dependent upon the frequency of the signal, the location, depth, the season of
the year and even time of day. The deep scattering layer is associated with the
volume of water just about the depth of the thermocline, and as the thermocline
changes depth according to the surface temperature and the time of day so also
does the deep scattering layer. Because many of the biological organisms in the
deep scattering layer are fish food, then the presence of the layer often means that
fish will be found nearby. The volume scatterers have target strengths which vary
with frequency. Thus, frequency response measurements can often indicate
types and sizes of organisms present in the acoustic field.

The other two types of reverberation are surface and bottom reverberation.
Surface reverberation varies with the frequency of the signal, wind speed, the
grazing angle, and wave height. Much of the scattering is caused by a layer of
bubbles trapped just below the surface. Bottom reverberation varies with the
roughness of the bottom, the type of the bottom, the frequency of the signal and
the grazing angle.

The other interference signal, noise, has many causes. At frequencies below
1 Hz, microsciesmic movements in the earth's crust cause random noise signals.
Further up the frequency scale, noise caused by animals, shipping, distant storms,
waves breaking on shores, pebble movements, and sea-surface movements
contribute to the noise spectrum up to about 100 kHz. Above 100 kHz the
predominant noise source is thermal noise, identical to Johnson noise in an
electrical resistance. In the deep ocean, these various noise sources have all been
well documented for various weather conditions (HORTON, 1957). However, for
locations near shores, in rivers and estuaries the noise encountered is usually
three of four times that found in the deep ocean.

8.3.3 Signal to Noise Ratio and Signal to Reverberation Ratio

Consider a target at range R metres, with an effective cross-section scattering
area equal to σ (metres)^2. The acoustic intensity of the echo signal at the
receiver is

$$E = \left(\frac{WA_T}{\lambda^2}\right)\left(\frac{10^{-0.2aR}}{R^4}\right)\left(\frac{\sigma}{4\pi}\right) \text{ Watts/m}^2, \text{ where}$$

the transmitter array of area A_T transmits W Watts acoustic energy into the water at a wavelength λ, and which suffers an absorption loss of a dB/m. This equation is written thus so that the individual terms can be identified when it is written in logarithmic form, i.e. $S = 10 \log_{10} E = $ (Source level $-$ propagation loss $+$ target strength) dB rel 1 W/m^2.

An estimate of the noise level can be obtained from a graph of the noise spectrum level (such as in URICK, 1967). The noise level is then given by

$$N = (\text{Noise spectrum level} + 10 \log_{10} B) - 10 \log_{10} \frac{4\pi A_R}{\lambda^2},$$

where B is the system bandwidth $(B > (\text{pulse length})^{-1})$ and A_R is the receiver array area. The last term represents the system gain due to the receiver directivity, the noise being omnidirectional. Below about 100 kHz the noise spectrum level probably depends upon sea-state, but above this frequency it is given by $(-256 + 20 \log_{10} (\text{frequency}))$ dB rel. 1 W/m^2 (MELLEN, 1952). The signal to noise ratio at the input to the receiver is therefore $S/N = $ [source level $-$ propagation loss $+$ target strength $+$ receiver directivity index $-$ noise spectrum level $- 10 \log_{10}$ (bandwidth)] dB, and if this exceeds the system detection threshold the target can be detected.

The equivalent plane-wave reverberation level from volume backscatterers at range R is $RL = $ [Source level $-$ propagation loss $+$ reverberation target strength/m$^3 + 10 \log_{10}$ (insonified volume)] dB rel. 1 W/m$^2 = SL - PL + 10 \log_{10} s_v + 10 \log_{10} (c\tau/2 \cdot \Psi R^2)$ dB rel. 1 W/m^2, where $c = $ velocity of sound, $\tau = $ pulse width, and Ψ is a function of the transmit/receive beam patterns. For surface and bottom reverberation, the equivalent plane-wave reverberation level is $RL = (SL - PL + 10 \log_{10} S_{s,b} + 10 \log_{10} (c\tau/2 \cdot \Phi R))$ dB rel. 1 W/m^2. Here again Φ is a function of the transmit/receive beam patterns. Values of Ψ and Φ for many array configurations are given by URICK (1967), where full derivations of the reverberation and noise equations are also given.

At 24 kHz, a typical deep ocean value of $10 \log_{10} s_v$ is -90 dB, but in the deep scattering layer it rises to about -70 dB. This coefficient depends upon the frequency, the density and sizes of the scatterers. The scattering coefficients $10 \log_{10} S_s$ and $10 \log_{10} S_b$ are also dependent upon these parameters and also upon the grazing angle.

Since both the wanted echo signals and the masking reverberation signals contain common SL and PL terms, an increase in the source level does not improve the signal to reverberation level, but a reduction in the pulse length τ will reduce the reverberation level. This is contrary to the case where noise alone masks the signal where τ is made as large as possible to reduce the bandwidth to a minimum and an increase in source level increases the system signal to noise ratio. Usually in a well designed sonar, the system just becomes reverberation limited at maximum range: at shorter ranges the system is noise limited.

8.3.4 Interference Signals in Air

The likely signal to noise ratio for an air sonar system must be estimated in order that the feasibility of using such a sonar can be assessed. In air, noise sources fall into two categories (1) thermal noise, kTB Watts where k is Boltz-mann's constant, T the temperature and B the bandwidth, and (2) the ambient noise created by insects, vehicles, wind, rain etc. At 5 kHz the ambient acoustic noise spectrum level has a lower limit of -149 dB rel. to 1 W/m² (Little, 1969). This ambient noise level falls by about 5 dB per octave increase in frequency. Random thermal motion of molecules in the atmosphere will produce an equiva-lent noise spectrum level of $4\pi kT/\lambda^2$ W/m², or -168 dB rel. to 1 W/m². This figure is some 20 dB below the lower limit for the ambient noise.

Electron shot noise in the receiver amplifier must also be included in any calculations. At audio frequencies, it is possible to design amplifiers with a total noise level referred to the input not much greater than kTB Watts, or -204 dB rel. to 1 W per unit bandwidth (Faulkner, 1968). This level may be compared with the ambient noise spectrum level of 149 dB rel. to 1 W/m² as measured in the air. Consider a receiver transducer with an area of $A_R = 1$ m², and a 10% efficiency: by assuming that the ambient noise is isotropic, the effective collecting area at 5 kHZ is $\lambda^2/4\pi A_R$ or 300 (mm)². Thus the ambient noise spectrum level referred to the amplifier input becomes -194 dB rel. to 1 W. This figure is some 10 dB above noise generated inside the receiver amplifier. In this example the ambient noise predominates over the other sources, but by changing either the frequency or the receiver array configuration the reverse may occur. Thus over the range of fre-quencies used for air sonars no firm general conclusion can be drawn about the dominant noise source.

Backscattering of acoustic energy in air is caused mainly by wind velocity changes and the thermal structure of the atmosphere, see Table 1. This backscat-tered energy produces volume reverberation, but as was observed in Section 8.3.2, one man's reverberation is another man's target. Air sonars intended for detecting solid objects, such as a mobility aid for a blind person, would regard this back-scattered energy as volume reverberation. However the backscattered energy is regarded as the target when the sonar becomes at atmospheric sounder to be used for meteorological purposes.

8.4 Factors Affecting Resolution in a Sonar System

8.4.1 General

The sonar equation predicts the performance of a system as far as the ability of a system to detect an echo against background noise and reverberation. Often other factors have to be considered so that targets may be separated in the sonar display. This section will consider the factors which limit range, bearing and Doppler resolution.

8.4.2 Range Resolution

The range resolution of a pulsed sonar system depends upon the pulse length, τ, of the transmitted pulse. The range resolution is improved when the pulse length is reduced and since $B\tau > 1$ (Gabor, 1946), a reduction in τ increases the system bandwidth, B, and hense worsens the noise level.

An echo-pulse received time t after the transmission of a pulse originates from a target at a range $0.5\,ct$, where c is the speed of sound. Because the pulse exists for a finite time τ, the receiver processes data originating in the range annulus $0.5\,c(t-\tau)$ to $0.5\,ct$. So that two targets may be resolved separately the echo from the nearer target must decay sufficiently so that a blank appears on the receiver display before the further target produces a point on the display. Consequently, the minimum detectable separation of targets, or the range resolution of the system, is $\Delta R = c\tau$ metres. From the $B\tau > 1$ inequality, this means $B > c/\Delta R$.

Often the system bandwidth is limited by the bandwidths of the transmit and receive transducers, each of which acts as a band-pass filter. If each has a bandwidth b, together in the system they will produce a bandwith $B = 0.64\,b$. Thus the system bandwidth is less than the bandwidth of the individual transducers and every effort is made to design high bandwidth transducers when good range resolution is required.

Apart from the pulsed carrier signal (with its matched filter approximated by a bandpass filter), the next most commonly used signal is the chirp signal. For the chirp signal, the full system bandwidth, B, is utilised by transmitting a pulse of carrier whose frequency is swept from f_1 Hz at the start of the pulse to $(f_1 - B)$ Hz at the end of the pulse. The pulse can be made relatively long in time so that for a fixed peak power, the total transmitted energy is increased, and the energy is spread more or less evenly over the whole of the available bandwidth. The received signal is passed through its matched filter: for the chirp signal this is a bandpass filter whose frequency-delay characteristic is the inverse of the frequency-time characteristic of the transmitted signal. Thus the higher frequency signals, near f_1 Hz, are delayed relative to the lower frequency signals, near $(f_1 - B)$ Hz. The result is that the pulse is compressed in time and the output from the matched filter has a much increased peak signal power. Since noise does not match the filter, there is also an increased signal to noise ratio. Often this technique is used where peak transmitter power is limited, either by technology or cavitation, while at the same time the long transmitter pulse does not degrade the system range resolution (SOMERS, 1970).

8.4.3 Bearing Resolution

The lateral resolution of a system is usually defined in terms of the beamwidth of the transducer arrays. Figure 3 shows the beam pattern of an untapered array of length d. Usually the beamwidth is taken as the width of the mainlobe where the power has fallen by half relative to the peak value of $\theta = 0$. For high resolution systems, d is usually much larger than the signal wavelength λ, and the beamwidth is approximately λ/d radians. However, as with the system range resolution, an allowance must be made for two closely spaced targets, and a blank must appear between the two target paints on the display. Hence the bearing resolution is usually taken as $2\lambda/d$ radians, which corresponds to the beamwidth between the first pair of zeros in the beam pattern of Fig. 3.

For an untapered line array of length d, the lateral resolution cannot be substantially less than d (TUCKER, 1968). However, if the array is focussed (equivalent to applying a quadratic phase taper across the array) (WELSBY, 1968), then the lateral resolution can be retained as equal to $2\lambda R/d$ metres for all ranges within the depth of focus. When the array is unfocussed the depth of focus extends from a distance F to infinity. If an omnidirectional source is placed at range F along the axis of the array, the elements at the edges of the array will lag in phase with respect to the array centre. One definition of F is that the resulting phase lag at the edges of the array is $\pi/4$ radians, and then $F = d^2/\lambda$ metres. Using a similar definition for the depth of focus, an array focussed at range F will have a depth of focus which extends from $0.5\,F$ to infinity (WELSBY et al., 1973).

8.4.4 Doppler Effect in Sonar

A target (or source) moving towards a sonar will change the received acoustic signal frequency by an amount proportional to the moving target's velocity component in a direction towards the sonar. For an active sonar, this frequency change is approximately $\Delta f = f_0\,\bar{u}/c$, where f_0 is the frequency projected towards the moving target, and $\bar{u}$ is the relative velocity component of the target and sonar. In water, the fractional change in frequency, $\Delta f/f_0$ is 1.33×10^{-3} for each metre per second relative velocity. The equivalent fractional bandwidth for a radar system is 6.67×10^{-9}. The relatively high fractional sonar bandwidths together with the low ping rate restricts the ability of a sonar to detect movements by using the Doppler effect. Consider a pulsed signal with a carrier signal frequency f_0, pulse width τ and a pulse repetition frequency f_s, see Fig. 4a. This waveform possesses the line frequency spectrum shown in Fig. 4b, which has an equation given by

$$\tau f_s \cdot \sum_{n=-\infty}^{+\infty} \left| \frac{\sin n\pi f_s \tau}{n\pi f_s \tau} \cdot \sin 2\pi (f_0 + n f_s) t \right|,$$

SCHWARTZ (1970). The lines in the spectrum are equally spaced f_s Hz apart in frequency. The first pair of zeros in the spectrum occur at $(f_0 \pm 1/\tau)$. Suppose a typical sonar has $f_0 = 50$ kHz, $\tau = 1$ ms, and range $= 1$ km ($f_s = 0.75$ Hz). The spread between the first pair of zeros in the frequency spectrum is 2 kHz and the spectral lines are spaced 0.75 Hz apart. A target with $\bar{u} = 1$ m/s will produce a Doppler shift of 66.7 Hz. In radar systems, Doppler data is obtained by noting the movement of the central line in the frequency spectrum from f_0 to $f_0 + \Delta f$, where Δf can have a value from 0 to $0.5 f_s$. Using the same technique, the sonar system whose parameters were quoted would only be able to detect Doppler shifts of ± 0.375 Hz which represents a ridiculously small maximum relative velocity of only ± 5.625 mm/s.

Consequently Doppler shifts in sonar are detected by noting movements of the complete mainlobe of the spectrum. The system pulse length must be sufficient that the relative velocity, $\bar{u}$, produces a frequency shift greater than $2/\tau$. If $\bar{u}$ has a minimum value of 1 m/s, then the system mentioned just previously would require a minimum pulse length of 30 ms. Thus a system with good Doppler resolution must necessarily have poor range resolution, and vice-versa. In the extreme some

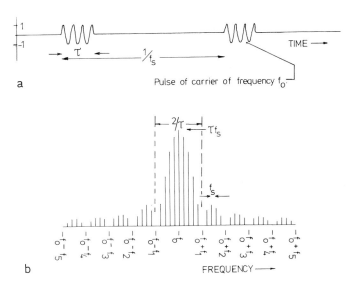

Fig. 4a and b. Frequency spectrum of a pulsed carrier waveform ($f_n = n/\tau$, bandwidth = $2/\tau$ Hz)

Doppler sonars dispense completely with pulses and transmit a continuous tone. The frequency spectrum, whose equation was given earlier, then reduces to a single line, and the system has no range resolution capability.

8.5 Applications

8.5.1 Echo Sounding

Early hydrographic surveying was carried out using a lead weight lowered by line from the deck of the survey vessel. The large errors introduced through the line being bowed by sub-surface currents and by drift of the ship were difficult to evaluate. Echo-sounders were developed from ASDIC equipments in the early 1920's. These were gradually improved and by 1930 production equipments were available with directional transducers which were directed with their acoustic beams vertical. The hydrographic surveyor was now able to obtain a line profile of the sea-bed immediately beneath the survey vessel. The paper records are still obtained by moving a stylus across the paper, the stylus marking the paper with an intensity proportional to the level of the received signals. The stylus movement corresponds to the range of the echoing targets, while the paper itself moves at rightangles to the stylus at a rate corresponding to the speed of the survey vessel, i.e. distance along the record represents distance travelled while distance across the record represents depth, see Fig. 5.

Until the advent of radio navigation aids in the middle of the 1940ies, the hydrographic surveyor relied upon theodolites ashore and the sextant at sea to fix the position of his vessel. Nowadays hydrographic surveying using echo-

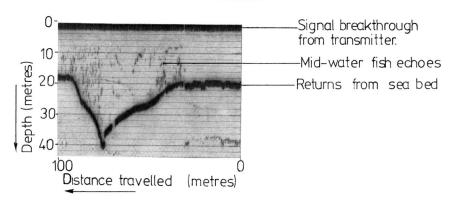

Fig. 5. Echo-sounder display

sounders and radio navigation aids has reached a high level of achievement (RITCHIE, 1970). However, with the development of very large bulk carrying vessels with deep draughts, the exact geometry of the sea-bed between sounding lines must now be known. Also from a scientific point of view, more knowledge is required both of the sea-bed, which is often unstable, and of its contours which can change over a relatively short period of time. Attempts have been made to survey complete areas using multiple echo-sounders, towed on a boom behind the survey vessel (FAHRENTHOLZ, 1963).

Although originally developed for hydrographic surveying, the simple echo-sounder can be used to detect fish shoals, and the fisheries industry soon began to adopt echo-sounders for this purpose. Marine biologists are now able to make estimates of fish stocks and distributions from echo-sounder records (DUNN, 1973).

8.5.2 Sonars with Fixed Beams

There are applications of sonars with fixed beams in biological studies. One application is that of counting migrating fish in rivers. In the U.S.A., the Bendix Corporation developed a sonar system for counting migrating salmon in rivers. A line of transducers is put across the river bed and the beams arranged to cover the whole of the river cross-section. Fish swimming through the beams are detected and counted. While this type of system is suitable in rivers with very large numbers of migrating fish, there are many false alarms. When there are smaller numbers of fish the relatively high false alarm rate is unacceptable. Where such conditions pertain, the type of system developed by BRAITHWAITE (1971) appears satisfactory. This system comprises four transducers, two mounted on either side of the river and the transducers are positioned so that two complete sound "curtains" are thrown horizontally across the river. Fish swimming through the beams are detected, the direction in which they are travelling can be determined from the order in which they enter the two beams, and they can be classified into size from their target returns.

On a much larger scale WESTON (1973) has described detection results obtained using a fixed bottom-laid sonar operating at about 1 kHz. Shoals of Cornish pilchards were observed and their movement along the beam could be seen on the displays at ranges out to tens of kilometres. Their movements could be associated with the tidal cycle and it was observed that the fish shoaled during daylight hours and then dispersed at night. Such sonars, possibly with mechanically or electronically steered beams could give valuable data concerning pelagic fish shoals, shoal sizes and movements over large areas.

8.5.3 Mechanically Rotated Sonar Array Systems

For fisheries purposes, fish need to be detected before they pass beneath the vessel. If the array is made directional, then by rotating it a map of acoustic targets can be built up and displayed on a cathode-ray tube in the same way as a radar. These so called mechanically scanned sonars suffer from the slow rate of data acquisition and presentation. For example, consider a sonar to have a maximum range of 1500 m and which is to scan a 120° sector with a 10° beam. It takes 2 sec for the acoustic energy to travel to maximum range and back to receiver, and the array must not rotate through more than one beam width in this time otherwise potential targets can be lost. In all it will take 24 sec to scan the 120° sector, and the display can only be up-dated once every 24 sec. Thus there is little or no time integration on the display when targets are moving. For such applications as civil engineering, where the targets are stationary, the use of very long persistance phosphors on the cathode-ray tube display permits time integration.

8.5.4 Side-scan Sonars

This type of sonar has an array beam pattern as indicated in Fig. 6. As the pulse of acoustic energy progresses away from the transducers it intersects with the bottom and a fraction of the acoustic signal is reflected back to the receiver. Here the receiver electronics dynamically compensate for spreading and absorption losses, and the resulting video signal intensity modulates the paper display.

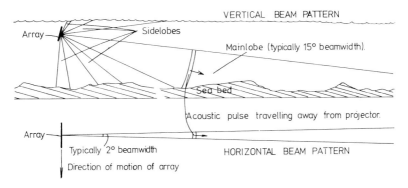

Fig. 6. Geometry of side-scan sonar beam patterns

This display is very similar to that used for the echo-sounder, distance along the paper represents distance travelled by the vessel, while the distance across the paper represents slant range of the targets displayed.

Different sea-bed features have different target strenghts, and rock formations, sand waves and other features such as shore-lines, anchorage chains etc. can often be distinguished on the display (CHESTERMAN et al., 1958). The records only give a subjective view of bottom features, there is no quantative measurement of the heights of the various features portrayed. Even so, the side-scan records provide much useful data for the marine surveyor, and the side-scan technique enables very large areas to be surveyed quickly. Because the equipment measures slant range, the record represents a geometrically distorted map of the sea-bed. These distortions can be removed by off-line processing (HOPKINS, 1970).

Ray bending causes additional distortions and range limitations and to overcome this problem the array is often towed in a hydrodynamic submerged body below the thermocline so that ray bending is minimised. The array is then almost completely stabilised against sea-surface motion, the noise picked up from the survey ship is reduced, and because the array is at a greater depth more power can be transmitted if required without exceeding the cavitation threshold.

Commercially available side-scan sonars operate in shallow water at frequencies between 30 kHz and 120 kHz, with ranges of up to about 2 km. One side-scan sonar developed by the Institute of Oceanographical Sciences operates at 6.4 kHz, with vertical and horizontal beam angles of 15° and 2.7°. Originally this sonar used a pulse length of 30 ms, and transmitter peak power of 50 kW, and under favorable conditions the maximum range was 22 km with a pulse repetition period of some 30 sec in water depths between 3000 m and 6000 m. By using a correlator receiver to obtain pulse compression it has been possible to use a 4 sec frequency modulated sweep of 100 Hz to obtain better discrimination against noise while at the same time reducing the transmitter power (SOMERS, 1970).

Although designed as a deep ocean geological and geophysical instrument, it has been used in shallow water (100 m) as a tool in fisheries research. On a single side-scan record shoals of fish cannot be distinguished from bottom features. However, by comparing two records of the same area, features which have moved can be assumed to be fish shoals (SOMERS, 1973). In shallow water operations of this type the sea may be surveyed at a rate equal to 200 km²/hr.

8.5.5 Sub-bottom Profiling Systems

Scientific investigation and commercial exploration of the earth's outer layers are carried out using seismic techniques. Seismic signals are often generated by explosions and the resulting signals analysed. At sea explosives are dropped into the sea and these are detonated when they reach the required depth. The resulting seismic signals are picked up by the hydrophones suspended from surface buoys; these signals are then radioed back to the survey vessel where they are recorded on magnetic tape and subsequently extensively analysed usually by digital computer techniques. Explosive sources are in some ways ideal in that they produce a very high acoustic power and the signals contain a wide spectrum of low frequencies. Unfortunately no two explosions are identical, and this increases the complexity of analysis.

An echo-sounder record often indicates that a certain amount of acoustic energy penetrates the sea-bed and produces sub-bottom returns even at relatively high frequencies. The depth of penetration increases as the frequency is reduced, but the use of low frequency sonar transmission usually means a low frequency bandwidth, hence poor range resolution, and for a fixed transducer size a wide angular beamwidth with the resulting poor lateral resolution. Thus the essential requirements for a shallow high resolution sub-bottom profiling system are low frequency, high power, wide bandwidth and narrow beamwidth signals. A sub-bottom profiling sonar operates in the same manner as an echo sounder with the same type of display. However, because the received signal level is low, unnecessary propagation losses in the water volume itself must be avoided and many profiling systems operate from hydrodynamically stabilised towed bodies so that the transducers can be positioned as close to the sea-bed as possible.

As an alternative to the more straightforward types of transmitter, a capacitance can be used to store a large quantity of electrical energy, and this can be discharged directly into the water causing a large spark. These so called sparker systems produce about 4 kilo-joules of energy, and they obtain penetration down to 300 m or so. Resolution is just sufficient to delineate geological hazards (such as gas pockets and active faults), diapiric structures and buried channels (TAYLOR, 1973).

Another form of acoustic source used in sub-bottom profiling work is the boomer or thumper. This is a dynamometric transducer which is constructed from two metal plates placed either side of a flat electromagnetic coil. The energy for the acoustic pulse is stored in a bank of capacitors and this energy is discharged into the coil. The resulting change in the magnetic field causes currents to flow in the two plates and the plates are forced apart to produce an impulse of acoustic energy in the water. Penetration depths down to about 1 km can be obtained using a stored charge of some 100 k Joules. The pulse from a boomer contains low frequencies, hence the good penetration, but it has a low frequency bandwidth, and therefore poorer resolution than the sparker type of source. Other sources include the air gun which discharges air under pressure into the water, and the gas gun where an explosive mixture of gases is ignited electrically.

Commercial sub-bottom profiling systems using normal electroacoustic transducers generate about 10 kW of transmitter power at acoustic frequencies near 4 kHz; they obtain penetrations down to 100 m and with a pulse-length of some 2 or 3 m the vertical resolution is of the order of 1 m. It is important that the width of the profiler beam pattern should be as narrow as possible because as the beam enters the sea-bed it diverges due to refraction. Also due to the finite width of the main-lobe a single point target in the sea-bed will produce a hyperbolic return on the resulting display, see Fig. 7. Other confusing returns are caused by sidelobes, so that the beam patterns of the profiler transducers should not only have a narrow main-lobe but also negligible sidelobe levels.

The non-linear property of water mentioned in Section 8.2.1 has been used to produce the narrow beam pattern with low sidelobe levels at low frequencies and wide frequency bandwidths. Essentially if two high energy acoustic beams of

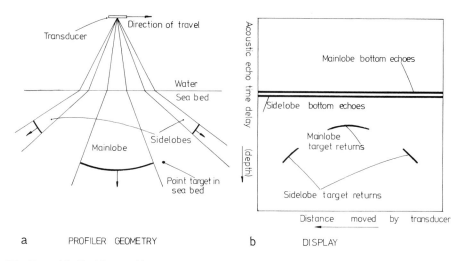

a   PROFILER GEOMETRY   b   DISPLAY

Fig. 7a and b. Problems with a sub-bottom profiler or an echo-sounder. Note how a point source produces a parabolic return on the display due to a finite beam-width and how the side lobes in the beam pattern cause ambiguities

primary frequencies f_1 and f_2 have a large intersecting volume, the nonlinear property of the water will cause interaction between the beams and various intermodulation frequencies $(f_1 - f_2)$ and $(f_1 + f_2)$ occur. When f_1 and f_2 are both of the order of hundreds of kHz, and nearly equal then $(f_1 - f_2)$ is a low frequency term. The increased absorption losses at the primary frequencies and at $(f_1 + f_2)$ mean that only the difference frequency $(f_1 - f_2)$ will propagate. If the various parameters are chosen correctly, the resulting difference frequency beam has a beamwidth of the same order as the primary frequencies, negligible sidelobe levels and a short pulse length, equal to that of the two primary frequencies (BERKTAY, 1973). Even though the conversion efficiency from the primary frequencies to the difference frequency is small, the large directivity index of the difference frequency beam means that the source level is high. Because $(f_1 - f_2) \ll f_1$, the non-linear acoustic array represents a considerable saving in array size over the corresponding low frequency array.

WALSH (1971) has described a successful sub-bottom profiling system which uses the non-linear acoustic principle. This sonar produced a 2° beam at 3.5 kHz from primary frequencies near 200 kHz, with a source level +8 dB rel. 1 W/m² at 1 m. range. The 2° beam had negligible side-lobes, and to increase the sensitivity of the system a chirp signal was used with a pulse compression receiver. WALSH (1971) also reports that when the same system is used as an echo-sounder, it produces excellent results in the deep ocean in water depths of up to 6 km.

8.5.6 Sonars with Sectors Scanned Electronically

For searching a sector, the simple mechanically scanned sonar requires a long time to scan the sector when a high angular resolution is required. One way of increasing the data rate is to use an array made in the shape of a cylinder, with the

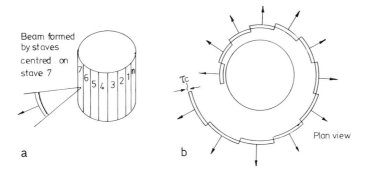

Fig. 8. (a) Multi-element cylindrical array for use to obtain electronic scanning over the full 360° sector, (b) the principle-firing

transducer elements fixed to the outside of the curved surface of the cylinder. Suppose the vertical staves are numbered 1 to n as indicated in Fig. 8a. By combining the outputs from staves 1 to m ($m < n$) with appropriate weighting to reduce sidelobes, then a beam is formed in a direction perpendicular to the centre elements in the arc 1 to m. The beam can be made to scan the complete 360° sector by selecting combinations of the staves electronically. The display is presented in the form of a planposition indicator similar to that commonly used in radar. This produces a dynamic display of the total underwater situation and it finds its application in fisheries work (JOHNSON and PROCTOR, 1973). Such a sonar is able to scan a complete 360° sector up to ranges of 1500 m in only 2 sec.

Where a high bearing resolution is required, it is possible to restrict the overall scanned sector to say 30°, and to use a large number of elements in a long line array to form a beam with a beamwidth of the order of 0.5°. The transmitter for such a system insonifies the whole sector and Fig. 9 illustrates the situation at the array when signals from a distant target at a bearing θ are received. There is a time delay of the arrival of the wavefront between adjacent elements in the array equal to $d/c \sin \theta$ seconds, where d is the spacing of the elements in the array and c the speed of sound. If a time delay, $(i-1)d/c \sin \theta$, is inserted in the signal path from the i th transducer element, the signals from all channels will be co-phasal, and they add up to produce a maximum response for a target at bearing θ. To accommodate signals arising from targets at other bearings, the time delay must be varied electronically. This may be accomplished by using a set of delay lines with multiple output taps. By summing appropriate outputs from the delay lines, the beam may be formed in the required direction, see Fig. 10. To form other beams the taps on the delay lines are switched electronically. The transmitter for such a system insonifies the whole sector with a pulse of acoustic energy. Provided the switching is fast, all directions are interrogated at least once during the pulse length τ, and no detectable targets are lost. The delay-line type of beamforming system can handle signals with a very wide bandwidth because the time delays are frequency independent.

Electrical delay networks which faithfully produce both the correct signal phase and amplitude information are available, but they are relatively expensive.

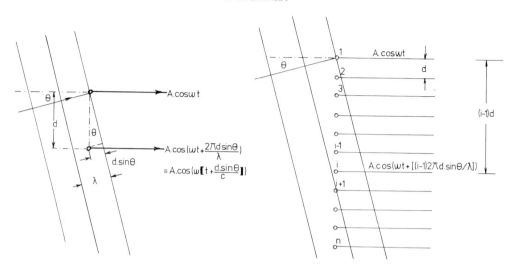

Fig. 9. Time-delay and phase relationships for a multielement line array intercepted by a plane wave originating from a source at bearing Θ

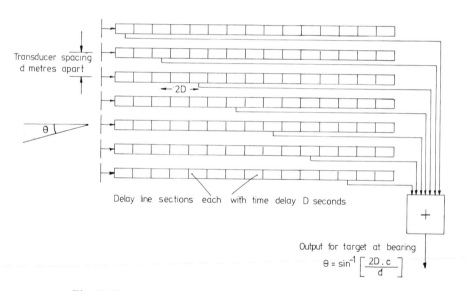

Fig. 10. The principle of electronic scanning by using delay lines

If, on the other hand, the amplitude information is dispensed with and the incoming signals are hardlimited, the delay lines can take the form of digital shift registers (ANDERSON, 1960). These can be produced very cheaply using modern integrated circuit technology.

A cathode-ray tube display is used and usually it is more convenient for $\sin\theta$ to be plotted against range in a so called B-scan. Most high resolution systems

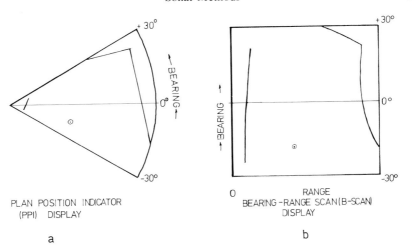

PLAN POSITION INDICATOR
(PPI) DISPLAY

a

BEARING-RANGE SCAN (B-SCAN)
DISPLAY

b

Fig. 11a and b. Relationships between target positions and geometries when presented on (a) a plan position indicator (PPI) display and (b) a bearing range (B-scan) display

have a relatively small scanned sector, so that $\sin\theta \simeq \theta$. The B-scan display presents a distorted acoustic map of the target field, but interpretation is quite simple. Figure 11 shows the relationship between a B-scan and the usual plan-position indicator type of display.

Referring again to Fig. 9, the distance between successive wavefronts is λ, the wavelength of the carrier signal and the time delay can be equated to a phase shift between the signals received on adjacent elements of $2\pi d/\lambda \sin\theta$ radians. Hence, phase shifts can be inserted between adjacent elements to make the signals co-phasal for targets at a bearing Θ. Because the required phase shift contains the term λ, the use of phase-shifts to produce off-axis beams is limited to narrow frequency band operations.

The n beams formed in a phased beam-former are Fourier transform coefficients. With modern computing techniques it is possible to perform the required n-point transform in real time on a digital computer using the Fast Fourier Transform algorithm. However, the hardware and software required for such a transform are still expensive, and analogue systems (as opposed to digital systems) have existed for many years for forming the discrete Fourier transform.

Those systems developed rely upon modulation techniques to produce the required phase shifts. If the signal from the i th transducer element in Fig. 9, $A_i \cos(\omega t + (i-1)2\pi d/\lambda \sin\Theta)$, is modulated with a carrier waveform $2\cos(i-1)\omega_s t$, the output signal will contain the lower sideband term $A_i \cos(\omega t + (i-1)(2\pi d/\lambda \sin\Theta - \omega_s t)$ which can be extracted by filtering. At time $t = 2\pi d \sin\Theta/\lambda\omega_s$, the unwanted phase term cancels. If all channels are treated in a similar manner, the resulting signals will be co-phasal once during the scanning period $2\pi/\omega_s$. Signals arising from other targets within the scanned sector on other bearings will produce a summed maximum response at some other time during the scanning period. This type of sonar is called a within pulse sector scanner and the beam so formed is swept across the sector as the value of $\omega_s t$ changes during the scanning period.

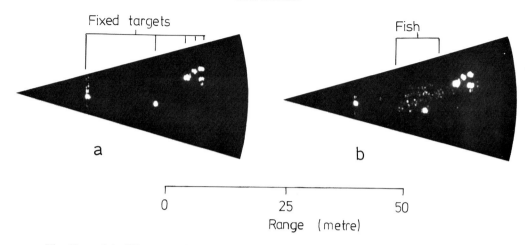

Fig. 12a and b. 30° sector, plan position indicator displays obtained from a within pulse sector scanning sonar. (a) Displays showing only fixed targets. (b) Display showing fixed targets and targets caused by an aggregation of fish

Since the phase term $\omega_s t$ repeats itself once every scanning period no detectable targets are lost, provided the accoustic pulse length τ is greater than $2\pi/\omega_s$ (MCCARTNEY, 1961; VOGLIS and COOKE, 1966). Figure 12 shows display photographs of a within pulse sector scanning sonar.

The main application of high resolution within-pulse sector scanning sonars has been for fisheries purposes. The Lowestoft Fisheries Laboratory of the U.K. Ministry of Agriculture, Fisheries and Food use a within-pulse sector scanning sonar which operates at 300 kHz, has a pulse length of about 0.1 ms, and a range of some 400 m. The sonar forms a scanning beam with a half-power beamwidth of 0.33°, and a total scanned sector of 30°. Because of the high angular resolution of the system, the array has to be stabilised against ship roll, pitch, yaw and heave motions, otherwise the resolution of the system is lost (MITSON and COOKE, 1971).

This particular ship mounted sonar has been used to investigate fish behavior, fish migration, fish reaction to gear, and to study the operation of fishing gear. When individual fish are required to be tracked, they are first caught and then released with a micro-miniature acoustic transponding tag attached (MITSON and STORETON-WEST, 1971). The very large signal received from the tag, which weighs only 4 g in sea water, makes it clearly distinguishable from all other background targets, and movements of a tagged fish relative to other features can easily be followed on the sonar display. This sonar has also been used to produce acoustic maps of the sea-bed. Wrecked ships have been discovered, and the effects of gravel excavation in the North Sea and English Channel observed. In principle, it is possible to use this system for hydrographic survey purposes. If the transducer beam is pointed sideways to the direction of motion of the boat, as in a side-scan sonar, and the sector is scanned vertically, then the resulting display will contain the quantative information missing from a side-scan sonar display. The heights of various sea-bed features can be measured at sufficient ranges, so that all features

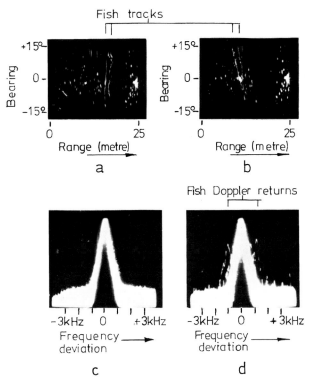

Fig. 13. (a and b) Photographs of the B-scan display of a within-pulse sector scanning sonar, showing tracks of salmon crossing the beam. (c) Spreading of the spectrum of the signal in a single frequency (1 MHz) continuous wave sonar by the doppler shifts due to turbulence etc., in the volume of the water. (d) Spectrum of a 1 MHz continuous wave sonar when fish are swimming across the beam. Extra doppler shifts are thought to be due to fish tail movements

between the hydrographer's echo-sounding survey lines can be mapped with precision at high speeds.

A 500 kHz within-pulse sector scanning sonar developed in the author's department has been used for short range work, up to 60 m, to study the response of individual fish to sound stimuli, to investigate the behaviour of fish around the cooling water intakes to a power station, and to count migrating salmon in a river estuary. The sonar has been used also to study the efficiency of a commercial salmon fishing industry in a Scottish river. Figure 13 shows two B-scan displays of the tracks of salmon in the River Spey in Scotland. To obtain these the sonar array was pointed across the river. The photographs show a large number of stationary targets, and they are time integrations of the display, the integration time being the time taken for the fish to swim across the sector. The fish tracks are therefore the continuous lines in Fig. 13a at about 15 m range, and in Fig. 13b there are again tracks from two fish at about 10 m range.

Because of the large number of background returns from rocks and other river bed features, the sonar operators found that they could concentrate on the sector

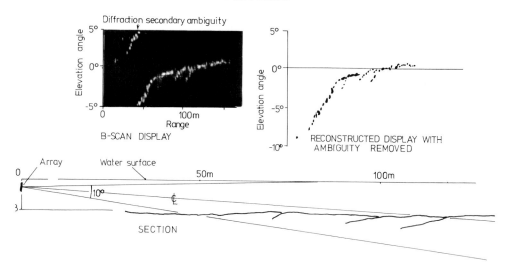

Fig. 14. Digital sonar display and interpretation when used in the vertical scanning mode

scanning sonar display for only short periods at a time. To overcome this operator fatigue, a 1 MHz continuous transmission Doppler sonar was used in conjunction with the sector scanning sonar. While the Doppler sonar gave neither
bearing nor range information, it did give an audio warning to the operator
whose full attention could then be focussed on the sector scanning sonar display.
The number of fish passing upstream could then be determined from this display.
Figure 13c shows the Doppler returns from the body of the water volume. There
are small Doppler shifts in the received frequency spectrum due to background
reverberation, turbulent eddies and the like. In Fig. 13d, there are two definite
additional Doppler returns at about 1 MHz ± 1.5 kHz. The within-pulse sector
scanning sonar displays of Fig. 13a and Fig. 13b indicate that the fish movements
were across the sonar beams, and these should not produce a Doppler shift
because the relative velocity, $\bar{u}$, of the targets and the sonar arrays is very small.
However, the Doppler sonar was extremely sensitive so that it is thought that the
detected Doppler shifts in Fig. 13d are due to the fish tail movements.

Another type of sonar which gives both range and bearing data from a single
widebeam pulsed frequency transmission is the so called digital sonar (NAIRN,
1968). This type of sonar is so named because all the processing is carried out by a
simple, cheap, purpose built serial digital computer. The sonar operates by hard
limiting all the incoming signals, which removes all the amplitude information,
and the need for dynamic range compensation of the received signals. Unlike all
the other sonars described here, bearing data is not obtained by measuring the
maximum response of the receiver array. Instead the digital sonar measures the
phase differences of the signals received on adjacent elements in a sectional receiver array. The hard limiting process allows only one target to be detected on
the display in one range annulus (about 200 mm for a digital sonar operating at
50 kHz). However, it is found that over a short time integration on the display
enables shapes of complex targets to be outlined. For example, the digital sonar

has demonstrated its ability to detect and display fishschools, fish movements, shore lines, and bottom features (CREASEY and BRAITHWAITE, 1969). Figure 14 illustrates how a digital sonar may be used as a hydrographic instrument. A true within-pulse sector scanning sonar produces sidelobes in the scanning pattern. If there are very large targets present, such as a transponding target, these sidelobes cause smearing across the display at all bearings at the particular range of the target. There are no scanning sidelobes associated with the digital sonar, so that the digital sonar is very suitable for use with transponding targets (CREASEY et al., 1973).

8.5.7 Remote Sensing Application of Sonar in Air

Air sonars are used in industrial applications and as a mobility aid for blind people. These applications are usually short range and cannot be considered as remote sensing instruments in either earth or bio-sciences. On the other hand, rocket grenades have been used for many years to obtain atmospheric temperature and wind profiles to heights in excess of 50 km (STRAND et al., 1956). In 1969, LITTLE discussed the application of the acoustics method for determining atmospheric profiles and structures by observing scattering produced by turbulence. Studies by various Soviet workers, for example KALISTRATOVA (1959), have resulted in an acoustic back-scattering equation based upon the statistical theory of turbulent mixing. Using this scattering equation, LITTLE (1969) calculated that a simple pulsed air sonar working at 5 kHz, with a 10 msec, 10 Watt pulse, and a $1 m^2$ transducer can have a signal to noise ratio as large as $+26$ dB when receiving energy backscattered from atmospheric thermal microstructure at a height of 1.5 km. BERAN (1970) reports estimates of the useful range of a lower frequency acoustic sounder as high as 10 km.

Meteorological applications using acoustics can be divided into three classes (1) those requiring the range of a particular scattering volume, (2) those requiring both the range and a measurement of the magnitude of the scattered energy, and (3) those employing a Doppler technique.

The monitoring of thermal inversion structures, which are important in the study of air pollution, can be achieved by measuring only the range of the scatterers. That this was possible was demonstrated by McALLISTER et al. (1968), who used a facsimile recorder, similar to an echo-sounder display, to produce a time record of the acoustic returns from low level scatterers. By employing a number of fixed acoustic sounders the topography of inversion phenomena can be mapped both in time and in space. The study of the motion of low level gravity waves can be achieved in the same manner.

By using a sounder working at a number of discrete frequencies, the humidity content of the atmosphere at different heights can be measured. This is achieved by making allowances for the variation of the target strength of turbulent scatterers as a function of frequency, proportional to $(frequency)^{1/3}$, and then measuring the absorption losses as a function of frequency and height. Since absorption losses are mainly due to humidity, the difference in losses at different frequencies gives a measure of the humidity (HARRIS, 1966).

The exploitation of the spectra of backscattered and forwardscattered signals is very a powerful remote sensing tool. Brown (1972) produced a sufficiently simple general solution to the scattering equation which can enable the received signals to be interpreted in terms of the mean air velocity, the mechanical turbulence and the temperature microstructure. He showed that the strength of the backscattered energy is due mainly to temperature fluctuations alone. For the case where propagation times are short compared to the lifetime of the mechanical turbulent eddies, Doppler shifts encountered on the forward path are cancelled by equal but opposite Doppler shifts on the reverse path. Thus in this situation, to a first order, the resulting Doppler shift will be that contributed only by scatterers in the terminal volume. This Doppler shift is proportional to the standard deviation of the eddy velocities in the terminal volume. The angle of arrival in a monostatic vertical sounder indicates the average air velocity between the sounder and the height of the scatterers.

References

Anderson, V. C.: Digital array phasing. Journal of the Acoustical Society of America **32**, 867–877 (1960).

Beran, D.: Application of acoustics in meteorology. University of Melbourne, Meteorology Department, Project EAR, Report No. 2 (1970).

Berktay, H. O.: Nonlinear acoustics. In: Griffiths, J. W. R., Stocklin, P. L., van Schoonevelt, C.: (Eds.): Signal Processing, pp. 311–326. London-New York: Academic Press 1973.

Braithwaite, H. B.: A sonar fish counter. J. Fish Biol. **3**, 73–82 (1971).

Brown, E. H.: Acoustic-doppler-radar scattering equation and general solution. J. Acoust. Soc. Amer. **5**, 2, 1391–1396 (1972).

Chesterman, W. D., Clynick, P. R., Stride, A. H. B.: An acoustic aid to sea bed survey. Acoustica **8**, 285–290 (1958).

Creasey, D. J., Braithwaite, H. B.: Experimental results of a sonar system with a digital signal processing unit. Appl. Acoust. **2**, 39–57 (1969).

Creasey, D. J., Gazey, B. K., Westwood, J. D.: The use of underwater acoustics in the control of diving operations. Proceedings of the Third World Congress of Underwater Activities (Flemming, N., Ed.). London (1973).

Dunn, W. I.: An echo-sounder for fish target strenght measurement. British Acoustical Society, Technical Meeting on Sonar in Fisheries, Lowestoft. Paper No. 73/24 (1973).

Fahrentholz, S.: Profile and area echograph for surveying and location of obstacles in waterways. J. Brit. Inst. Radio Eng. **26, 2**, 181–187 (1963).

Faulkner, E. A.: The design of low-noise audio-frequency amplifiers. Radio Electr. Eng. **36, 1**, 17–30 (1968).

Gabor, D.: Theory of communication. J. Inst. Electr. Eng. **1946**, Part III, 429–457.

Hall, L.: The origin of ultrasonic absorption in water. Physcis Rev. **73**, 775 (1948).

Harris, C. M.: Absorption of sound in air versus humidity and temperature. J. Acoust. Soc. Amer. **39**, 1125–1132 (1966).

Hopkins, J. C.: Cathode-ray tube display and correction of side-scan sonar signals. Proceedings of the Institute of Electronic and Radio Engineers, Conference on Electronic Engineering in Ocean Technology, Swansea. Conference Publication No. **19**, 151–158 (1970).

Horton, J. W.: Fundamentals of Sonar. United States Naval Institute, Annapolis (1957).

Johnson, H. M., Proctor, L. W.: Advanced design of a fishing sonar. Paper presented at Oceanology International 1973, Society of Underwater Technology (1973).

Kallistrotova, M. A.: Procedure for investigating sound scattering in the atmosphere. Soviet Physics-Acoustics **5**, 512–514, (Translated from Akusticheskii Zhurn. **5, 4**, 496–498) (1959).

KAY, L.: A plausible explanation of the bat's echolocation acuity. Animal Behaviour **10**, 34–41 (1962).

LITTLE, C. G.: Acoustic methods for remote probing of the lower atmosphere. Proc. Inst. Electr. Electron. Eng. **57**, 4, 571–578 (1969).

MARQUET, W. M.: Submerged navigation and submersible instrumentation. Technical Progress Report, Woods Hole Oceanographic Institution, WHO 1-73-37, (MAXWELL, A. E., Ed.) 4–19 (1973).

MCALLISTER, L. G., POLLARD, J. R., MAHONEY, A. R., SHAW, P. J. R.: Acoustic sounding—a new approach to the study of atmospheric structure. Weapons Research Establishment Technical Note, CPD(T) 160, Salisbury, South Australia (1968).

MCCARTNEY, B. S.: An improved electronic sector-scanning sonar receiver. J. Brit. Inst. Radio **22**, 6, 481–488 (1961).

MELLEN, R.: The thermal-noise limit in the detection of underwater acoustic signals. J. Acoust. Soc. Amer. **24**, 478–481 (1952).

MITSON, R. B., COOKE, J. C.: Shipboard installation and trials of an electronic sector scanning sonar. Radio Electron. Eng. **41**, 8, 339–350 (1971).

MITSON, R. B., STORETON-WEST, T.: A transponding acoustic fish tag. Radio Electron. Eng. **41**, 11, 483–489 (1971).

NAIRN, D.: Clipped-digital technique for sequential processing of sonar signals. J. Acoust. Soc. Amer. **44**, 5, 1267–1277 (1968).

RANDALL, R. H.: An introduction to acoustics. Reading. Pal Alto, London: Addison-Wesley 1951.

RAYLEIGH, Lord.: Theory of Sound. New York: Dover 1945.

RITCHIE, G. S.: Problems in bathymetric surveying presented by modern trends in shipbuilding. Radio Electron. Eng. **40**, 5, 219–224 (1970).

SCHWARTZ, M.: Information transmission, modulation and noise, 2nd Ed. New York: McGraw-Hill 1970.

SOMERS, M. L.: Signal processing in project GLORIA. Proceedings of the Institute of Electronic and Radio Engineers, Conference on Electronic Engineering in Ocean Technology, Swansea. Conference Publication No. **19**, 109–120 (1970).

SOMERS, M. L.: Some recent results with a long range side-scan sonar. In: GRIFFITHS, J. W. R., STOCKLIN, P. (Eds.): Signal Processing, pp. 757–767. London-New York: Academic Press 1973.

STRAND, W. G., NORDBERG, W., WALSH, J. R.: Atmospheric temperature and winds between 30 and 80 km. J. Geophys. Res. **61**, 45–56 (1956).

TAYLOR, D. M.: What the "Decca-Profiler" can do. Ocean Industry, October 51–53 (1973).

TUCKER, D. G.: Near field effects in electronic scanning sonar. J. Sound Vibration. **8**, 3, 390–394 (1968).

TUCKER, D. G., GAZEY, B. K.: Applied underwater acoustics. Oxford: Pergamon Press 1966.

URICK, R. J.: Principles of Underwater Sound for Engineers. New York: McGraw-Hill 1967.

VOGLIS, G. M., AND COOKE, J. C.: Underwater applications of an advanced acoustic scanning equipment. Ultrasonics **4**, 1–9 (1966).

WALSH, G. M.: Practical application of finite amplitude techniques to narrow beam depth measurements. Proc. Symp. Linear Acoust. Birmingham, Brit. Acoust. Soc. **1971**, 197–208.

WELSBY, V. G.: Electronic scanning of focussed arrays. J. Sound Vibration **8**, 3, 390–394 (1968).

WELSBY, V. G., CREASEY, D. J., BARNICKLE, N. J.: Narrow beam focussed array for electronically scanned sonar, some experimental results. J. Sound and Vibration **30**, 2, 237, 248 (1973).

WESTERVELT, P. J.: Parametric acoustic array. J. Acoust. Soc. Amer. **35**, 535–537 (1963).

WESTON, D. E.: Fish traces on a long-range sonar display. British Acoustical Society, Technical Meeting on Sonar in Fisheries, Lowestoft, Paper No. 73/21 (1973).

WOOD, A. B.: A textbook of sound. London: Bell 1949.

9. Digital Picture Processing

PH. HARTL

9.1 Introduction

9.1.1 General Remarks

From an analysis point of view an image can be considered as a two-dimensional area of picture elements (pixels), each of which is characterized by a gray level. In other words, an image is a brightness function $f(x,y)$ of two space variables x and y. The gray levels are assumed to represent the radiant energy of some electromagnetic spectral range. But they could also be a measure of some other features, such as distance, velocity or polarization etc.

If the image is not in black and white (B/W image), but in color, then it can be considered as a combination of 3 B/W pictures, each one representing the radiant energy of a subdivision of the spectrum. Therefore, three functions $f_r(x,y)$, $f_b(x,y), f_g(x,y)$, representing the brightness functions of the visible red, blue, and green spectral regions, for example, would then be necessary for image analysis. We will deal with color pictures only in Section 9.4.6. With this exception we will assume the images to be B/W.

We will deal with automatic pictures processing. This plays an important role in practice, because there is (ROSENFELD, 1970; ANDREWS et al., 1970; BRESSANIN et al., 1973; TOLIMSON, 1972):

1. a growing demand on real-time decision making on the basis of image information;

2. a fast increasing amount of image data to be processed;

3. only a limited number of skilled interpreters available; they should be used for non-conventional image processes and not for standard ones;

4. the request for quantitative analysis, in order to extract from the pictures as much information as possible.

For image digitization we can assume, that the pixels lie on the integer values of x and y. The grey levels shall be quantized into 2^k steps, therefore characterizing the image by k bits per pixel. k might be typically 6 bits or more, thus 64 different gray levels or more may be distinguished within an image.

A typical B/W image consists of $N^2 = 4000 \times 4000$ pixels. The information content of an image is thus in the order of $N^2 k = 1.6 \cdot 10^7 \cdot 6 \approx 10^8$ bits or more. This is very large and is the reason for problems, which arise in storage capacity and process time, if a large quantiy of images has to be analysed (ALLEN et al., 1973).

9.1.2 Relation between Image and Real Scene

The actually recorded image $f(x, y)$ is, mathematically spoken, related to a true scene $s(u, v)$ by a transformation T

$$f(x, y) = T[s(u, v)]. \tag{1}$$

$s(u, v)$ can be determined from the image by executing the inverse operation T^{-1}

$$s(u, v) = T^{-1}[f(x, y)]. \tag{2}$$

This requires the knowledge of the transformation T. T is a function of

1. the imaging system, i.e. the sensor performance;
2. the platform attitude and position as well as of the illumination conditions, which determine under what geometric and radiometric conditions the mapping was performed by the sensor;
3. the terrain geometry
4. the medium between object and sensor, i.e. of the atmospheric conditions.

T is for this reasons the product of several transformations. Some of them are wanted, some of them are, unfortunately, not avoidable.

One particularly wanted transformation is the ideal sensor related projection. This may be a perspective or an orthographic, or a panoramic, or some other projection without any distortions. In addition, also the generation of standard cartographic products such as Mercator or Polar Stereographic projections may be part of the wanted image transformations. These transformations are the nominal ones. They are axactly determined. Without distortions by the unavoidable transformations, two control points per image would allow to determine scale. orientation and, therefore, the true position of all pixels and of the corresponding objects. In addition, a gray scale for brightness calibration would determine the exact intensity distribution $s(u, v)$.

But additional unavoidable transformation cause problems of distortions and degradations. Some typical examples of degrading transformations are listed in Fig. 1. There are two different classes, the one dealing with geometric or spatial transformations and the other with photometric or radiometric ones. Geometric transformations are concerned with the relationship between the coordinates of the objects and the actual pixel locations. Radiometric transformations leave the geometry of the pixels unchanged, but modify the gray levels. Some of the transformations are simple and can be considered to be linear, or quadratic. Some distortions are, however, complicated, for example the pin cushion distortion, and can be determined and treated best by means of calibration images or control points. Fortunately, these latter ones are typical for some sensor types and their distortions change only slowly, so that the corrections can be done by table lookup (WEBBER, 1973).

The geometric errors can be measured in terms of rms errors in meters on ground. In case of a normal distribution with zero mean the rms number corresponds to the 68% value. Typical rms values for the ERTS images are roughly 200 m for sensor alignment, 100 m for ephemeris position, 30 m for exposure time and 700 m for attitude errors. So totally, the external errors are in the order of

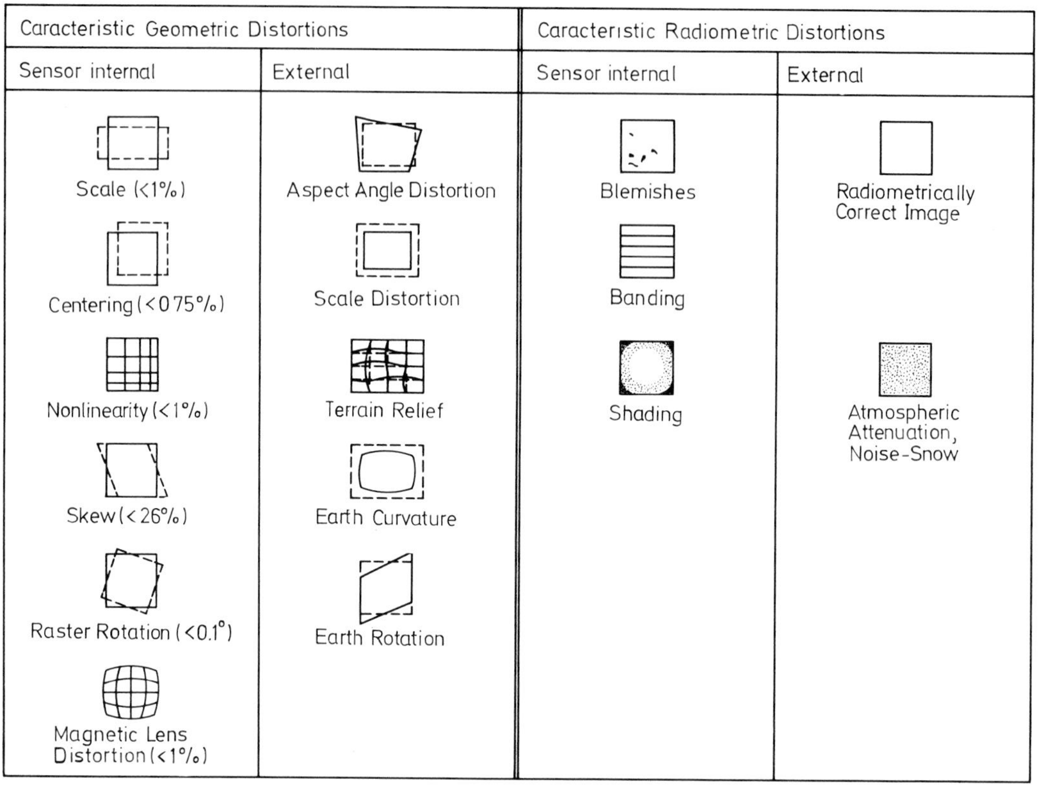

Caracteristic Geometric Distortions		Caracteristic Radiometric Distortions	
Sensor internal	External	Sensor internal	External
Scale (<1%)	Aspect Angle Distortion	Blemishes	Radiometrically Correct Image
Centering (<0 75%)	Scale Distortion	Banding	
Nonlinearity (<1%)	Terrain Relief	Shading	Atmospheric Attenuation, Noise-Snow
Skew (<26%)	Earth Curvature		
Raster Rotation (<0.1°)	Earth Rotation		
Magnetic Lens Distortion (<1%)			

Fig. 1. Characteristic geometric and radiometric distortions. Figures in parenthesis are representative for ERTS-RBV images. (BERNSTEIN and SILVERMAN, 1971; NASA, 1971)

700 m rms. Special internal errors for the ERTS-RBV are given in Fig. 1. They result in a total of 1355 m rms. For the ERTS-MSS the internal errors combine to an rms positional mapping error of 26 m and are very small as compared to the external errors.

Exact correction of a picture, i.e. compensation of the image degradations and establishing an image, which maps the scene in accordance with the wanted transformations, is called image restoration. This procedure is necessary with regards to geometry for cartographic maps (MALILA et al., 1972; BERNSTEIN, 1973).

Opposite to that, an image registration transforms multiple images of the same scene to match each other and to make them uniquely adressable: The locations of the pixels are geometrically identical, but have some distortions with respect to the nominal positions. To put images into registry becomes necessary for common use of the color composites of an image, or for change detection. Some of the problems involved in image registration and restoration will be dealt with in Paragraph 3.

Similar to the geometric aspect, we can also distinguish in the radiometric case between absolute and relative accuracy. If the radiance of a pixel is related to the radiance of another pixel in the image, then we are talking about relative radiometry. If the brightness of the pixel is, however, related to the radiance into the sensor, then we are talking about absolute radiance.

As calibration is usually done by means of a reference light source, which may be known only with a limited accuracy of, say 1–3%, the absolute accuracy is also limited.

The radiometric errors are a function of spatial frequency: Electrical as well as optical and photographic systems respond with a decreasing output level for an increasing frequency. In other words, the modulation transfer function (MTF), which is the ratio of the brightness variations in the image to those in the original, varies with spatial frequency and introduces, therefore, a radiometric error. For the ERTS images the corresponding radiometric errors on a 70 mm film are 9% for 40 cycles per mm, 20% for 5 cycles per mm, and 1% for 10 cycles per mm, including the effects of the sensor, the electronics and the photosystem. For lower spatial frequencies, such as 1 cycle per mm on a 70 mm film, the radiometric accuracy is independent of the target size (NASA, 1971). Shading is another radiometric error: At different spatial locations of the face the sensor might have a different response. Also different signal levels might be of influence.

Many other error sources—atmospheric, sun angle etc.—must be considered to reduce the system radiance error. But for the various error sources different models must be applied, which are too specific as to be discussed in this chapter. Tools for image correction are given by the image enhancement and filtering processes (Paragraph 4) and by application of hardware like the electron beam recorder, (2.2.3 and 2.2) (MALILA and NALEPKA, 1973; COUSIN et al., 1972; BAKIS et al., 1972).

9.1.3 Enhancement and Filtering

Aside from restoration and registration image processes, are also used for image enhancement and image filtering. By means of these latter procedures, the interesting part of the picture information shall be demonstrated in a more pronounced way and the background information shall be suppressed or at least deemphasized. Procedures for these purposes are, of course, dependent on the particular interest of the special user. They shall help the analyst and are used for feature extraction and classification, as will be shown in Paragraph 5. Both, image enhancement and image filtering apply processes, some of which are better understood in the spatial domain of the image signal, others are better dealt with in the frequency domain, as will be seen in Paragraph 4. In order to understand the background herefore, these dual procedures will be discussed from a mathematical point of view in the Appendix 1 (ANDREWS et al., 1970; ROSENFELD, 1970).

9.2 The Elements of an Image Data Processing and Analysis System

An image data processing and analysis system can be broken down into 5 subsystems for (1) data acquisition, (2) data handling, (3) data preprocessing, (4) data analysis, and (5) data storage and retrieval.

9.2.1 Data Acquisition Subsystem

Image data can be collected by various types of sensors, including photographic and tv-cameras, mechanical scanners, radar and lidar systems, etc. A general purpose image data center must, therefore, be able to accept and process very different data formats. The delivered raw data may have different support media, like photographic film, analog or digital tapes, or they may be telemetry signals. The sizes of the images and the arrangements of the data as well as the data encoding may vary considerably.

In addition to the image data themselves, auxiliary information also has to be acquired, which is related to the sensor characteristics, the locations, to time, and to geometric and radiometric corrections. The raw data have to be annotated with these information after the quality check. This annotation assures easy access to the image data, when stored in the data bank and allows the image data corrections, when ever it becomes necessary. Figure 2 (see pp. 366 and 367) shows an ERTS-precision processed output film with the auxiliary data necessary for image analysis.

The acquired data will be stored in the historical file, which is part of the data storage and retrieval subsystem. A master copy is produced from those raw data, which will be processed further.

9.2.2 Data Handling Subsystem

9.2.2.1 Concept

From the master copy the data are converted to formats, which are especially suitable for the subsequent processes. Paper prints and enlargements as well as transparencies will be produced from the photographic images. Also digitization of the picture information and recording on computer compatible magnetic tapes (CCT) is often requested, in order to allow automatic image processing. From the analog magnetic tapes the data have to be digitized and stored an CCT's.

The digital signals, if stored on high density digital tapes (HDDT) must also be recorded on CCT's. These are the only ones, which can be easily addressed and all computers, equipped with tape recorder facilities can handle them. Typical data characteristics of CCT's, HDDT's, and analog tapes are listed in Table 1.

Data will be displayed for visual inspection during the automatic process and final products should be delivered often in form of hard copies. Conversion of optical signals into digital ones and vice versa is necessary.

Table 2 lists the output formats of the ERTS Image Processing System.

9.2.2.2 Image Input and Output Devices

Various techniques are applied for the optical-digital-optical conversion. Four widely used basic concepts are (1) the rotating drum microdensitometer, (2) the flying-spot scanner, (3) the electron beam recorder, and (4) the laserscanner recorder.

Table 1. Typical data characteristics of computer compatible tapes (CCT's), high density digital tapes (HDDT's) and analog tapes

Type of tape	No. of tracks	Bit density per track	Transfer rate per track	Capacity of tape
CCT	6 (8)	$2.8 \cdot 10^3$ bit/cm^2	$320 \cdot 10^3$ bit/s	$2.4 \cdot 10^8$ bit ($3.2 \cdot 10^8$) for 12.7 mm $\times$ 720 m
HDDT	28	$8.7 \cdot 10^4$ bit/cm^2	$2.4 \cdot 10^6$ bit/sec	$6 \cdot 10^{10}$ bit for 25.4 mm $\times$ 2800 m
Analog tapes		$6 \cdot 10^4$ bit/cm^2	$15 \cdot 10^6$ bit/sec	$6.65 \cdot 10^{10}$ bit for 50.8 mm $\times$ 2160 m

Table 2. Output formats of ERTS image processing system. (R. JOHNSON and R. BUITEN, 1971)

Product	Sensor	Format	Notes
Bulk-master images	RBV and MSS	B and W Film of all spectral channel at 1:1000000 scale	– 70 mm (1:3370000 scale is 1st bulk product-enlarged to 1:10^6. – For all users for screening, direct analysis, and correlative use.
Bulk-color master images	RBV and MSS	Color composites of 3 (3 or 4 for MSS) spectral channels at 1:10^6 scale	– RBV images to be in precision (70 meters, rms) registration. – RBV and MSS registration to geo. coordinates (70 meters, rms). – Cartographic applications, improved false color and multispectral interpretation of RBV image data.
Precision-color images	RBV and MSS	Color composites of 3 (3 of 4 for MSS) spectral channels at 1:10^6 scale	
Special-digitized image video data	Bulk and precision-RBV and MSS	Computer-compatible tape – 9-track, 800 bits/inch	– For user applications, requires single and multichannel video analysis and enhancement.

The principle of the rotating drum microdensitometer is shown in Fig. 3. The transparency of the image to be digitized is fixed on a rotating drum. A light source, a lamp or a laser, scans along the drum axis. One rotation of the drum corresponds to one image line, one scan of the light source completes one complete image scan and lasts as long as one drum rotation times the number of image lines.

By the transparent image the light beam is modulated. The photomultiplier detects the brightness variations as electrical signals, which are time varying proportional to the image brightness. Analog to digital conversion of these signals concludes this digitization process.

PH. HARTL

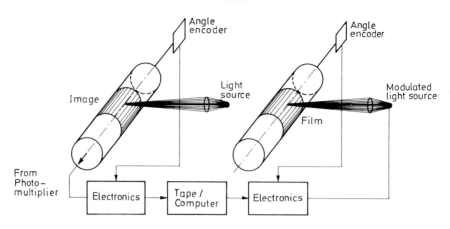

Fig. 3. Rotating drum microdensitometer and recorder

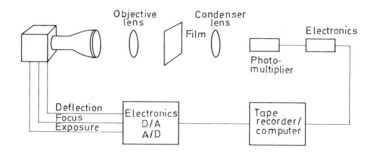

Fig. 4. CRT-reader/-writer (flying spot scan system)

Inversely, a second drum is used for the write mode. Here, a photosensitive film is fixed on the drum. The light source is modulated by the digital to analog converted values of the pixels. Light source and drum scans or rotates, respectively, in the same manner as described for the read mode.

Instead of a lamp, the light source can also use a laser beam. There is the advantage in this case, that, because of the dense light concentration a high conversion speed can be achieved and fine corn film renders high resolution, in this case.

The flying spot scan system is based on a cathode ray tube (CRT); Fig. 4. No mechanical moving parts are involved. The electron beam scans instead, and generates the moving light spot on the screen. The transparent image lies in between the CRT-screen and a photomultiplier. Therefore, also here the light becomes modulated by the brightness function of the image. This is detected by the photomultiplier and converted into digital signals, similar to the procedure for the rotating drum. In the write mode, a photosensitive film is exposed to the light spot of the CRT-screen, and the light spot generating electron beam is controlled by the digital to analog converted signals.

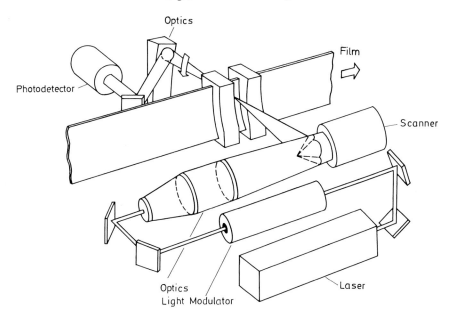

Fig. 5. Principle of the laser-scanner-recorder. (Source: RCA)

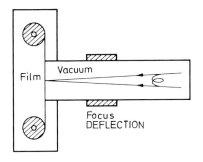

Fig. 6. Electron beam recorder

The laser beam recorder is, from its principle, very similar to the rotating drum equipment. The mechanical arrangement is, however, such, that geometric distortions are widely eliminated. The laser beam is rotating and not the film. This one moves along the rotation axis; Fig. 5.

The electron beam recorder is a high quality image recorder. The principle is shown in Fig. 6. The electron beam is focused in all 3 dimensions on the electronic sensitive film. The film chamber is evacuated.

The laser beam recorder provides the highest accuarcy and the best resolution. About 40 000 pixels can be resolved per scan and the distortion is as low as 0,005%.

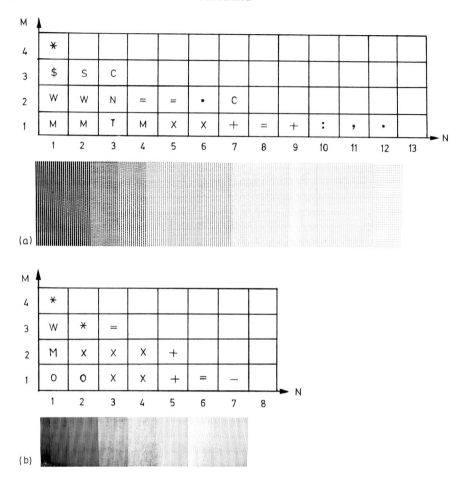

Fig. 7 a and b. Generation of gray shades with line printers; two examples: (a) $N=13$ shades; (b) $N=8$ shades. $M=$number of characters. (Source: General Electric, IBM)

The electron beam recorder resolves 10000 pixels per scan and provides an accuracy equivalent to 0,05% distortion.

For image writing a straightforward method is the alphanumeric printout. The gray levels are represented by printing sets of alphanumeric characters, examples of which are given in Fig. 7.

9.2.3 Data Preprocessing Subsystem

Referring to the ERTS-image preprocessing practice the products of this subsystem can be classified into three categories, which take into account the different interests of the users. The quick look bulk images will be distributed widely. The second type, the precision film imagery, will have eliminated geometric error

Table 3. ERTS-performance data. (R.JOHNSON and R.BUITEN, 1971)

	Parameter	MSS	RBV
Bulk processing	Throughput	10000 Images per week	
	radiometric accuracy	$\pm3\%$	$\pm10\%$
	registration accuracy	115 meters	325 meters
	mapping accuracy	880 meters	930 meters
Precision processing	Throughput	455 Images per week	
	radiometric accuracy	$\pm4\%$	$\pm10\%$
	registration accuracy	120 meters	110 meters
	mapping accuracy	85 meters	70 meters

to a very large extent, and finally a third category of products is destined for quantitative processes and analysis.

The bulk image data will be corrected for errors, which are originated by the spacecraft, attitude, and earth rotation, since these errors are systematic and do not change significantly within short times. The auxiliary data which are necessary for this process are, of course, annotated to the tapes. The preprogrammed corrections, just mentioned, and the bulk image production are performed with the help of an electron beam recorder. This is a well adapted equipment, since the beam can be controlled in its coordinates as well as in its intensity by the signals, which are computed as error correcting signals. 10000 images can be produced with sufficient quality by this way; see Table 3.

The remaining errors, which are left for correction to precision film imagery, are determined by automatic correlation with reseau marks and recognizeable ground objects in the image (see Paragraph 3), and by calibration methods of radiometric errors. The precise image processing renders accuracies, as given in Table 3.

The third category of preprocessing techniques deals with the processes for filtering and enhancement, which are needed for classification and analysis. The various possibilities are discussed in Paragraph 4.

The classification processes extract finally the information from the picture— saying, for example, which areas or which percentage of the whole scene will be cornfields, etc. They are decision making. They need the user's decision about the criteria, which features shall define a specific class.

9.2.4 Data Storage and Retrieval Subsystem

For data storage three different types of file are used, (1) the historical file, (2) the working file, and (3) the user's file.

The historical file stores the original documentation. Only part of it will be processed and analysed immediately. The working file is used for products determined for actual utilization. Not only the raw image data, but also additional information, necessary for further processing, is stored here. Catalogs of material parameters, with regards to radiometric data, and maps are typical examples for that.

The third file is the user's file. It stores information of general interest, for example the results about the agricultural or forestrial usage of certain areas, achieved by remote sensing data (SKINNER and GONZALEZ, 1973).

9.3 Geometric Corrections

Three different techniques are in practical use, image gridding, registering and mapping. Before dealing with them, some general remarks about the linear, quadratic and higher order geometric transformations are necessary.

9.3.1 Linear Transformations

Referring to Eq.(1) and Fig. 1 a, geometric distortion can be defined by a transformation of the coordinates. Many of these transformations are simple and can be specified for an image region by a pair of linear equations:

$$x = Au + Bv + C$$

$$y = Du + Ev + F \tag{3}$$

$$f(x, y) = f(Au + Bv + C, \ Du + Ev + F).$$

The capital letters are parameters to be determined for each special case. A pure displacement or translation is given for

$$A = E = 1, \quad B = D = 0.$$

A pure rotation is determined by $A^2 + B^2 = D^2 + E^2 = 1$.
A change of scale is equivalent to $A/B = D/E; \ C = F = 0$.

For these special cases the transformation parameter can, therefore, be determined very easily (NAGY, 1972).

9.3.2 Quadratic and Higher Order Transformations

More complicated transformations can be approximated by means of quadratic equations or polynomials of grade N. The latter case, including the linear and quadratic as special cases, is given by the following equations:

$$x = \sum_{p=0}^{N} \sum_{q=0}^{N-p} A_{pq} u^p v^q,$$

$$y = \sum_{p=0}^{N} \sum_{q=0}^{N-p} B_{pq} u^p v^q. \tag{4}$$

The determination of the parameter values increases considerably for higher values of N. There is sometimes the choice to use either one pair of higher order transformation equations or several linear equation systems. Under the latter condition the image is subdivided into more sectors, where for each sector one pair of linear equations can be applied. In general, this is to be preferred as compared to the higher order possibility; it saves computer time.

9.3.3 Reference Marks

The most bothersome distortions cannot even be dealt with Eq. (4). Among these complicated transformations the pincushion distortions and the optical aberrations are representative. For them it is recommended to use reseau or fiducial marks. They are located very accurately on the faceplate of the sensor. The differences between the known locations and the apparent ones in the image give a map of sensor internal errors. As these sensor errors change slowly, it is sufficient to determine these fiducial coordinates only from time to time.

Besides of these references marks just described, there are two other types, which are used for external geometrical distortions. These are the pass points and the ground control points. Pass points are arbitrarily selected locations for image registration purposes. Ground control points (GCP) are used for mapping operations and are locations of known positions.

9.3.4 Image Gridding

This is the simplest form of a geometric correction. The image itself is left without any modifications. Instead of correcting the locations of the pixels some standard map information is superimposed over the image. Geographical coordinate lines, coastal lines or even border lines can be used for these purposes. This makes orientation in the image easy, the points of the grid lines are determined clearly and the pixels in between the lines can be determined by interpolation.

This technique is applied in weather satellite maps, for example. It is an adequate method for low accuracy requirements. It can be implemented easily, i.e. with little computertime, as only relatively few coordinate points for the gridding must be calculated. It can, however, become difficult, to compare 2 images of this type, taken with two different sensors or at different times.

9.3.5 Image Registration

One image is taken as a reference or master copy. The others are to be matched to it. In order to save process time the image registration is usually done in two steps: (1) Global preprocessing is performed by means of a few initial pass points, distributed over the whole image. By this procedure overall scale and rotational differences shall be eliminated. (2) Then local transformations are done in using a larger number of pass points.

The automatic matching of the points is not as easy as the manual operation. In order to understand the problem, one has to consider, that data are stored as digits on a tape and the pattern of the pass point must be recognized. It is only known with tolerances, where this pattern will be located; it lies within a "window". For all possible positions of the reference mark a "similarity-degree" must be calculated. The coordinates, for which the similarity between a search area of the image to be registered and the window area of the reference image is highest, are destined to indicate true position. The following subparagraph will make this more clear. Besides, also the problem of defining the similarity will become evident there.

9.3.5.1 Shadow Casting Method

IBM has developed a strategy for the reseau mark detection and location for the ERTS-RBV-camera. 81 fiducial marks are located at the faceplate of this camera. These marks are of cross shape with arms, 4 pixels wide and 32 pixels long; Fig. 8. A one percent uncertainty of the location corresponds in the RBV camera to 41 pixels in the X and Y direction. The search area of 128×128 pixels is, therefore selected, being the nearest binary number to $2 \times 41 + 32 = 114$ (BERN-STEIN and SILVERMAN, 1971).

Starting with a 1% uncertainty of the location, a block of 128×128 elements is centered around a reseau mark's location. For both, the row and the column the sums S_r and S_c respectively are calculated:

$$S_r = \sum_{n=1}^{128} f_{m,n},$$
$$S_c = \sum_{m=1}^{128} f_{m,n}. \tag{5}$$

$f_{m,n}$ being the grey level of the mth row and the nth column. For both, the row and the column we receive a sequence $\{S_r\}$, $\{S_c\}$, respectively. This operation is called "shadow casting". As the reseau marks correspond to small values with regard to grey levels—they should be dark—the sums around the center of the reseau marks should be minimum. Indeed, one of the conditions of the IBM program is, that the data values of the reseau arms are much less than those of the surrounding areas, for example at least 8 counts less. The actual minimum S_r of the row sums and S_c of the column sums are assumed to indicate the mark center. Then the sums $S_{r-5} \ldots S_r \ldots S_{r+5}$ and $S_{c-5} \ldots S_c \ldots S_{c+5}$ are used as fitting points for quadratic function of the estimated value $\hat{S}_r, \hat{S}_c$ respectively. This assumes, that a quadratic transformation is sufficient, to describe the distortions of the camera in a small sector around the reseau mark.

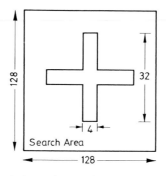

Fig. 8. Reseau-mark pixel search area. (BERNSTEIN and SILVERMAN, 1971)

For column sum, the following equation holds:

$$\hat{S}_c = A + Bx + Cx^2 .$$

The parameters A, B, C are determined by means of the fitting points:

$$S_{c-5} = A - B5 + C25; \quad S_c = A; \quad S_{c+5} = A + B5 + C25;$$

therefore

$$A = S_c; \quad B = \frac{S_{c+5} - S_{c-5}}{10}; \quad C = \frac{S_{c+5} + S_{c-5} - 2S_c}{50}.$$

The estimated value of S as a function of x can, therefore, be written:

$$\hat{S}_c = \frac{S_{c+5} + S_{c-5} - 2S_c}{50} x^2 + \frac{S_{c+5} - S_{c-5}}{10} x + S_c . \tag{6}$$

The corresponding equation for the row values is given by just replacing the subscript c by r.

Those sums, $S_{c-4} \ldots S_{c+4}$, which exceed the quadratic function, are detected by a computer dynamic threshold. More than 2 sums must be smaller than the detection threshold, given by the estimated value $\hat{S}_c$ and $\hat{S}_r$ respectively. If this is true, then the center of the reseau mark is assumed to be the one which was found to have the minimum S_c, S_r.

9.3.5.2 Ground Control Point Correlation

Errors external to the sensor, such as platform attitude errors, earth curvature influence and earth rotation can be determined by use of a suitable number of ground control points in case of restoration, or by use of pass points in case of registration. Two images of the same scene which do not differ, in magnification and rotation are used, the one being the reference for the other. The best translational fitting has to be looked for. As shown in Fig. 9 a search area S and a window W are defined. S is an array of $L \times L$ pixels, which may assume one of J gray levels, or in other words: The gray level $S(i,j)$ of the pixel with the co-ordinates (i,j) assumes values between the following limits:

$$0 \leqq S(i,j) \leqq J-1 ;$$

$$1 \leqq i,j \quad \leqq M . \tag{7}$$

The window W should be an array of $M \times M$ pixels, M being smaller than L. W is a subimage of that image, which should be put into registry. The gray levels $W(l,m)$ should lie in the same range as those of S:

$$0 \leqq W(l,m) \leqq J-1 . \tag{8}$$

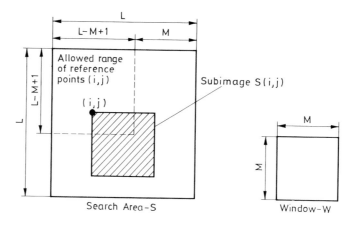

Fig. 9. Search area and window for ground control point correlation. (BERNSTEIN and SILVERMAN, 1971)

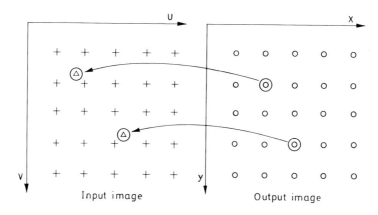

Fig. 10. Geometric correction. (BERNSTEIN and SILVERMAN, 1971)

As can be seen from Fig. 10 subimages, fully contained in the window, are determined by the following condition:

$$1 \leqq l, m \leqq M,$$

$$1 \leqq i, j \leqq L - M + 1. \tag{9}$$

We define the coordinates (i, j) of the upper left corner of the subimage as the indicator of the window's location. For registration the window (i, j) has to be placed to all locations, for which it is fully in the search space. For all positions, the similarity of window and subimage has to be calculated. The maximum of similarity being achieved at window position (i^*, j^*) determines the correct position of the window. Similarity can be defined by various algorithms. Most widely used is correlation. The normalized correlation R_N is defined by

$$R_N^2(i,j) = \frac{[\sum_{l=1}^{M} \sum_{m=1}^{M} W(l,m) S(i+l-1, j+m-1)]^2}{[\sum_{l=1}^{M} \sum_{m=1}^{M} W^2(l,m)] [\sum_{l=1}^{M} \sum_{m=1}^{M} S^2(i+l-1, j+m-1)]}. \tag{10}$$

For all window locations allowed, R_N is to be calculated by this formula. All R_N-values will lie between zero and one. If $R_N(i,j)$ is assumed to be the third dimension to the search area, then a correlation surface can be drawn, the peak of which determines the maximum similarity.

Correlation is complicated and costly. M^2 points of the window are to be compared with $(L-M+1)^2$ points of the search area thus $M^2(L-M+1)^2$ correlations must be performed. Therefore BERNSTEIN and SILVERMAN (1971) proposed a different approach, which is called "sequential similarity detection".

In this procedure the positions of the windows are chosen at random over the entire search area, and are widely distributed. This renders, in general, a great deal of new information for each test. For determination of similarity the error between the windowing pairs is calculated, being defined by:

$$E(i,j) \equiv \sum_{l=1}^{M} \sum_{m=1}^{M} |S(i+l-1, j+m-1) - W(l,m)|. \tag{11}$$

In the ideal case a minimum of zero is guaranteed for

$$W = S(i^* + l - 1, j^* + m - 1).$$

This computation can be done much easier than the one for Eq. (10). Besides: For points, far from the mimimum, the gray level pairs of the corresponding pixels do not match on the average and the difference accumulates rapidly. As soon as the errors of the components of Eq. (11) accumulate for a location (i,j) to a value above a threshold T, the operations for this (i,j)—window position are stopped. By this way, computer time can be reduced considerably, as only for those positions the differences are completely calculated, for which values of E are below threshold. Among them the one (i^*,j^*) location is then given as E-minimum.

9.3.6 Geometric Correction Concepts

Figure 10 shows the principle of mapping the input image into the output image. It is assumed that no high frequency distortions and also no local ones are to be taken into account. Then the transformation can be performed by means of polynomial mapping according to Eqs. (3) and (4). Coordinates of an output pixel do generally not coincide with the coordinates of any input pixel, but correspond with some location in between pixels. Interpolation procedures have to be performed by two steps in order to achieve the correct gray level: Each pixel of the output image is retransformed into the image. The pixels surrounding it there and their gray levels are used to determine by interpolation, which gray level is to be attributed to the transformed pixel.

A simplified version of this concept is called point shift algorithm.

9.3.6.1 Point Shift Algorithm

Here it is assumed that input and output image are similar enough to each other as to guarantee that pixels error locations are at most one-half pixel spacing in each axis.

Therefore, it is possible to assign to each transformed pixel a nearest neighbor. If input pixel with coordinates (m,n) is the one closest to the actually mapped location of the (i,j)th output pixel, then the gray level $g(m,n)$ is assigned to that output pixel (i,j).

9.4 Image Enhancement and Filtering Processes

We will now consider the images with regards to their use for an analyst. He receives the images, which may still have some degradations because of noise, attenuation as a function of spatial frequency, low contrast between the points of interest and the background, etc. It is in this paragraph, where we deal with the possibilities, which are available to overcome these problems (Bressanin et al., 1973; Kritikos, 1971).

9.4.1 Enhancement Processes

One way to show some features in a more pronounced way is to enhance the contrast. For this purpose the gray level distribution can be manipulated. Very often, the original gray level histogram of an image shows that either the bright or the dark levels are preferred. This means that the range available for the gray levels is not used in its full span and, therefore, the various contrasts of the image scenes are not as large as they could be. A redistribution of the gray levels can bring improvement. The gray level transformation can be done in various ways;

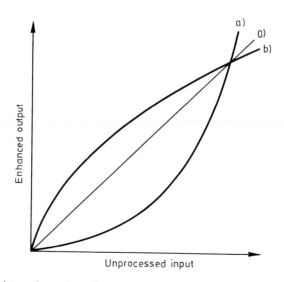

Fig. 11. Gray level transformations for contrast enhancement: *a* Exponential and *b* logarithmic correction

Fig. 11. The gray levels can be reassigned on the basis of an exponential characteristic, curve *a*, which compresses the dark levels and expands the bright ones.
Noise can be suppressed by this way, because generally the noise level is low as
compared to the signal level. The gray levels can be changed in accordance to a
logarithmic function, where the bright levels are compressed and the dark ones
expanded, curve *b*.

A more sophisticated method is the redistribution of the gray levels in such a
way that the resultant histogram is almost rectangular. One example in this
direction is shown in Fig. 12.

The method of density level stretching expands those gray levels linearly over
the full scale, which are between some boundaries *A* and *B*; Fig. 13 a. A more
interesting procedure is multi density level stretching; Fig. 13 b. The effect ist
that everything in the picture is in high contrast. This shows on the one side the
features, which otherwise would be hidden. But it is, on the other side, very
sensitive to noise. So, only common use with the original image is recommended.

Density slicing is another widely used enhancement process. As shown in
Fig. 13 c, the continuous brightness of the image signals becomes quantized. The
simplest way is to use only one binary step, i.e. to fix some brightness level *T* as
threshold value and to map signal in white, if it corresponds to a level above and
black for a level below *T*; Fig. 13 d.

A more interesting way is to use *N* different thresholds and thus quantize the
gray levels into *N* steps. A human observer can distinguish between 60–100 gray
levels. In density slicing only the order of ten levels is applied. This makes identification of areas with changing gray levels often much easier. Figure 14 a demonstrates the feasibility. Density slicing is a first step towards classification techniques. Each quantized level can characterize a tolerance limit for a given feature.
For example climatic zones can be defined via specific temperatur ranges in a
thermal IR-image (BRISTOR, 1970).

9.4.2 Image Smoothing

In order to smooth the gray level changes the actual brightness of the pixel
is replaced by an average value, for example of the 8 neighboring pixels and itself
in the two-dimensional case, and by the 2 neighboring pixels and itself in the
one-dimensional case. In this simplest case of a low pass filter, which de-
emphasizes the higher spatial frequencies, the transformation is given by the
Eqs. (39) and (40) of the appendix, with $M = 1$ and $N = 1$ and

$$h_i = \tfrac{1}{3}, \qquad h_{ik} = \tfrac{1}{9}:$$

One-dimensional low pass filter

$$f(x=m) = \tfrac{1}{3}s_{m-1} + \tfrac{1}{3}s_m + \tfrac{1}{3}s_{m+1}. \tag{12}$$

Two-dimensional low pass filter

$$f(x=m, y=n) = \tfrac{1}{9}[s_{m-1,n-1} + s_{m-1,n} + s_{m-1,n+1} + s_{m,n-1} + s_{m,n}$$

$$+ s_{m,n+1} + s_{m+1,n-1} + s_{m+1,n} + s_{m+1,n+1}]. \tag{13}$$

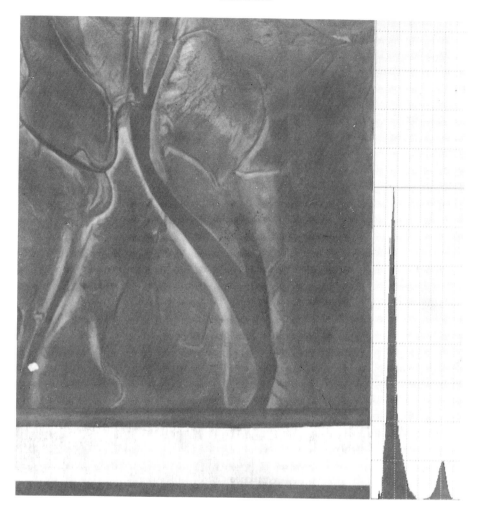

Fig. 12a and b. Equalization of gray level histogram (Triendl, 1972). (a) Air photo and gray level histogram. (b) Equal distribution of gray levels

In general, the weighting factors need not to be equal and also neighborhood, which will be involved in the averaging process, can be larger.

The digital modulation transfer function MTF of Eq.(12) can be calculated by means of Eq.(42):

$$H(f_a) = \tfrac{1}{3}[1 + \cos 2\pi f_a - j \sin 2\pi f_a + \cos 2\pi f_a + j \sin 2\pi f_a]$$

$$= \tfrac{1}{3} + \tfrac{2}{3} \cos 2\pi f_a . \tag{14}$$

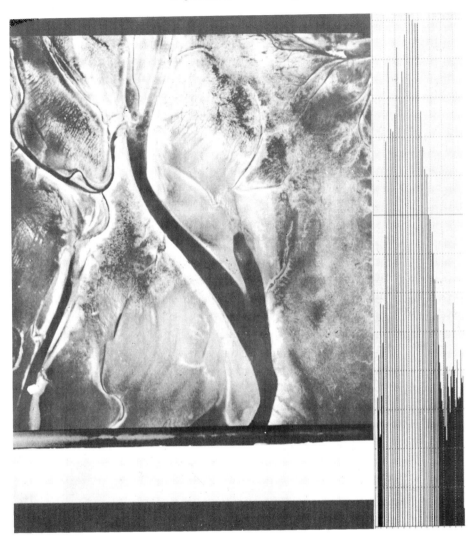

Fig. 12 b

This is shown in Fig. 15 as curve *a*. Curve *b* of the same figure shows the MTF for a five point averaging filter, for which the following equation can easily be calculated (SELZER, 1968).

$$H(f_a) = \tfrac{1}{5}[1 + 2\cos 2\pi f_a + 2\cos 4\pi f_a] . \tag{15}$$

It is obvious from Fig. 15 that these filters have really low pass character, i.e. they do not allow higher frequencies to be transferred.

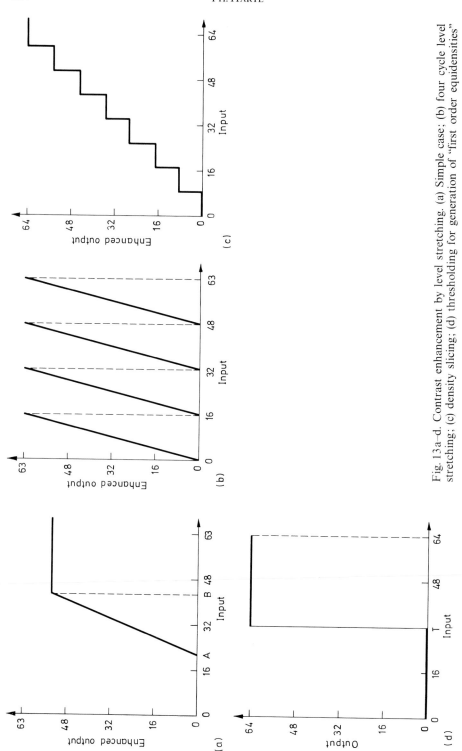

Fig. 13a–d. Contrast enhancement by level stretching. (a) Simple case; (b) four cycle level stretching; (c) density slicing; (d) thresholding for generation of "first order equidensities"

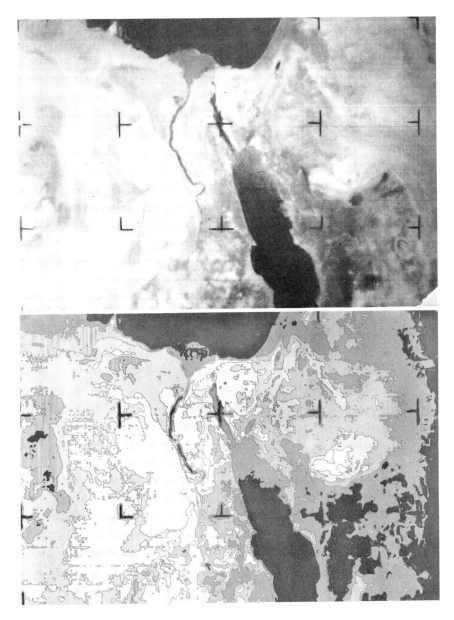

Fig. 14a. Density slicing. (KRITIKOS, 1971). Above: NIMBUS picture. Below: Six level density slice plus border lines

Fig. 14b. Image enhancement by pseudoplastic mapping (KRITIKOS, 1971). Left: Color infrared satellite photo. Right: Pseudoplastic mapping of the above image

9.4.3 Image Sharpening

Low pass filtering is of interest, if large low contrast features contain random noise. The latter one can be suppressed partly, as it has preferably high frequency components. If the low-pass filter is subtracted from the original filter, i.e. if the gray values of the low-pass filtered image are subtracted pixel by pixel from the unfiltered ones, then the complementary image is produced, the high pass filtered picture.

Overemphasizing of contrasts is sometimes necessary, in order to give the observer some hint that changes exist in the gray levels. But direct contrast

Fig. 14b

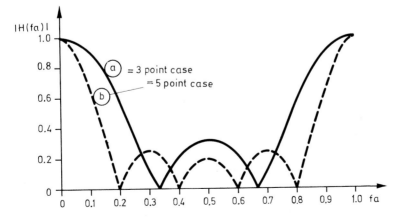

Fig. 15. MTF for equal weight from pass filters

enhancement may cause, under practical conditions, saturation in the image. A high pass filter, however, removes the background, which is slowly varying, and shows the faster varying feature. Various interesting examples for practical application of the high pass filter and other enhancement processes in the bio-medical application are given in by SELZER (1968). Also differentiation in any direction x can be applied for accentuation of changes

$$\frac{\partial f_{i,j}}{\partial x} = f_{i,j} - f_{i,j-1} .$$ (16)

If sharpening is wanted in every direction, then the derivative should be taken in the gradient direction, i.e. in the direction of fastest change in gray level. The corresponding derivative is equal to $\sqrt{\left(\frac{\partial f_{i,j}}{\partial x}\right)^2 + \left(\frac{\partial f_{i,j}}{\partial y}\right)^2}$ for any pair of orthogonal directions x, y.

The Laplacian operator

$$\frac{\partial^2 f_{i,j}}{\partial x^2} + \frac{\partial^2 f_{i,j}}{\partial y^2} = f_{i,j} - \tfrac{1}{4}(f_{i-1,j} + f_{i+1,j} + f_{i,j-1} + f_{i,j+1}) .$$ (17)

is independent of the direction. Here the new brightness in (i,j) becomes equal to the difference between the brightness of the pixel (i,j) and the average brightness of the surrounding points in the 2 orthogonal directions x, y. This operator en-hances the brightness changes very sharply, but does this also with noise. Reduc-tion of the noise content in the image before performing the enhancement process should, therefore, be considered.

Finally a combination of the original image or a low pass filtered image with the Laplacian processed picture is interesting. By this way, the positive values of the differentiation and the negative ones correspond to an increase or decrease of the brightness and generate a shadowing effect. The result is a pseudoplastic image. Although looking like an altitude variation in a relief the gray levels are, of course, still just brightness indicators and not even quantitatively fixed brightness levels. Figure 14b shows an example for this technique (GRIFFITH, 1973; SCHÄRF, 1973).

9.4.4 Matched Filters

These processes operate by cross-correlating the picture with a matrix of weights. The weights are proportional to the values of the gray levels the interest-ing image pattern would have. The ground control point correlation of Paragraph 3.5.2 applies this technique, where the weight matrix is the digital window image and the correlation process is given by Eq. (10). This technique is also called matched filter process or feature selective filtering. It can be applied in the spatial as well as in the spectral domain. SELZER uses a very simple feature selective filter, which we shortly describe in the following (SELZER, 1968).

Figure 16 shows the feature selective filter matrix which is used to enhance lines, that are linear or nearly linear over short distances. It is only binary weight-

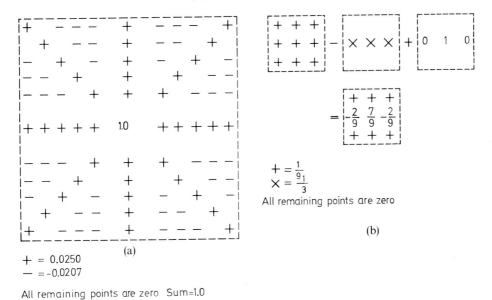

+ = 0.0250
— = -0.0207

All remaining points are zero   Sum=1.0

Fig. 16a and b. Weighting factors for feature-selective filters that (a) enhance lines that are linear or nearly linear over a short distance (SELZER, 1968); (b) remove scan line noise by adding the difference between the scene brightness average and the line brightness average to the pixel value to be corrected

ing, i.e. the levels of the pixel are either multiplied by ($+0.0250$) or by (-0.0207). This makes the output of the filter large, if the filter is centered on a straight line of anyone of the vertical, horizontal, or diagonal directions. If only vertical lines should be enhanced, then only the filter weights of the vertical line in the center of the weighting matrix is made positive, all others negative. In case of horizontal lines to be enhanced, the horizontal line through the center is the only positive weighted one.

A filter for finding a square in the original image could be implemented by means of a weighting matrix with a positive core of a size according to the square dimensions and a negative frame. The resulting correlation surface has its maximum at the center of the square.

9.4.5 Ratio Mapping and Generalization of Image Combination Procedures

For amplifying differences in two images, it is very useful to substract the gray levels point by point form each other, in order to receive the image-differences. This is a change detection method: The difference of the same scene, but taken in different spectral ranges, or at different times, becomes visual in the image difference.

Another method on a set of images from the same scene is ratio mapping. Here, the gray levels of various images are related to any one of them taken as reference. The reference can also be the average image of the whole set. If an image is to be subject of ratio mapping the value of the gray level is divided by the value of the gray level of the corresponding pixel of the reference image. This ratio is widely independent of the variable illumination effects, i.e. shading, atmospheric influences, sun angle, scan angle, cloud shadows, etc.

There are also other processes of ratio mapping, for example for contrast mapping $f - g/f + g$, if f is the gray level in one image and g the gray level in the other one, the reference image, taken for corresponding pixels (BILLINGSLEY and GOETZ, 1973).

In general, one can describe many enhancement processes, which can be performed for 2 images, by the following equation:

$$f_{\text{out}} = \frac{Af + Bg + C}{Df + Eg + F} (Gf + H)^I + J.$$

(18)

The capital letters are parameters, which can assume various values, depending on what method should be implemented. Ratio mapping is equivalent to $A = E \neq 0$, all other parameters equal to zero. $A = -B$ and $D = E = A$, but $C = F = G = H = J$ will render contrast mapping. Also offset, exponentiating etc. can be done.

9.4.6 Pseudocolor Transformation

Especially for human viewing operations in the color domain have proven useful. One reason is that the eye is more sensitive to changes in hue than to changes in brightness. Therefore, pseudo color transformations have been applied, which convert brightness to hue. In other words: Different shades of a B/W image are transformed to different colors.

Inversely, from color images some colortones can be extracted and used to transform this filtered image into a new pseudo color image. This is a method to enhance faint color differences (HELBIG, 1972; HELBIG, 1973; LAMAR and MERIFIELD, 1973).

9.5 Feature Extraction and Classification

There are four different effects, which can be used for feature recognition:

1. Spectral effects, i.e. color tones, or more generally frequency dependent reflectance and emittance features (signatures).

2. Spatial effects, i.e. patterns, such as shape, texture, size, or topology.

3. Polarization effects, which are especially applicable in the microwave region and.

4. Temporal effects, such as variations in intensity of scattered light.

Most important are spectral and spatial effects and we will only deal with them. In the spectral domain computer techniques are superior to human interpretation. In the spatial domain human interpretation qualities are not yet achieved by automated processes. The most promising way is to make use of both capabilities, those of the human viewer and those of the computer, which is possible with an interactive computer system. The quantitative analysis is performed then by the computer, the human intelligence and experience teaches the computer, what features are most suitable (NICHOLS, 1973; MOWER, 1972).

9.5.1 Spectral Features

The basic idea can be understood from Fig. 17. This shows the remittance of three different materials as a function of wavelength, the signatures. In the various wavelengths the remittance values of the materials differ more or less. Taking the wavelength λ_1 and λ_2, we can identify the materials as spots in a two-dimensional space. If we would have taken N different wavelength, then the corresponding space would be N-dimensional, in which the materials could be identified in form of coordinates, which correspond to the related remittance values. But let us go back to the simplest two dimensional case of Fig. 17b. The material can be discriminated, because they are on different places. In reality, the materials will not always have the same remittance. They will vary because of many factors: Plants may have been planted at different times, on different ground, moisture might be different, soil might be prepared better or worse, illumination, etc. might be different. Therefore, the material will identify not only a point but a point and its surroundings as typical for it in the feature space. Some neighboring point of a different material can, under these circumstances, have a feature, which is not unique. Then it is necessary to look for another feature, i.e. another dimension or wavelength, for which a discrimination is possible. At this wavelength the difference in remittance should be large as compared to the variance. One additonal help in this sense is, as described by LANDGREBE (1971), to use the dependence of the reflectance from the season: Fig. 17c. It might well be, that some materials differ sufficiently only at certain times.

9.5.2 Spatial Features

Textures are local prospects of statistical nature such as picture of random dots, brick walls, weaves etc. Some examples of synthetic textures are shown in Fig. 18.

It is obvious that enhancement and filtering processes are basic tools for extracting spatial features. By applying some averaging process to each one of the extracted features one can achieve some parameter, which can be used for classification.

The following two subparagraphs deal with two methods for texture-context feature extraction.

9.5.2.1 Some Approach for Spatial Feature Classification

As an example, the original image can be filtered for features such as lines, edges, dots, etc. of given spatial frequencies, in different directions, etc. Then, each one of the filtered images can be submitted to low pass filtering. The averaging process extracts features with regard to textures. This is obvious because of tex-

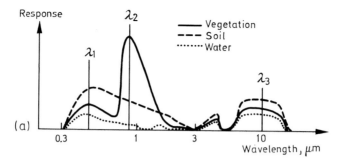

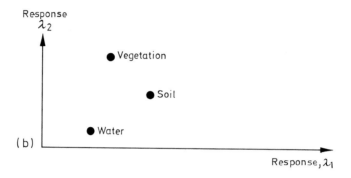

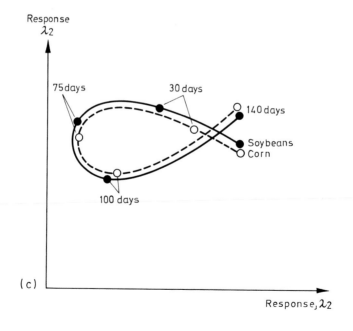

Fig. 17a–c. Signatures. (a) Relative response as a function of wavelength; (b) feature space; (c) temporal change. (Landgrebe, 1971)

Fig. 18. Examples of synthetic textures (TRIENDL, 1973). Left: Original synthetic texture.
Right: Low pass filtered image

ture being defined by averaging processes. With these filtered outputs as components of the feature space classification can be conducted.

E. E. TRIENDL (1972) has applied this procedure. The principle block diagram
is shown in Fig. 19a. In an example, he has applied two local filter processes to an
image (Fig. 19b) with 240×500 picture elements of a recultivation forest. Filter A
was a low pass filter for 3×3 pixels and gray scale expansion. Filter B was a high
frequency filter for 3×3 pixels. Averaging was done over 11×11 pixels. The features gained by these processes are related to brightness and roughness, respectively. Classification was done on the basis of a clustering algorithm. The result is
shown in Fig. 19c. Problems exist at the boundaries due to the averaging process.

9.5.2.2 Texture-context Feature Extraction Algorithm

An interesting approach for extracting a set of textural features, which are
quickly computable, was developed by HARALICK (1972). The features which are
considered are all functions of distance and angle. Various matrices are defined
for this reason, each having matrix elements, which are gray level dependence
frequencies P_{ij}. Each number P_{ij} indicates, how often in the image 2 pixels occur,
which have a mutual distance d and one of the pixel has gray level i, while the
other one has gray level j. With N different gray levels this determines an
$N \times N$ matrix $P(d)$ for a fixed d.

In order to take into consideration the angular dependence between pixels 4
different directions are investigated, namely $0°, 45°, 90°, 135°$, as shown in Fig. 20.

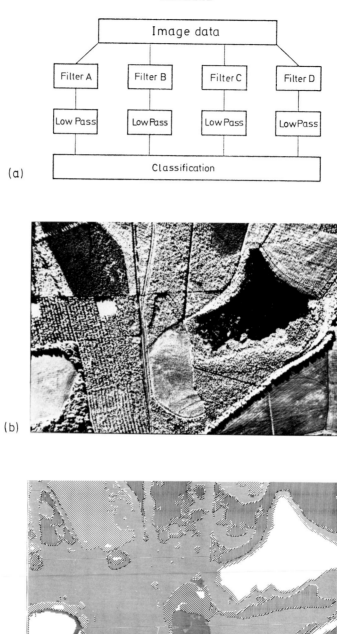

(a)

(b)

(c)

Fig. 19 a–c. A procedure for feature extraction (Triendl, 1972). (a) Block diagram; (b) original image; (c) processed image

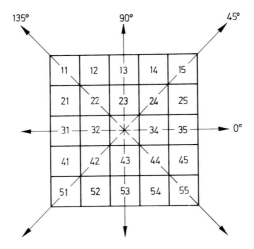

Fig. 20. Numbering of the neighboring pixels of the central pixel

The nearest neighbors $d=1$ of the central pixel * are 32 and 34 for the $0°$ case, 22 and 44 for the $135°$ case etc.

For given distance four different matrices with the elements $P_{ij}(d, 0°)$, $P_{ij}(d, 45°)$, $P_{ij}(d, 90°)$, $P_{ij}(d, 135°)$ are defined. From any matrix of this kind, textural features can be calculated.

First of all, one can use them for distinguishing between homogeneous and inhomogeneous features. A homogeneous image has only few dominant gray level transitions. The angular second moment features f_1 as defined by Eq.(19) is a measure of this homogeneity:

$$f_1 = \sum_{i=1}^{N} \sum_{j=1}^{N} \left(\frac{P_{ij}}{R}\right)^2 . \tag{19}$$

It is the sum of the squared terms of the gray level spatial dependence matrix normalized by the total number R of pixel pairs.

The contrast feature f_2 is a measure of the amount of boundaries present in the image. It is defined as a difference moment of the P matrix

$$f_2 = \sum_{n=0}^{N-1} n^2 \sum_{|i-j|=n} \left(\frac{P_{ij}}{R}\right) . \tag{20}$$

The correlation feature f_3 measures the gray level dependencies by means of the equation

$$f_3 = \frac{\sum_{i=1}^{N} \sum_{j=1}^{N} \frac{i \cdot j \cdot P_{ij}}{R} - \mu_x \mu_y}{\sigma_x \sigma_y} \tag{21}$$

μ_x, μ_y being the means and
σ_x, σ_y being the standard deviations of the marginal distributions P_x, P_y, obtained by summing the rows and the columns of P/R.

f_3 is actually a two-dimensional autocorrelation function of the image for a particular distance and angle lag.

If each of these formulas f_1, f_2, f_3 are calculated for 4 different angles and 3 different distances, then in total 36 different features are available (HARALICK, 1972).

Numerical example see Appendix 2.

HARALICK et al. (1973) demonstrate that easily computable textural features have a wide applicability for image classification.

9.5.3 The Decision Process

9.5.3.1 Supervised Technique

We will discuss the decision process first on the basis of Fig. 21. One simple procedure is to start with the determination of the centers of the material classes, their mean values. Then the locus of points equidistant from each pair of these mean values can be calculated. The whole space is subdivided by this procedure and attributed to the various classes (materials). These boundaries are then valid for the subsequent gathered samples, too.

In other words, the establishment of the boundaries was made for some initial samples, the training samples, in such a way as to allow for the mass of data to classify them with hopefully a minimum of misinterpretations. Errors in the decision process may be unavoidable in a few cases, because in a great number of samples exceptionally large deviations in feature may exist for the one or the other sample. This is a consequence of the statistical character of the feature.

This classification technique is called supervised technique. The data of the training samples can be gathered in this technique in two different ways, from the (1) signature data bank and (2) by the extrapolation method.

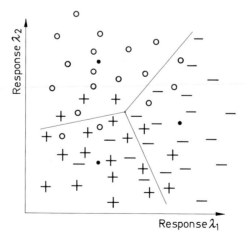

Fig. 21. Example of a decision rule

The signature data bank shall have stored the feature data, gathered systematically by many measurements. The user can apply the data for all materials he supposes to encounter and put them as coordinates in the N-dimensional feature space. Then he can fix the decision lines.

Although this is a straightforward approach, it may not be easy in reality to procede this way even then, if all data necessary are already gathered in the data bank. Changes in illumination, view angle, influences of climatic conditions etc. might cause influences on the signatures, which deteriorate the probability of success.

In the extrapolation method the remote sensing mission is always started with first gathering training samples from the material categories (classes), which might be encountered during the mission. With these data collected from limited areas, the subdivision of the feature space occurs. The distribution of each one of the classes with regards to the feature space for the mass of the data is extrapolated. This procedure—although requiring some extra time for the collection of the training samples—has the advantage, that the variations of the feature data are not large, because the data collection is done within short times, in a limited area, with the same set of data.

9.5.3.2 Unsupervised Technique

Besides of the supervised technique there is the unsupervised technique. This is based on the clustering technique. Without *a priori* knowledge about the features to be encountered the remotely sensed data are grouped into a number of classes. This technique makes use of the fact that different classes will differ in the features and therefore cluster in different locations, i.e. around different centers of the feature space. The physical identification of the classes is to be done by help of ground truth measurements after the decisions about the subspaces. The *a posteriori* ground truth information can be gained with less effort as compared to the supervised technique, because the selection of the typical sample areas can be optimized on the basis of the cluster process results: These show the areas of identical features. For the ground truth it is, therefore, possible to select just that representative area of the class, which is geographically closest.

So, the main difference between supervised and unsupervised technique is this: The supervised technique starts with classes, which are of informational value— the features are known—and tries to separate the clusters. The non-supervised technique starts with separable clusters and checks afterwards, if the features of the clusters are typical for specific materials and for which ones.

9.5.4 The Decision Criteria

Each pixel of an image must be cataloged into the class (category) it belongs to, on the basis of its features. We assume N features to be available for the classification process. That is, an N-dimensional measurement vector or pattern d is given per pixel, whose components are the features extracted from the sensor data on the basis of textural or spectral information.

M different classes are taken into account. We assume for each class c_i, $i=1...M$, to have a multivariate N-dimensional probability density function, i.e. a class is only determined in a statistical manner. In other words, the classes c_i are not fixed as single vectors, but given as subspaces: The M subspaces have to be defined.

The subdivision of the N-dimensional feature space into subspaces in an "optimal manner" is the task of the classification. There are various optimal decision criteria and proper selection depends on the users intention. Well known and widely used are the Bayes criteria, which minimizes the "average cost (loss) L" for given a priori probabilities, and the maximum likelihood criterion. This is the special case that misclassifications have all the same costs and correct classification has no cost.

The average cost L is defined by the equation

$$L = \sum_{j=1}^{M} \sum_{i=1}^{M} L_{ij} P(c_i|d). \qquad (22)$$

$P(c_i|d)$ is the probability of c_i being decided, given the vector d. L_{ij} is the cost associated with the decision c_i, if the measurement actually belongs to c_j.

If L_{ij} is chosen to be $\qquad L_{ij}=0 \quad$ for $\quad i=j$,

$$L_{ij}=1 \quad \text{for} \quad i \neq j,$$

then we have the maximum likelihood case.

L_{ij} weighs the individual classifications and misclassifications. As these factors have different importance for the various applications their numbers, and therefore the L-matrix, must be determined by the user. For some users, it might be extremely important not to make a misclassification for one specific class. For example it might be necessary to detect all places where a specific disease occurs. But it migth be less important to have erroneously assumed that it exists on a place, where it turns out by ground truth, that this is not true.

9.5.4.1 The Bayes Criterion

The measurement vector d should be assigned to the class c_i if and only if the conditional probability $P(c_i|d)$ is greater or equal to the conditional probability $P(c_j|d)$ for any class j (BROONER, 1972; CRANE and RICHARDSON, 1972).

$$P(c_i|d) \geq P(c_j|d). \qquad (23)$$

This conditional "a posteriori probabilities" can be calculated from the "a priori probabilities" by means of the Bayes rule:

$$P(c_i|d) \cdot P(d) = P(d|c_i) \cdot P(c_i)$$

$$P(c_i|d) = \frac{P(d|c_i) \cdot P(c_i)}{P(d)}. \qquad (24)$$

Therefore, d is assigned to class c_i, if and only if

$$\frac{P(d|c_i) \cdot P(c_i)}{P(d)} \geqq \frac{P(d|c_j) \, P(c_j)}{P(d)}.$$

Taking into account that $P(d) \neq 0$, we can rewrite this equation in the following form:

$$P(d|c_i) \cdot P(c_i) \geqq P(d|c_j) \cdot P(c_j). \tag{25}$$

This decision rule minimizes the average number of misclassifications.

The probability density functions $P(d|c_i)$ and $P(c_i)$ must be known for the decision. The estimations for the probability of the class c_i can be gained by ground truth, or in lack of data the assumption can be made that all categories have equal likelihood. For the $P(d|c_i)$ estimate a training set is used. Besides, often the assumption is made that $P(d|c_i)$ is multivariate normal and can be written in the following way:

$$P(d/c_i) = \frac{\exp[-\tfrac{1}{2}(d-\mu_i)^t Q_i^{-1}(-\mu_i)]}{(2\pi)^{N/2}|Q_i|^{\frac{1}{2}}} \tag{26}$$

$Q_i = E[(d-\mu_i)(d-\mu_i)^t] =$ covariance matrix of d from c_i; $|Q_i| =$ matrix of Q_i; $\mu_i = E(d) =$ mean of d; $t =$ transpose.

Both values can be estimated from the training set:

$$\hat{Q}_i = \frac{1}{K} \sum_{k=1}^{K} (d_k - \mu_i)(d_k - \mu_i)^t$$

$$\mu_i = \frac{1}{K} \sum_{k=1}^{K} d_k$$

where $d_1, d_2, \ldots, d_k$ are the vectors of the training set for category c_i.

For equal probability of all classes the decision rule requests to select that class c_i, for which $P(d|c_i)$ becomes maximum. Instead of $P(c|d_i)$ we can also maximize with respect to its logarithm $\ln P(d|c_i)$. In the normal distribution case of Eq.(26) this means that

$$\ln\{P(d|c_i)\} = -\tfrac{1}{2}\{(d-\mu_i)^t Q_i^{-1}(d-\mu_i) + \ln Q_i + N \ln \tfrac{\pi}{2}\}$$

should become a maximum, or

$$(d-\mu_i)^t Q_i^{-1}(d-\mu_i) + \ln|Q_i| = \text{Minimum}. \tag{27}$$

This is called the quadratic decision rule, as it requires a quadratic function of $(d-\mu_i)$ to be minimized. The calculation of $(d-\mu_i)^t Q_i (d-\mu_i)$ can be expressed in a different way (KRIEGLER et al., 1973):

$$Q_i = (\sigma) \cdot (\varrho) \cdot (\sigma) \tag{28}$$

where (σ) is a diagonal matrix of the standard deviation and (ϱ) is the correlation matrix with all 1's on the diagonal and values of 0 to 1 off the diagonal.

$$Q_i^{-1} = \left(\frac{1}{\sigma}\right) \cdot (\varrho)^{-1} \cdot \left(\frac{1}{\sigma}\right) \tag{29}$$

is the inverse of equation.

Thus

$$(d - \mu_i)^t Q_i^{-1} (d - \mu_i) = \left(\frac{d - \mu_i}{\sigma}\right)^t \cdot (\varrho)^{-1} \cdot \left(\frac{d - \mu_i}{\sigma}\right).$$

Assuming that the terms $|d - \mu_i/\sigma|$ of interest are smaller than a large number, say 8, we can neglect in the calculation the still larger values and save computer time. Besides, the correlation matrix is symmetric. Therefore, we can simplify the quadratic form as follows

$$\left(\frac{d - \mu_i}{\sigma}\right)^t \cdot (B)^t \cdot (B) \cdot \left(\frac{d - \mu_i}{\sigma}\right) = (u)^t(u) \tag{30}$$

(B) is the upper triangular matrix formed by the decomposition of $(\varrho)^{-1}$.

$$(u) = (B)\left(\frac{d - \mu_i}{\sigma}\right)$$

and the final matrix is

$$(u)^t(u) = \sum_{i=1}^n u_i^2 \tag{31}$$

where u_i are the elements of the (u) vector.

KRIEGLER et al. (1973) report that this procedure exists in hardwired form in the MIDAS system, where 2×10^5 maximum likelihood decisions are performed per second. 8 signals and 9 classes are accepted by the machine.

The quadratic decision rule has been applied very often and very successfully, as is published by numerous authors. This is not as obvious as it seems to be, because the assumptions about the normal distribution of the data for each class are erroneous. Besides, the values of d for neighboring pixels are not independent from each other, as is assumed. Also the uncertainties about $P(c_i)$ and about the loss function are factors, which could implement misclassifications.

9.5.4.2 Sequential Techniques

Sequential techniques are appropriate, if the features are extracted in sequence. L_k shall indicate the likelihood ratio

$$L_k = P_k(y|c_i)/P_k(y|c_j)$$

at stage k. Two stopping boundaries A and B are defined, which divide the feature space into three regions: (1) region A for class c_i, (2) region B associated with class c_j and (3) a region of indifference (SU and CUMMINGS, 1972).

If $L_k \geq A$, d is decided to be of class c_i.

If $L_k < B$, d is decided to be of class c_j.

If $B < L_k < A$,   additional feature has to be extracted before the decision can be made. If no additional features are available, then finally a maximum likelihood decision has to be performed by changing the boundaries to $A = B$.

9.5.4.3 Potential Function

The relationship of any point in the feature space to a measurement vector of the training sample is defined in this concept by a potential function $p(k)$. This is in some analogy to the electrostatic potential. It states that any measurement vector is closely related to a sample vector of the training set, if the distance Δ between them is small. For larger distances Δ this relationship rapidly decreases. In distance Δ_k the potential of the training sample k is given by

$$p(k) = \frac{1}{1 + a\Delta_k^2} \tag{32}$$

a is a parameter, which can be chosen; it defines the rate of decrease of the potential.

The mean potential function p_i of a class c_i is given by the equation

$$p_i = \frac{1}{K_i} \sum_{k=1}^{K_i} p(i, k). \tag{33}$$

Here, K_i is the number of sample points in the training set for class c_i. A vector d is decided to be a member of class c_i if for this class the mean potential is a maximum (TOLIMSON, 1972).

9.5.5 Unsupervised Classification

In the previous paragraphs we have assumed the distributions of the measurement vectors for the classes to be known. In the unsupervised classification this is not any more requested. Therefore different decision criteria must be chosen, such as clustering. This is based on the fact, that measurement vectors of the same class will be close together in the feature space, i.e. they will cluster around a center. By some way, therefore cluster centers for the various classes have to be established and each measurement vector must be assigned to the center nearest by. If a new vector is added to a center, then the cluster center has to be determined again. As soon, as all vectors are assigned, the cluster centers may be different from the ones originally started with. If this is so, then in form of an iterative process the assignment of the measurement vectors is repeated, until the changes become finally negligible.

Not only the coordinates of the cluster centers but also the number of clusters can be changed. If it turns out, that only a very limited number of vectors is contained in one cluster, then this cluster might be suppressed and the vectors reassigned to the remaining clusters. Inversely, if some vectors are too far away from any center, then this vectors might become new clustering centers (Su, 1972; Steiner, 1972).

9.5.5.1 Sequential Clustering Technique

A small number K of N dimensional vectors d_i, $i=1 \ldots K$ is tested first, if they constitute a class. The mean vector $\bar{d}$ is calculated as well as the distance Δd_K between the sample vector and the mean vector:

$$\bar{d} = \frac{1}{K} \sum_{i=1}^{K} d_i,$$

$$(\Delta d_K)^2 = \frac{1}{M} \sum_{i=1}^{K} (d_i - \bar{d})^2.$$

$$(34)$$

If $\quad \dfrac{(\Delta d_K)^2}{(\bar{d})^2} \leq T,$

$$(35)$$

then the samples are assumed to form a class. T is a threshold to be chosen, $\bar{d}$ is the cluster center. A new sample is accepted a member of the same category, if it satisfies also the above given condition. Otherwise it is discarded and a new sample is then taken for repeat.

After test of all measurement vectors, the discarded samples are submitted the same procedures as just mentioned. A small number of measurement vectors are taken to form a second cluster and all samples discarded so far are now checked for the second category and accepted or, once more, discarded. This procedure is repeated, until all samples are grouped into clusters.

Finally, the classes are printed out including the mean and variance.

9.5.6 Recognizing of Bridges, Rivers, Lakes, and Islands

A procedure for recognizing these kinds of objects was developed by R. Bajcsy and M. Tavakoli (1973). It uses a "world model" and applies a set of features, i.e. extracts from the image textures, shapes, size, signatures, etc. The world model states that

1. Water has brightness levels of smaller or equal to 10% of the maximal value in ERTS-MSS-band 3, whereas for land it is larger than 10%. The texture is homogeneous.

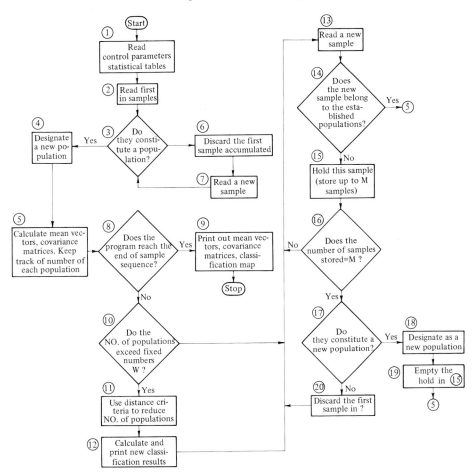

Fig. 22. Flowchart of statistical sequential clustering algorithm. (Su and Cummings, 1972)

2. Rivers have gray levels of water, the texture is homogeneous, the boundaries are open, the contrast is large. Rivers are below bridges and are surrounded by land.

3. Lakes have also gray levels of water, are of homogeneous texture. They have large contrast, have boundaries mostly closed, but open to rivers. They are surrounded by land and surround islands themselves.

4. Bridges have gray levels of land, homogeneous texture, are of thin shape, elongated and smoothly curved. They are connected to land with their two shorter sides, and surrounded with water on their longer sides.

5. Islands have gray level of land, continuous texture, closed boundaries, and are completely surrounded by water. They must not be connected with land.

The program can find all bridges, identify lakes, and rivers, and all islands of an image with some limitations, as shown by the authors, by means of several concrete examples.

9.6 Appendix 1: The Image Signal in Spatial and Frequency Domain

(ANDREWS et al., 1970; ROSENFELD, 1970; SELZER, 1968)

9.6.1 Signals in the Spatial Domain

Starting with the spatial domain, the signal can be considered as being composed of the spatially distributed energy radiated from all points of the scene $s(u, v)$ to be mapped. As long as the image process can be considered to be linear, the image $f(x, y)$ is given by the equation

$$f(x, y) = \iint_{-\infty}^{+\infty} s(u, v)k(u, v, x, y)dudv , \qquad (36)$$

where $k(u, v, x, y)$ is the function, which describes the effect of the sensor generating the picture. It is the impulse-response or point-spread function of the sensor. The image would be equal to the point-spread function, it the scene would be a single point source of light.

If the impulse-response function is invariant to shifts in the coordinates of the point source, we can write

$$h(u, v, x, y) = h(x - u, y - v) \qquad (37)$$

and Eq. (36) can be rewritten in the form

$$f(x, y) = \iint_{-\infty}^{+\infty} h(x - u, y - v)s(u, v)dudv$$

$$f(x, y) = h(x, y) * s(x, y) . \qquad (38)$$

This pair of equation expresses, that the image function $f(x, y)$ is the convolution of the scene function $s(u, v)$ with the point-spread function. It is the overlapping of all signals radiating points of the scene, each of which is weighted by the function $h(x - u, y - v)$. For the digitized image with $M \times N$ pixels we modify Eq. (38) in the following manner:

$$f(x = m, y = n) = \sum_{k=-M}^{M} \sum_{l=-N}^{N} h_{k,l} s_{m-k,n-l} \qquad (39)$$

or of the simpler one dimensional case

$$f(x = m) = \sum_{k=-M}^{+M} h_k s_{m-k} . \qquad (40)$$

The weighting function is here the set $s = \{s_{-M}, \ldots, s_{-1}, s_0, s_1, \ldots, s_M\}$ of values $s_i, i = -M \ldots + M$, which correspond to the influences, the gray levels i pixel apart have on the gray level of the image point O. Examples are shown in paragraph 4.

9.6.2 Signals in the Frequency Domain

The signal can be considered in the frequency domain, too. This is based on the Fourier Transform (FT): Any signal which is absolutely integrable can be decomposed into sinusoids of various frequencies, amplitudes and phases. For an image system we measure the frequencies in cycles per pixel. A frequency of 0.2 indicates a sinusoid with a period of 5 pixels.

The Fourier Transform $H(f_a, f_b)$ of the point-spread function h is given by the equation

$$H(f_a, f_b) = \iint_{-\infty}^{+\infty} h(u, v) e^{-j2\pi(uf_a + vf_b)} \cdot du\, dv . \tag{41}$$

For the one-dimensional discrete case, the Fourier Transform of h is given by

$$H(f_a) = \sum_{k=-\infty}^{+\infty} h_k e^{-2\pi j f_a k} \tag{42}$$

and for the two-dimensional case it is defined by

$$H(f_a, f_b) = \sum_{k=-\infty}^{+\infty} \sum_{l=-\infty}^{+\infty} h_{k,l} \cdot e^{-2\pi j (f_a k + f_b k)} , \tag{43}$$

with $j = \sqrt{-1}$, $f_a, f_b \ldots$ frequencies in cycles per sample.

The same type of transformation holds for the other functions f, s. Fourier Transforms are expressed by the corresponding capital letters.

For the one-dimensional case, let us now show that the transfer function, given in the spatial domain by a convolution, is determined in the frequency domain as a product. The FT of $f_n(x)$ from Eq.(40) is given by

$$F(f_a) = \sum_{n=-\infty}^{+\infty} f_n e^{-j2\pi f_a \cdot n} \tag{44}$$

and in substituting Eq.(40) into (44)

$$F(f_a) = \sum_{n=-\infty}^{+\infty} \left[\sum_{k=-M}^{+M} h_k s_{n-k} \right] \cdot e^{-j2\pi f_a \cdot n}$$

and after rearrangement

$$F(f_a) = \sum_{k=-K}^{+K} h_k \sum_{n=-\infty}^{+\infty} s_{n-k} \cdot e^{-j2\pi f_a \cdot n} .$$

By change of the variable

$$F(f_a) = \sum_{h=-K}^{+K} h_k \sum_{m=-\infty}^{+\infty} s_m \cdot e^{-j2\pi f_a(m+k)}$$

$$= \sum_{k=-K}^{+K} h_k \cdot e^{2\pi f_a k} \sum_{m=-\infty}^{+\infty} h_m e^{-j2\pi f_a m}$$

we achieve the result:

$$F(f_a) = G(f_a) \cdot H(f_a) . \tag{45}$$

This can, of course, be generalized to the two- and multidimensional case. It shows that a convolution procedure of the spatial domain corresponds to the simpler procedure of a multiplication in the frequency domain. In the frequency domain the optical transfer function (OTF), defined as the relative amplitude and phase response to the sinusoidal sensor input, is used. This corresponds to the point-spread function in the spatial domain. The amplitude response is called the modulation transfer function (MTF). The system MTF curve is the cascaded response of the subsystem.

9.7 Appendix 2: Numerical Example for Texture-context Features

Given the matrix of 4 gray level values of a 4×4 image

$$\begin{pmatrix} 0 & 1 & 1 & 0 \\ 1 & 1 & 0 & 0 \\ 2 & 2 & 0 & 2 \\ 2 & 3 & 3 & 0 \end{pmatrix}$$

the matrix of P_{ij} values which indicate how often a pixel with gray level i has a pixel with gray level j as neighbor is of the following form

$$\begin{pmatrix} P_{00} & P_{01} & P_{02} & P_{03} \\ P_{10} & P_{11} & P_{12} & P_{13} \\ P_{20} & P_{21} & P_{22} & P_{23} \\ P_{30} & P_{31} & P_{32} & P_{33} \end{pmatrix}.$$

Neighborhood must be further defined by distance d and the angle:

$$P(d=1.0°) = \begin{pmatrix} 2 & 3 & 2 & 1 \\ 3 & 4 & 0 & 0 \\ 2 & 0 & 2 & 1 \\ 1 & 0 & 1 & 2 \end{pmatrix}; \qquad P(d=1.45°) = \begin{pmatrix} 4 & 0 & 1 & 1 \\ 0 & 4 & 1 & 0 \\ 1 & 1 & 2 & 1 \\ 1 & 0 & 1 & 0 \end{pmatrix};$$

$$P(d=1.90°) = \begin{pmatrix} 4 & 2 & 2 & 1 \\ 2 & 2 & 2 & 0 \\ 2 & 2 & 2 & 1 \\ 1 & 0 & 1 & 0 \end{pmatrix}; \qquad P(d=1.135°) = \begin{pmatrix} 2 & 4 & 1 & 0 \\ 4 & 0 & 1 & 0 \\ 1 & 1 & 0 & 2 \\ 0 & 0 & 2 & 0 \end{pmatrix}.$$

Then the angular second moment feature can be calculated.

$$f_1(0°) = \sum_{i=0}^{3} \sum_{j=0}^{3} \left(\frac{P_{ij}}{R} \right)^2,$$

where $R =$ number of pixel pairs, which adds up to $R = 24$, because the two outer columns of the image have neighbor pixels in the inner side ($4+4=8$), whereas the inner columns have pixel neighbors in both sides ($8+8=16$). Taking the squares of the elements $P_{ij}(d=1.0°)$ we receive:

$$f_1(0°) = \frac{1}{24^2}(2^2 + 3^2 + 2^2 + 1^2 + 3^2 + 4^2 + 0^2 + 0^2 + 2^2 + 0^2 + 2^2 + 1^2$$

$$+ 1^2 + 0^2 + 1^2 + 2^2) = \left(\frac{1}{24}\right)^2 \cdot 58 = 0.1007.$$

The contrast feature is

$$f_2(90°) = \sum_{n=0}^{3} n^2 \sum_{|i-j|=n} \left(\frac{P_{ij}}{R}\right),$$

where again $R = 24$ because of the 4×4 image matrix. Taking the elements of the matrix $P(d=1.90°)$ and multiplying each one with the corresponding gray level difference $(i-j)^2$, we receive the following equation:

$$f_2(90°) = \frac{1}{24}[0^2 \cdot 4 + 1^2 \cdot 2 + 2^2 \cdot 2 + 3^2 \cdot 1$$

$$1^2 \cdot 2 + 0^2 \cdot 2 + 1^2 \cdot 2 + 2^2 \cdot 0$$

$$2^2 \cdot 2 + 1^2 \cdot 2 + 0^2 \cdot 2 + 1^2 \cdot 1$$

$$3^2 \cdot 1 + 2^2 \cdot 0 + 1^2 \cdot 1 + 0^2 \cdot 0]$$

$$= \frac{1}{24}[1^2 \cdot 10 + 2^2 \cdot 4 + 3^2 \cdot 2] = \frac{44}{24} = 1.8408.$$

The correlation feature is

$$f_3(45°) = \frac{\sum_{i=0}^{3} \sum_{j=0}^{3} \frac{i \cdot j \cdot P_{ij}}{R} - \mu_x \mu_y}{\sigma_x \sigma_y}.$$

Taking the elements of the $P(d=1.45°)$-matrix the number of pixel pairs is $R = 18$, and

$$\sum_{i=0}^{3} \sum_{j=0}^{3} ijP_{ij} = 0 \cdot 0 \cdot 4 + 0 \cdot 1 \cdot 0 + 0 \cdot 2 \cdot 1 + 0 \cdot 3 \cdot 1$$

$$1 \cdot 0 \cdot 0 + 1 \cdot 1 \cdot 4 + 1 \cdot 2 \cdot 1 + 1 \cdot 3 \cdot 0$$

$$2 \cdot 0 \cdot 1 + 2 \cdot 1 \cdot 1 + 2 \cdot 2 \cdot 2 + 2 \cdot 3 \cdot 1$$

$$3 \cdot 0 \cdot 1 + 3 \cdot 1 \cdot 0 + 3 \cdot 2 \cdot 1 + 3 \cdot 3 \cdot 0$$

$$= 4 + 2 + 2 + 8 + 6 + 6 = 28.$$

We calculate the means μ_x, μ_y by using $P_i = \sum_{j=0}^{3} P_{ij}$ and $P_j = \sum_{i=0}^{3} P_{ij}$:

$$\mu_x = \frac{1}{R}\sum_{i=0}^{3} iP_i = \frac{1}{R}(0 \cdot 6 + 1 \cdot 5 + 2 \cdot 5 + 3 \cdot 2) = 21 \cdot \frac{1}{R},$$

$$\mu_y = \frac{1}{R}\sum_{j=0}^{3} jP_j = \frac{1}{R}(0 \cdot 6 + 1 \cdot 5 + 2 \cdot 5 + 3 \cdot 2) = 21 \cdot \frac{1}{R}.$$

The standard deviations σ_x, σ_y are given by the equations:

$$\sigma_x^2 = \sum_{i=0}^{3} \sum_{j=0}^{3} i^2 \frac{P_{ij}}{R} - \mu_x^2 = \frac{1}{R}$$

$$[0^2 \cdot 6 + 1^2 \cdot 5 + 2^2 \cdot 5 + 3^2 \cdot 2] - \mu_x^2 \; ;$$

$$\sigma_y^2 = \sum_{i=0}^{3} \sum_{j=0}^{3} j^2 \frac{P_{ij}}{R} - \mu_y^2 = \frac{1}{R}$$

$$[0^2 \cdot 6 + 1^2 \cdot 5 + 2^2 \cdot 5 + 3^2 \cdot 2] - \mu_y^2 \; ;$$

$$\sigma_x^2 = \sigma_y^2 = \frac{1}{R} 43 - \left(\frac{1}{R} \cdot 21 \right)^2 .$$

Finally, f_3 is calculated by:

$$f_3(45°) = \frac{\dfrac{1}{18} \cdot 28 - \dfrac{1}{18^2} \cdot 21^2}{\dfrac{1}{18} \cdot 43 - \dfrac{1}{18^2} \cdot 21^2} = \frac{28 - \dfrac{441}{18}}{43 - \dfrac{441}{18}} = 0.1795 .$$

References

ALLEN, G. R., BONRUD, L. O., COSGROVE, J. J., STONE, R. M.: The design and use of special purpose processors for the machine processing of remotely sensed data. Proc. Conference on Machine Processing of Remotely sensed data, Oct. 16–18, Purdue Univ. TEEE Catalog No. 73 CHO 834—2 GE, pp. 1A-25–1A-42 (1973).

ANDREWS, H. C., TESCHER, A. G., KRUGER, R. P.: Image processing by digital computer. IEEE Spectrum, pp. 20–32, July 1970.

BAJCSY, R., TAVAKOLI, M.: A computer recognition of bridges, islands, rivers and lakes from satellite pictures. Conference on machine processing of Remotely Sensed Data, Purdue University. Lafayette, ind. Oct. 16–18, pp. 2A-54–2A-68 (1973).

BAKIS, R., WESLEY, M. A., WILL, P. M.: Digital correction of geometric and radiometric errors in ERTS data. Information Processing 71-North-Holland Publishing Co., Nr. 1138–1143 (1972).

BERNSTEIN, R.: Results of precision processing (scene correction) of ERTS-1 images using digital image processing techniques. ERTS-1 Symposium, March 5–9. Goddard Space Flight Center NASA SP-327 (1973).

BERNSTEIN, R., SILVERMAN, H.: Digital techniques for earth resource image data processing, AIAA paper No. 71–978, AIAA 8th Annual Meeting and Technical Display, Wash., D.C., Oct. 25–28, 1971.

BILLINGSLEY, F. C., GOETZ, A. F.: Computer techniques used for some enhancements of ERTS Images Symposium on significant results obtained from the earth resources technology satellite-1, Vol. I: Technical Presentations, Section B. NASA SP-327. Paper I-9, March 5–9, 1973.

BRESSANIN, G., ERICKTON, J. et al.: Data processing systems for earth resources surveys ESRO CR-295, Contractor Report, September 1973.

BRISTOR, C. L.: Earth resources data processing as viewed from an environmenttal satellite data processing experience case. AIAA Earth Resources overservations and information systems meeting, Annapolis, Ma. March 2–4, 1970.

BROONER, W. G.: Alias spectral parameters affecting automated image interpretation using Bayesiau probability techniques. Processing of the 8th International Symposium on Remote Sensing of Environment, pp. 1929–1949, Oct. 1972.

COUSIN, S. B., ANDERSON, A. C., PARIS, J. F., POTTER, J. F.: Significant techniques in the processing and interpretation of ERTS-1 data. Proc. of 8th International Symposium on Remote Sensing of Environment, Ann Arbor, Mich., pp. 1151–1158, Oct. 1972.

CRANE, R. B., RICHARDSON, W.: Performance evaluation of multispectral scanner classification methods, Proceedings of the Eighth International Symposium on Remote Sensing of Environment, pp. 815–820, Ocotober 1972.

GRIFFITH, A. K.: Edge detection in simple scenes using a priori information. IEEE Transactions on computers, C-22, No. 4, pp. 371–381, April 1973.

HARALICK, R. M.: Data processing at the University of Kansas, Paper No. 38, 4th Annual Earth Resources Program Review, Vol. II (University Programs), NASA MSC, Houston, Texas, APP. II (1972).

HARALICK, R. M., SHANMUGAM, K. S.: Combined spectral and spatial processing of ERTS imagery data. Symposium on significant results obtained from the Earth Resources Technology Satellite-1. Vol. I: Technical Presentations Section B. Paper I 16. New Carrollton. Md., March 5–9, 1973.

HARALICK, R. M., SHANMUGAM, K. S., DINSTEIN, I.: Textural features for image classification. IEEE transactions on Systems, Man and Cybernetics, Vol. SMC-3, No. 6, pp. 610–621, Nov. 1973.

HELBIG, H. S.: A new method for evaluating and mapping colors in aerial photographs. Eighth international Symposium on Remote Sensing of Environment, Ann Arbor, Mich., October 2–6, 1972.

HELBIG, H. S.: Fast automated analysis and classification of color pictures by signature and pattern recognition using a color scanner. Conference on machine Processing of Remotely Sensed Data, Purdue University, Lafayette, Ind., Oct. 16–18, 1973.

HORTON, C. R., DOBBINS, L. W.: Image Data Transmission. An assessment of practical systems. Ground Data Processing system for ERTS spacecraft imagery. RCA recording systems, Camden, N.J., January 1973.

JOHNSON, R., BUITEN, R.: Design of the ERTS image processing system. AIAA 8th Annual meeting and technical display Washington D.C., Oct. 25–28, 1971.

KRIEGLER, F., MARSHALL, R., LAMPERT, S., GORDON, M., CONNELL, C., KISTLER, R.: Multivariate interactive digital analysis system (MIDAS): A new fast multispectral recognition system. Proc. Conference on Machine Processing of Remote Sensed Data. Purdue University IEEE Catalog No. 73 CHO, 834-2 GE, pp. 2 B-O–2 B-13, Oct. 16–18, 1973.

KRITIKOS, G.: Einige Verfahren der digitalen Bildverarbeitung. Bildmessung und Luftbildwesen, pp. 242–252. Karlsruhe: Herbert-Wichmann, pp. 242–252, November 1971.

LAMAR, J., MERIFIELD, P. M.: Pseudocolor Transformation of ERTS imagery. Symposium on significant results obtained from the Earth Resources Technology Satellite-I, Vol. I: Technical Presentations, Section B. Paper I 12, New Carrollton, Md., NASA SP-32, March 5–9, 1973.

LANDGREBE, D.: Systems Approach to the use of Remote Sensing International workshop on Earth Resources Survey Systems, Vol. 1, pp. 139–154, NASA SP-283, May 3–14, 1971.

MALILA, W. A., HIEBER, R. H., McCLEER, A. P.: Correlation of ERTS MSS Data and earth coordinate systems. Conference on machine processing of remotely sensed Data, Purdue University, Lafayette, Ind., pp. 2 B-O–2 B-13, October 16–18, 1972.

MALILA, W. A., NALEPKA, R. F.: Atmospheric effects in ERTS-1 data and advanced information extraction techniques, Symposium on significant results obtained from the earth resources technology satellite-1, Vol. B, NASA SP-327, pp. 1097–1121, March 5–9, 1973.

MOWER, R. D.: The interpretation of multispectral imagery. An analysis of automated vs. human interpretation techniques, Ann Arbor, pp. 789–814, 1972.

NAGY, G.: Digital image-processing activities in remote sensing for earth resources. Proceedings of the IEEE, Vol. 60, No. 10, pp. 1177–1200, October 1972.

NASA: ERTS Data User's Handbook, GSFC, Document 71 SD 4249, 1971.

Nichols, J. D.: Combining human and Computer Interpretation capabilities to analyze ERTS imagery. Symposium on significant results obtained from the earth resources technology satellite-1, Vol. I. NASA SP-327, Paper I 14, March 5–9, 1973.

Rosenfeld, A.: Picture Processing by computer. New York-London: Academic Press 1970.

Salvato, P.: Iterative techniques to estimate signature vectors for mixture processing of multispectral data. Conference on machine processing of remotely sensed data. Purdue University, Lafayette, Ind., Oct. 16–18, 1973.

Schärf, R.: Erzeugung linienhafter Bildmuster aus Grantonbildern mit Hilfe der Kontrastgradienten. Forschungsbericht aus der Wehrtechnik, BMVg-FBWT 73–10, 1973.

Selzer, R. H.: The use of computers to improve biomedical image quality. Proceedings AFIPS, Fall Joint Computer Conference, pp. 817–845, 1968.

Skinner, C. W., Gonzalez, R. C.: On the management and processing of earth resources information. Conference ou machine processing of remotely sensed data. Purdue University, Lafayette, Ind., Oct. 16–18, 1973.

Steiner, D.: Computer-processing and classification of multi-variate information from remote sensing imagery. Proc. of the eighth International Symposium on Remote Sensing of Environment, pp. 895–907, Oct. 1972.

Su, M. M., Cummings, R. E.: An unsupervised classification technique for multispectral remote sensing data, Proceedings of the 8th International Symposium on Remote Sensing of Environment, Ann Arbor, Michigan, October 1972.

Tolimson, R. F. (Ed.): Geographical Data Handling, Vol. I. Symposium on geographical information systems, Ottawa, Aug. 1972.

Triendl, E. E.: Skeletonization of noisy handdrawn symbols using parallel operations. Pattern Recognition, Vol. 2, pp. 215–226. Pergamon Press 1970.

Triendl, E. E.: Automatic terrain mapping by texture recognition. Proc. 7th internat. Symposium on remote sensing of environment Sept. 1972.

Triendl, E. E.: Texturerkennung und Texturreproduktion, Kybernetik **13**, 1–5 (1973).

Webber, W. F.: Techniques for image registration. Conference on machine processing of remotely sensed data, Purdue University, Lafayette, Ind., pp 1 B-1–1 B-23, Oct. 16–18, 1973.

Subject Index

Page numbers in italics refer to headings, page numbers in bold face refer to figure legends and tables

Color Plates

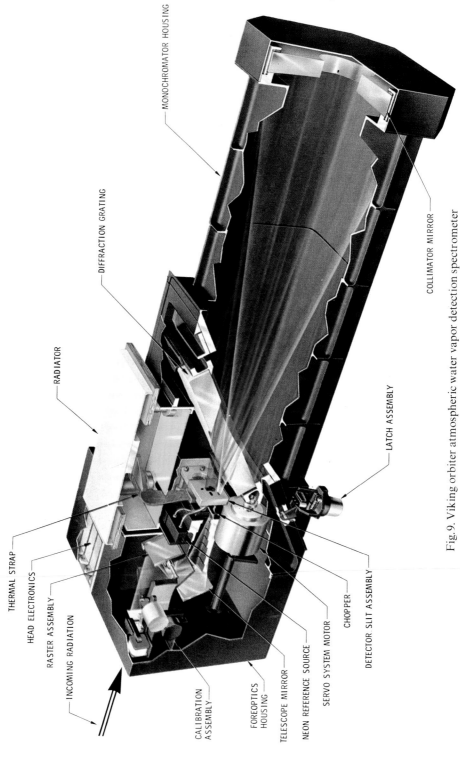

MONOCHROMATOR HOUSING

DIFFRACTION GRATING

COLLIMATOR MIRROR

RADIATOR

LATCH ASSEMBLY

THERMAL STRAP

HEAD ELECTRONICS

RASTER ASSEMBLY

INCOMING RADIATION

CALIBRATION ASSEMBLY

FOREOPTICS HOUSING

TELESCOPE MIRROR

NEON REFERENCE SOURCE

SERVO SYSTEM MOTOR

CHOPPER

DETECTOR SLIT ASSEMBLY

Fig. 9. Viking orbiter atmospheric water vapor detection spectrometer

Fig. 11 of paper by P. W. SCHAPER: Infrared Sensing Methods 361

Fig. 11. ERTS multi spectral photograph of Los Angeles, California, basin

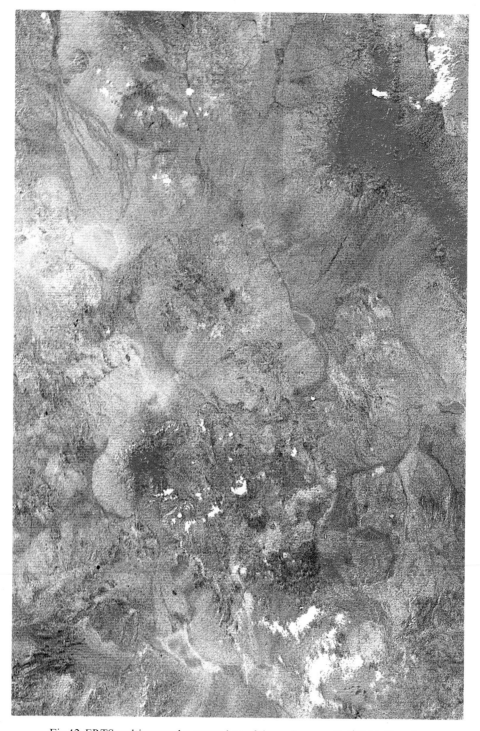

Fig. 12. ERTS multi spectral scanner data of the region near Goldfield, Nevada

Fig. 13 of paper by P.W.SCHAPER: Infrared Sensing Methods   363

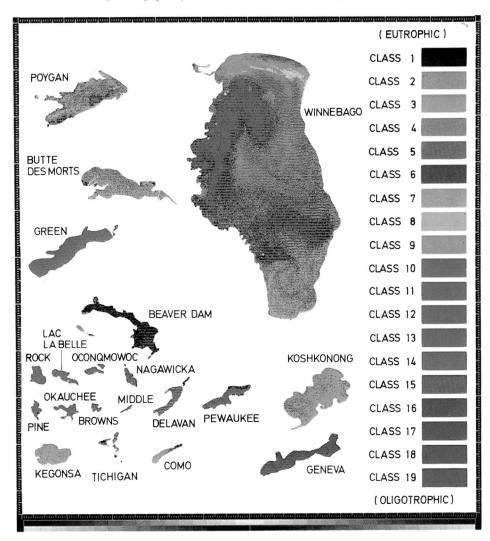

Fig. 13. Lake water quality classification from ERTS multi spectral scanner data

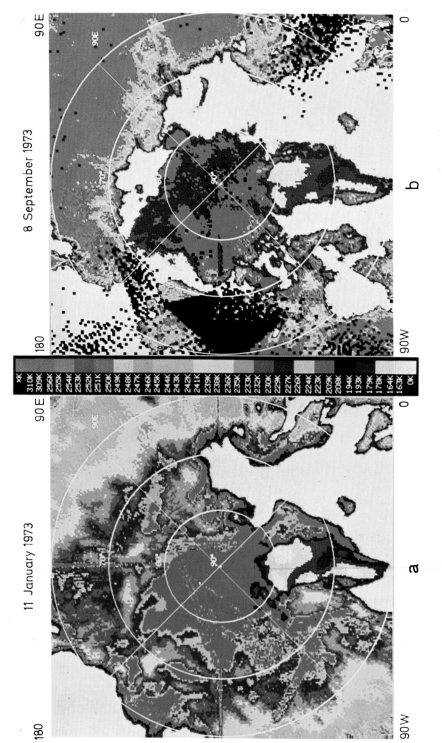

Fig. 23a and b. False color polar projection maps of 1.55 cm microwave radiometer data (North pole region) obtained from the Nimbus 5 Scanning radiometer on (a) January 11, 1973; (b) September 8, 1973. (GLOERSEN et al., 1974b)

Fig. 26 of paper by E. SCHANDA: Passive Microwave Sensing 365

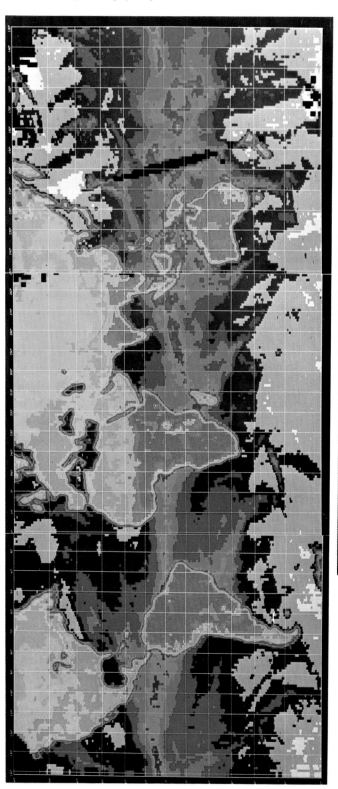

12 –16 January 1973

Fig. 26. Radio brightness map of the world. Nimbus 5 Electrically Scanned Microwave Radiometer ($\lambda = 1.55$ cm). (WEBSTER et al., 1975)

Fig. 2. ERTS-precision-processed output film image format. Actually a $1:10^6$-scale ortho- ▶
photo in a universal transverse mercator (UTM) projection. (JOHNSON and BUITEN, 1971;
NASA, 1971)

1. Bulk annotation block and gray scale, copied from input image.
2. Issuing agency and satellite number.
3. 30-minute longitude coordinates identified along top edge.
4. Internal 30-min-lat-long ticks (when requested).
5. Image I.D. number.
6. Color block.
7. 30-min-latitude coordinates* identified along left edge of tick mark frame.
8. Color-registration mark, 2 places.
9. Cartographic and positional data.
10. Bottom line:
    — Date, month, year of picture exposure
    — Format center latitude and longitude, indicating in degrees and minutes
    — Latitude and longitude of the nadir in degrees and minutes, i.e. position of satellite projected to earth surface
    — Sensor and spectral band identification and exposure times
    — Sun elevation angle and azimuthangle
    — Satellite heading, measured clockwise from true north, orbit revolution, ground recording station
    — Image processed normal or abnormal, predicted or definite ephemeris used
    — Identification of agency and project (ERTS)
    — Encoded information for satellite control center.
11. Image scale and bar scales.
12. 50000 meter UTM easting coordinates* along bottom edge of tick mark frame. Full coordinate and UTM zone given for westmost tick and at utm 2 ne boundaries.
13. Acquisition data and format center, bottom line for other data.
14. 50000 meter utm northing coordinates* along right edge of tick mark frame, first coordinate value identified in full.

----

* Tick marks in tick mark frame: longitude ticks on outer border, with even-degree marks accentuated. UTM ticks on inner border with 1000000 meter marks accentuated).

Fig. 2 of paper by PH. HARTL: Digital Picture Processing   367

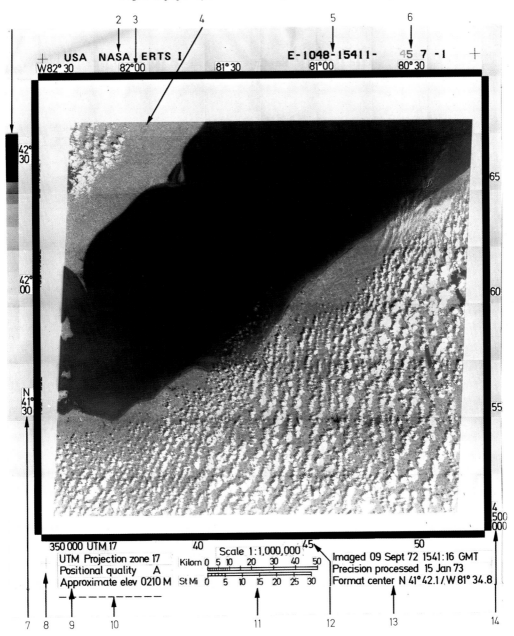

Fig. 2

Ecological Studies
Analysis and Synthesis

Editors: W.D.Billings, F.Golley,
O.L.Lange, J.S.Olson

Distribution rights for the
whole series for
U.K., Commonwealth and the
Traditional British Markets
(excluding Canada):
Chapman & Hall Ltd., London

Springer-Verlag
Berlin
Heidelberg
New York

Oecologia

In Cooperation with the
International Association for Ecology (Intecol)

Editor-in-Chief:H. Remmert
Editorial Board: L.C. Birch, L.C. Bliss,
M. Evenari, D.M. Gates, J.J. Gilbert, J. Jacobs,
T. Kira, O.L. Lange, H. Löffler, D. Neumann,
I. Phillipson, H. Remmert, F. Schaller,
K.E.F. Watt, W. Wieser, C.T. de Wit, H. Ziegler

OECOLOGIA reflects the dynamically
growing interest in ecology. Emphasis is
placed on the functional interrelationship of
organisms and environment rather than on
morphological adaption. The journal publishes
original articles, short communications, and
symposium reports on all aspects of modern
ecology, with particular reference to autecol-
ogy, physiological ecology, population
dynamics, production biology, demography,
epidemiology, behavioral ecology, food cycles,
theoretical ecology, including population
genetics.

Sample copies as well as subscription and
back-volume information available upon
request.

Please address:

Springer-Verlag, Werbeabteilung 4021
D-1000 Berlin 33, Heidelberger Platz 3

or

Springer-Verlag, New York Inc.
Promotion Department
175 Fifth Avenue, New York, N Y 10010

Springer-Verlag
Berlin
Heidelberg
New York